NUREG-1757
Vol. 1, Rev. 2

Consolidated Decommissioning Guidance

Decommissioning Process for Materials Licensees

Final Report

Manuscript Completed: September 2006
Date Published: September 2006

Prepared by
K.L. Banovac, J.T. Buckley, R.L. Johnson,
G.M. McCann, J.D. Parrott, D.W. Schmidt,
J.C. Shepherd, T.B. Smith, P.A. Sobel,
B.A. Watson, D.A. Widmayer, T.H. Youngblood

Division of Waste Management and Environmental Protection
Office of Nuclear Material Safety and Safeguards
U.S. Nuclear Regulatory Commission
Washington, DC 20555-0001

ABSTRACT

As part of its redesign of the materials license program, the U.S. Nuclear Regulatory Commission (NRC), Office of Nuclear Material Safety and Safeguards (NMSS) has consolidated and updated numerous decommissioning guidance documents into a three-volume NUREG. Specifically, the three volumes address the following topics:

(1) "Decommissioning Process for Materials Licensees";
(2) "Characterization, Survey, and Determination of Radiological Criteria"; and
(3) "Financial Assurance, Recordkeeping, and Timeliness."

This three-volume NUREG series replaces NUREG-1727 (NMSS Decommissioning Standard Review Plan) and NUREG/BR-0241 (NMSS Handbook for Decommissioning Fuel Cycle and Materials Licensees). NUREG-1757 is intended for use by NRC staff, licensees, and others.

Volume 1 of this NUREG series, entitled "Consolidated Decommissioning Guidance: Decommissioning Process for Materials Licensees," takes a risk-informed, performance-based approach to the information needed to support an application for decommissioning a materials license and compliance with the radiological criteria for license termination in 10 CFR Part 20, Subpart E. The approaches to license termination described in this guidance will help to identify the information (subject matter and level of detail) needed to terminate a license by considering the specific circumstances of the wide range of radioactive materials users licensed by NRC. Licensees should use this guidance in preparing license amendment requests. NRC staff will use this guidance in reviewing these amendment requests.

Volume 1 is intended to be applicable only to the decommissioning of materials facilities licensed under 10 CFR Parts 30, 40, 70, and 72 and to the ancillary surface facilities that support radioactive waste disposal activities licensed under 10 CFR Parts 60, 61, and 63. However, parts of this volume are applicable to reactor licensees, as described in the Foreword to this volume.

PAPERWORK REDUCTION ACT STATEMENT

The information collections contained in this NUREG are covered by the requirements of 10 CFR Parts 19, 20, 30, 33, 34, 35, 36, 39, 40, 51, 60, 61, 63, 70, 72, and 150 which were approved by the Office of Management and Budget, approval numbers 3150–0044, 0014, 0017, 0015, 0007, 0010, 0158, 0130, 0020, 0021, 0127, 0135, 0199, 0009, 0132, and 0032.

PUBLIC PROTECTION NOTIFICATION

The NRC may not conduct or sponsor, and a person is not required to respond to, a request for information or an information collection requirement unless the requesting document displays a currently valid OMB control number.

CONTENTS

CONTENTS

CONTENTS

PART II: DECOMMISSIONING PLANS

CONTENTS

FIGURES

TABLES

CONTENTS

APPENDICES

FOREWORD

> NRC staff suggests that licensees contact NRC or the appropriate Agreement State authority to assure understanding of what actions should be taken to initiate and complete decommissioning at facilities.

In September 2003, U.S. Nuclear Regulatory Commission (NRC) staff in the Office of Nuclear Material Safety and Safeguards (NMSS)[1] consolidated and updated the policies and guidance of its decommissioning program in a three-volume NUREG series, NUREG-1757, "Consolidated Decommissioning Guidance." This NUREG series provides guidance on: planning and implementing license termination under the NRC's License Termination Rule (LTR), in the Code of Federal Regulations (CFR), Title 10, Part 20, Subpart E; complying with the radiological criteria for license termination; and complying with the requirements for financial assurance and recordkeeping for decommissioning and timeliness in decommissioning of materials facilities. The staff periodically updates NUREG-1757, so that it reflects current NRC decommissioning policy.

In September 2005, the staff issued, for public comment, draft Supplement 1 to NUREG-1757, which contained proposed updates to the three volumes of NUREG-1757. Draft Supplement 1 included new and revised decommissioning guidance that addresses some of the LTR implementation issues, which were analyzed by the staff in two Commission papers (SECY-03-0069, Results of the LTR Analysis; and SECY-04-0035, Results of the LTR Analysis of the Use of Intentional Mixing of Contaminated Soil). These issues include restricted use and institutional controls, onsite disposal of radioactive materials under 10 CFR 20.2002, selection and justification of exposure scenarios based on reasonably foreseeable future land use (realistic scenarios), intentional mixing of contaminated soil, and removal of material after license termination. The staff also developed new and revised guidance on other issues, including engineered barriers.

The staff received stakeholder comments on Draft Supplement 1 and prepared responses to these comments. The stakeholder comments are located on NRC's decommissioning Web site, at http://www.nrc.gov/what-we-do/regulatory/decommissioning/reg-guides-comm.html, and the NRC staff responses are located on the same Web site and also in the Agencywide Documents Access and Management System at ML062370521. Supplement 1 has not been finalized as a separate document; instead, updated sections from Supplement 1 have been placed into the appropriate locations in revisions of Volumes 1 and 2 of NUREG-1757. The staff plans to revise Volume 3 of this NUREG series at a later date, and that revision will incorporate the Supplement 1 guidance that is related to Volume 3.

---

[1] As of September 2006, NRC is planning to reorganize NMSS and the Office of State and Tribal Programs (STP) to create two new offices: the Office of Federal and State Materials and Environmental Management Programs, which will focus on materials programs; and the new NMSS, which will focus on fuel cycle programs. This reorganization is scheduled to take effect on October 1, 2006. This document contains references to NMSS and STP. These references will be updated in future revisions of this document.

FOREWORD

NRC is currently moving toward increasing the use of risk information in its regulation of nuclear materials and nuclear waste management, including the decommissioning of nuclear facilities. NRC's risk-informed regulatory approach to the decommissioning of nuclear facilities represents a philosophy whereby risk insights are considered together with other factors to better focus the attention and resources of both the licensee and NRC on the more risk-significant aspects of the decommissioning process and on the elements of the facility and the site that will most affect risk to members of the public following decommissioning. This results in a more effective and efficient regulatory process.

The term "risk-informed," as used here, refers to the results and findings that come from risk assessments. A risk assessment is a systematic method for addressing risk. The end results of such assessments (e.g., the calculation of predicted doses from decommissioned sites) may relate directly or indirectly to public health effects. NRC staff has developed this guidance to implement the risk-informed approach and intends that the guidance be implemented in a risk-informed manner.

The primary decommissioning guidance documents used by licensees and NRC staff are NUREG-1757 and NUREG-1700, "Standard Review Plan for Evaluating Nuclear Power Reactor License Termination Plans." Table 1 below describes the general applicability of these documents. NUREG-1537, "Guidelines for Preparing and Reviewing Applications for the Licensing of Non-Power Reactors," contains guidance for non-power reactor licensees and NRC staff, which includes a section on decommissioning and license termination for non-power reactors.

Table 1. **Contents and Applicability of Key Decommissioning Guidance Documents**

Volume and Status [1]	Title	Licensees to Which the Guidance Applies
NUREG-1757, Vol. 1, Rev. 2; September 2006	Consolidated Decommissioning Guidance: Decommissioning Process for Materials Licensees	Fuel cycle, fuel storage, and materials licensees.[2] Limited applicability to reactor licensees (see text below).
NUREG-1757, Vol. 2, Rev. 1; September 2006	Consolidated Decommissioning Guidance: Characterization, Survey, and Determination of Radiological Criteria	All licensees that are subject to the LTR (fuel cycle, fuel storage, materials, and reactor licensees).
NUREG-1757, Vol. 3; September 2003	Consolidated NMSS Decommissioning Guidance: Financial Assurance, Recordkeeping, and Timeliness	Fuel cycle, fuel storage, and materials licensees.[2]
NUREG-1700, Rev. 1; April 2003	Standard Review Plan for Evaluating Nuclear Power Reactor License Termination Plans	Power reactor licensees.

1 Versions listed are current as of September 2006. Please refer to the NRC's Public Electronic Reading Room at http://www.nrc.gov/reading-rm/doc-collections/nuregs to obtain the most up-to-date version.

2 Licensees regulated under 10 CFR Parts 30, 40, 60, 61, 63, 70, and 72 (for 10 CFR Parts 60, 61, and 63, only the ancillary surface facilities that support radioactive waste disposal activities). Because uranium recovery facilities are not subject to 10 CFR Part 20, Subpart E, refer to NUREG-1620, Rev. 1, Section 5, for decommissioning guidance for uranium recovery facilities that are subject to 10 CFR 40, Appendix A.

The current document, NUREG-1757, Volume 1, Revision 2, was intended to be applicable only to materials licensees. However, parts of this Volume are also applicable to reactor licensees, and the most relevant sections are listed in Table 2.

Table 2. **Sections of NUREG-1757, Volume 1, That Are Applicable to Reactor Licensees**

Section	Title
Chapter 6	Radiological Criteria for Decommissioning
Section 15.4	Decommissioning Surveys
Section 15.7	National Environmental Policy Act Compliance
Section 15.11.1	Current NRC Approach to Releases of Solid Materials
Section 15.12	Onsite Disposal of Radioactive Materials Under 10 CFR 20.2002
Section 15.13	Use of Intentional Mixing of Contaminated Soil
Section 17.7	Restricted Use
Section 17.8	Alternate Criteria
Appendix B	Screening Values
Appendix H	EPA/NRC Memorandum of Understanding
Appendix M	Overview of the Restricted Use and Alternate Criteria Provisions of 10 CFR Part 20, Subpart E

Revision 1 of Volume 1 was published in September 2003. As mentioned above, the current document, Revision 2 of Volume 1, incorporates changes based on finalizing the guidance of draft Supplement 1. Table 3 describes the most significant changes to the guidance in this volume.

Table 3.   Summary of Major Changes to Volume 1, Revision 2

Subject	Affected Sections
Restricted Use and Institutional Controls	Section 17.7 Section 17.8 Appendix M
Onsite Disposal of Radioactive Materials under 10 CFR 20.2002	NEW Section 15.12
Intentional Mixing of Contaminated Soil	NEW Section 15.13 Section 17.1.3
Removal of Material after License Termination	NEW Section 15.11.1
Other Issues and Changes	Section 5.2

NUREG-1757 is intended for use by applicants, licensees, NRC license reviewers, and other NRC personnel. It is also available to Agreement States and the public.

This NUREG is not a substitute for NRC regulations, and compliance with it is not required. The NUREG describes approaches that are acceptable to NRC staff. However, methods and solutions different than those in this NUREG will be acceptable, if they provide a basis for concluding that the decommissioning actions are in compliance with NRC regulations.

Larry W. Camper, Director
Division of Waste Management and Environmental Protection
Office of Nuclear Material Safety and Safeguards

ACKNOWLEDGMENTS

The writing team thanks the individuals listed below for assisting in the development and review of this revision of the report. All participants provided valuable insights, observations, and recommendations.

The team thanks Justine Cowan, Loleta Dixon, and Agi Seaton of Computer Sciences Corporation.

The Participants

Nuclear Regulatory Commission Staff

Cameron, Francis X.
Cameron, Jamnes L.
Flanders, Scott C.
Gillen, Daniel M.
Hull, John T.
Isaac, Patrick J.
Jensen, E. Neil
Leslie, Bret W.
McConnell, Keith I.
Orlando, Dominick A.
Ott, William R.
Persinko, Andrew
Smith, Brooke G.
Spitzberg, Blair B.
Treby, Stuart A.

Organization of Agreement States

Cortez, Ruben (Texas Department of State Health Services)
Galloway, Gwyn (Utah Department of Environmental Quality)
Helmer, Stephen (Ohio Department of Health)
Young, Robert N. (Tennessee Department of Environment and Conservation)

Conference of Radiation Control Program Directors
Hsu, Stephen (California Department of Health Services)

ABBREVIATIONS

> The following terms are defined for the purposes of this three-volume NUREG report.

ACAP	Alternative Cover Assessment Program
ADAMS	Agencywide Documents Access and Management System
AEA	Atomic Energy Act (of 1954, as amended)
AEC	U.S. Atomic Energy Commission (became Energy Resource Development Agency and Nuclear Regulatory Commission)
ALARA	As low as is reasonably achievable
ALCD	Alternative Landfill Cover Demonstration
ANSI	American National Standards Institute
APF	Assigned Protection Factors
ASME	American Society of Mechanical Engineers
ASTM	American Society for Testing and Materials
Bq	becquerel
BRT	Bankruptcy Review Team
BTP	Branch Technical Position
CAM	Continuous Air Monitor
CATX	Categorical Exclusion
CEDE	Committed Effective Dose Equivalent
CEQ	Council on Environmental Quality
CERCLA	Comprehensive Environmental Response, Compensation, and Liability Act
CFR	Code of Federal Regulations
Ci	curie
cpm	counts per minute
DCD	Decommissioning Directorate (Nuclear Regulatory Commission)
DCGLs	Derived Concentration Guideline Levels
DFP	Decommissioning Funding Plan
DOE	U.S. Department of Energy
DOT	U.S. Department of Transportation

ABBREVIATIONS

DP	Decommissioning Plan
dpm	disintegrations per minute
DQA	Data Quality Assessment
DQO	Data Quality Objective
DWMEP	Division of Waste Management and Environmental Protection (Nuclear Regulatory Commission)
EA	Environmental Assessment
Eh	redox potential
EIS	Environmental Impact Statement
EMC	Elevated Measurement Comparison
EML	DOE Environmental Measurements Laboratory (formerly the Health and Safety Laboratory)
EPA	U.S. Environmental Protection Agency
EPAD	Environmental and Performance Assessment Directorate (Nuclear Regulatory Commission)
EPA/NRC MOU	Memorandum of Understanding between the Environmental Protection Agency and the Nuclear Regulatory Commission dated October 9, 2002
ER	Environmental Report
FEP	Feature, Event, and/or Process
FFIEC	Federal Financial Institutions Examination Council
FHLM	Federal Home Loan Mortgage Corporation
FNMA	Federal National Mortgage Association
FONSI	Finding of No Significant Impact
FR	*Federal Register*
FSS	Final Status Survey
FSSP	Final Status Survey Plan
FSSR	Final Status Survey Report
FUSRAP	Formerly Utilized Sites Remedial Action Program
GEIS	Generic Environmental Impact Statement
GNMA	Government National Mortgage Association

GPO	Government Printing Office
HEPA	high-efficiency particulate air
HSA	Historical Site Assessment
IC	Institutional Control
ICRP	International Commission on Radiological Protection
IMC	Inspection Manual Chapter
IMNS	Division of Industrial and Medical Nuclear Safety (Nuclear Regulatory Commission)
IP	Inspection Procedure
IROFS	Items Relied on for Safety
ISA	Integrated Safety Analysis
ISCORS	Interagency Steering Committee on Radiation Standards
ISFSI	Independent Spent Fuel Storage Installation
ISO	International Organization for Standardization
LA	License Amendment
LA/RC	legal agreement and restrictive covenant
LBGR	Lower Bound [of the] Gray Region
LLD	lower limit of detection
LPDR	Local Public Document Room
LTC	long-term control
LTP	License Termination Plan
LTR	License Termination Rule
MARLAP	Multi-Agency Radiological Laboratory Analytical Protocols Manual
MARRSIM	Multi-Agency Radiological Survey and Site Investigation Manual (NUREG-1575)
mCi	millicurie
MCL	Maximum Contaminant Level
MDA	Minimum Detectable Activity
MDC	Minimum Detectable Concentration
MIP	Master Inspection Plan

ABBREVIATIONS

MOU	Memorandum of Understanding
mrem	millirem
mSv	millisievert
NAIC	National Association of Insurance Commissioners
NAS	National Academy of Sciences
NCRP	National Council on Radiation Protection and Measurements
NCS	Nuclear Criticality Safety
NCSA	Nuclear Criticality Safety Analysis
NEPA	National Environmental Policy Act
NIST	National Institute of Standards and Technology
NMMSS	Nuclear Materials Management and Safeguards System
NMSS	Office of Nuclear Material Safety and Safeguards (Nuclear Regulatory Commission)[2]
NOAA	National Oceanic and Atmospheric Administration
NORM	Naturally Occurring Radioactive Material
NRC	U.S. Nuclear Regulatory Commission
OC	Office of Controller
OCC	Office of the Comptroller of the Currency
OE	Office of Enforcement (Nuclear Regulatory Commission)
OGC	Office of General Counsel (Nuclear Regulatory Commission)
OSHA	U.S. Occupational Safety and Health Administration
PCBs	Polychlorinated Biphenyls
pCi	picocurie
PDF	Probability Density Function
PDR	Public Document Room
P&GD	Policy and Guidance Directive

---

[2] As of September 2006, NRC is planning to reorganize NMSS and STP to create two new offices: the Office of Federal and State Materials and Environmental Management Programs, which will focus on materials programs; and the new NMSS, which will focus on fuel cycle programs. This reorganization is scheduled to take effect on October 1, 2006. This document contains references to NMSS and STP. These references will be updated in future revisions of this document.

pH	hydrogen power
PM	Project Manager
PMF	probable maximum flood
PMP	probable maximum precipitation
PPE	personal protective equipment
PSR	Partial Site Release
QA	Quality Assurance
QAPP	Quality Assurance Project Plan
QA/QC	Quality Assurance and Quality Control
RAI	Request for Additional Information
RCRA	Resource Conservation and Recovery Act
REMP	Radiological Environmental Monitoring Program
RF	Resuspension Factor
RG	Regulatory Guide (also known as Reg Guide)
RIS	Regulatory Issue Summary
ROD	Record of Decision
RSO	Radiation Safety Officer
RSSI	Radiation Site Survey and Investigation [Process]
RWP	Radiation Work Permit
SCP	Site Characterization Plan
SCR	Site Characterization Report
SDMP	Site Decommissioning Management Plan
SDWA	Safe Drinking Water Act
SER	Safety Evaluation Report
SOPs	Standard Operating Procedures
SRP	[NMSS Decommissioning] Standard Review Plan (NUREG–1727)
SSAB	site-specific advisory board

ABBREVIATIONS

STP	[Office of] State and Tribal Programs (Nuclear Regulatory Commission)[3]
Sv	sievert
TAR	Technical Assistance Request
TDS	Total Dissolved Solids
TEDE	Total Effective Dose Equivalent
TENORM	Technologically Enhanced Naturally Occurring Radioactive Material
TI	Transport Index
TLD	Thermoluminescent Dosimeter
TOC	Total Organic Carbon
TODE	Total Organ Dose Equivalent
TRU	Transuranic(s) [radionuclides]
UECA	Uniform Environmental Covenants Act
UMTRA	Uranium Mill Tailings Remedial Action
UMTRCA	Uranium Mill Tailings Radiation Control Act
USACE	U.S. Army Corps of Engineers
U.S.C.	U.S. Code
USDA	U.S. Department of Agriculture
USGS	U.S. Geological Survey
WAC	waste acceptance criteria
WRS	Wilcoxon Rank Sum [test]

---

[3] As of September 2006, NRC is planning to reorganize NMSS and STP to create two new offices: the Office of Federal and State Materials and Environmental Management Programs, which will focus on materials programs; and the new NMSS, which will focus on fuel cycle programs. This reorganization is scheduled to take effect on October 1, 2006. This document contains references to NMSS and STP. These references will be updated in future revisions of this document.

GLOSSARY

The following terms are defined for the purposes of this three-volume NUREG report.

Acceptance Review. The evaluation the NRC staff performs upon receipt of a license amendment request to determine if the information provided in the document is sufficient to begin the technical review.

Activity. The rate of disintegration (transformation) or decay of radioactive material. The units of activity are the curie (Ci) and the becquerel (Bq) (see 10 CFR 20.1003).

Affected parties. Representatives of a broad cross-section of individuals and institutions in the community or vicinity of a site that may be affected by the decommissioning of the site.

ALARA. Acronym for "as low as is reasonably achievable," which means making every reasonable effort to maintain exposures to radiation as far below the dose limits as is practical, consistent with the purpose for which the licensed activity is undertaken, and taking into account the state of technology, the economics of improvements in relation to the state of technology, the economics of improvements in relation to the benefits to the public health and safety, and other societal and socioeconomic considerations, and in relation to utilization of nuclear energy and licensed materials in the public interest (see 10 CFR 20.1003).

Alternate Criteria. Dose criteria for residual radioactivity that are greater than the dose criteria described in 10 CFR 20.1402 and 20.1403, as allowed in 10 CFR 20.1404. Alternate criteria must be approved by the Commission.

Aquifer. A geologic formation, group of formations, or part of a formation capable of yielding a significant amount of ground water to wells or springs.

Background Radiation. Radiation from cosmic sources, naturally occurring radioactive material, including radon (except as a decay product of source or special nuclear material) and global fallout as it exists in the environment from the testing of nuclear explosive devices or from past nuclear accidents such as Chernobyl that contribute to background radiation and are not under the control of the licensee. Background radiation does not include radiation from source, byproduct, or special nuclear materials regulated by NRC (see 10 CFR 20.1003).

Broad Scope Licenses. A type of specific license authorizing receipt, acquisition, ownership, possession, use, and transfer of any chemical or physical form of the byproduct material specified in the license, but not exceeding quantities specified in the license. The requirements for specific domestic licenses of broad scope for byproduct material are found in 10 CFR Part 33. Examples of broad scope licensees are facilities such as large universities and large research and development facilities.

Byproduct Material. (1) Any radioactive material (except special nuclear material) yielded in, or made radioactive by, exposure to the radiation incident to the process of producing or utilizing special nuclear material; and (2) the tailings or wastes produced by the extraction or concentration of uranium or thorium from ore processed primarily for its source material content, including discrete surface wastes resulting from uranium solution extraction processes (see 10 CFR 20.1003).

Categorical Exclusion (CATX). A category of regulatory actions which do not individually or cumulatively have a significant effect on the human environment and which the Commission has found to have no such effect in accordance with procedures set out in 10 CFR 51.22 and for which, therefore, neither an environmental assessment nor an environmental impact statement is required (see 10 CFR 51.14(a)).

Certification Amount of Financial Assurance. See *prescribed amount of financial assurance.*

Certification of Financial Assurance. The document submitted to certify that financial assurance has been provided as required by regulation.

Characterization survey. A type of survey that includes facility or site sampling, monitoring, and analysis activities to determine the extent and nature of residual radioactivity. Characterization surveys provide the basis for acquiring necessary technical information to develop, analyze, and select appropriate cleanup techniques.

Cleanup. See *decontamination.*

Closeout Inspection. An inspection performed by NRC, or its contractor, to determine if a licensee has adequately decommissioned its facility. Typically, a closeout inspection is performed after the licensee has demonstrated that its facility is suitable for release in accordance with NRC requirements.

Confirmatory Survey. A survey conducted by NRC, or its contractor, to verify the results of the licensee's final status survey. Typically, confirmatory surveys consist of measurements at a fraction of the locations previously surveyed by the licensee, to determine whether the licensee's results are valid and reproducible.

Critical Group. The group of individuals reasonably expected to receive the greatest exposure to residual radioactivity for any applicable set of circumstances (see 10 CFR 20.1003).

DandD code. The Decontamination and Decommissioning (DandD) software package, developed by NRC, that addresses compliance with the dose criteria of 10 CFR 20, Subpart E. Specifically, DandD embodies NRC's guidance on screening dose assessments to allow licensees to perform simple estimates of the annual dose from residual radioactivity in soils and on building surfaces.

Decommission. To remove a facility or site safely from service and reduce residual radioactivity to a level that permits (1) release of the property for unrestricted use and termination of the license or (2) release of the property under restricted conditions and termination of the license (see 10 CFR 20.1003).

Decommission Funding Plan (DFP). A document that contains a site-specific cost estimate for decommissioning, describes the method for assuring funds for decommissioning, describes the means for adjusting both the cost estimate and funding level over the life of the facility, and contains the certification of financial assurance and the signed originals of the financial instruments provided as financial assurance.

Decommissioning Groups. For the purposes of this guidance document, the categories of decommissioning activities that depend on the type of operation and the residual radioactivity.

Decommissioning Plan (DP). A detailed description of the activities that the licensee intends to use to assess the radiological status of its facility, to remove radioactivity attributable to licensed operations at its facility to levels that permit release of the site in accordance with NRC's regulations and termination of the license, and to demonstrate that the facility meets NRC's requirements for release. A DP typically consists of several interrelated components, including (1) site characterization information; (2) a remediation plan that has several components, including a description of remediation tasks, a health and safety plan, and a quality assurance plan; (3) site-specific cost estimates for the decommissioning; and (4) a final status survey plan (see 10 CFR 30.36(g)(4).

Decontamination. The removal of undesired residual radioactivity from facilities, soils, or equipment prior to the release of a site or facility and termination of a license. Also known as remediation, remedial action, and cleanup.

Derived Concentration Guideline Levels (DCGLs). Radionuclide-specific concentration limits used by the licensee during decommissioning to achieve the regulatory dose standard that permits the release of the property and termination of the license. The DCGL applicable to the average concentration over a survey unit is called the $DCGL_W$. The DCGL applicable to limited areas of elevated concentrations within a survey unit is called the $DCGL_{EMC}$.

Dose (or *radiation dose*). A generic term that means absorbed dose, dose equivalent, effective dose equivalent, committed dose equivalent, committed effective dose equivalent, or total effective dose equivalent, as defined in other paragraphs of 10 CFR 20.1003 (see 10 CFR 20.1003). In this NUREG report, dose generally refers to *total effective dose equivalent (TEDE)*.

Durable institutional controls. A legally enforceable mechanism for restricting land uses to meet the radiological criteria for license termination (10 CFR 20, Subpart E). Durable institutional controls are reliable and sustainable for the time period needed.

Effluent. Material discharged into the environment from licensed operations.

GLOSSARY

Environmental Assessment. A concise public document for which the Commission is responsible that serves to (1) briefly provide sufficient evidence and analysis for determining whether to prepare an environmental impact statement or a finding of no significant impact, (2) aid the Commission's compliance with NEPA when no environmental impact statement is necessary, and (3) facilitate preparation of an environmental impact statement when one is necessary (see 10 CFR 51.14(a)).

Environmental Impact Statement. A detailed written document that ensures the policies and goals defined in the NEPA are considered in the actions of the Federal government. It discusses significant impacts and reasonable alternatives to the proposed action.

Environmental Monitoring. The process of sampling and analyzing environmental media in and around a facility (1) to confirm compliance with performance objectives and (2) to detect radioactive material entering the environment to facilitate timely remedial action.

Environmental Report (ER). A document submitted to the NRC by an applicant for a license amendment request (see 10 CFR 51.14(a)). The ER is used by NRC staff to prepare environmental assessments and environmental impact statements. The requirements for ERs are specified in 10 CFR 51.45–51.69.

Exposure Pathway. The route by which radioactivity travels through the environment to eventually cause radiation exposure to a person or group.

Exposure Scenario. A description of the future land uses, human activities, and behavior of the natural system as related to a future human receptor's interaction with (and therefore exposure to) residual radioactivity. In particular, the exposure scenario describes where humans may be exposed to residual radioactivity in the environment, what exposure group habits determine exposure, and how residual radioactivity moves through the environment.

External Dose. That portion of the dose equivalent received from radiation sources outside the body (see 10 CFR 20.1003).

Final Status Survey (FSS). Measurements and sampling to describe the radiological conditions of a site or facility, following completion of decontamination activities (if any) and in preparation for release of the site or facility.

Final Status Survey Plan (FSSP). The description of the final status survey design.

Final Status Survey Report (FSSR). The results of the final status survey conducted by a licensee to demonstrate the radiological status of its facility. The FSSR is submitted to NRC for review and approval.

Financial Assurance. A guarantee or other financial arrangement provided by a licensee that funds for decommissioning will be available when needed. This is in addition to the licensee's regulatory obligation to decommission its facilities.

Financial Assurance Mechanism. Financial instruments used to provide financial assurance for decommissioning.

Floodplain. The lowland and relatively flat areas adjoining inland and coastal waters including flood-prone areas of offshore islands. Areas subject to a one percent or greater chance of flooding in any given year are included (see 10 CFR 72.3).

Footprint. The portion of a site undergoing decommissioning, which is comprised of all of the areas of soil containing residual radioactivity, where intentional mixing is proposed to meet the release criteria.

General Licenses. Licenses that are effective without the filing of applications with NRC or the issuance of licensing documents to particular persons. The requirements for general licenses are found in 10 CFR Parts 30 and 31. Examples of items for which general licenses are issued are gauges and smoke detectors.

Ground Water. Water contained in pores or fractures in either the unsaturated or saturated zones below ground level.

Historical Site Assessment (HSA). The identification of potential, likely, or known sources of radioactive material and radioactive contamination based on existing or derived information for the purpose of classifying a facility or site, or parts thereof, as impacted or non-impacted (see 10 CFR 50.2).

Hydraulic Conductivity. The volume of water that will move through a medium in a unit of time under a unit hydraulic gradient through a unit area measured perpendicular to the direction of flow.

Hydrology. Study of the properties, distribution, and circulation of water on the surface of the land, in the soil and underlying rocks, and in the atmosphere.

Impact. The positive or negative effect of an action (past, present, or future) on the natural environment (land use, air quality, water resources, geological resources, ecological resources, aesthetic and scenic resources) and the human environment (infrastructure, economics, social, and cultural).

Impacted Areas. The areas with some reasonable potential for residual radioactivity in excess of natural background or fallout levels (see 10 CFR 50.2).

Inactive Outdoor Area. The outdoor portion of a site not used for licensed activities or materials for 24 months or more.

Infiltration. The process of water entering the soil at the ground surface. Infiltration becomes percolation when water has moved below the depth at which it can be removed (to return to the atmosphere) by evaporation or transpiration.

Institutional Controls. Measures to control access to a site and minimize disturbances to engineered measures established by the licensee to control the residual radioactivity. Institutional controls include administrative mechanisms (e.g., land use restrictions) and may include, but are not limited to, physical controls (e.g., signs, markers, landscaping, and fences).

Karst. A type of topography that is formed over limestone, dolomite, or gypsum by dissolution, characterized by sinkholes, caves, and underground drainage.

Leak Test. A test for leakage of radioactivity from sealed radioactive sources. These tests are made when the sealed source is received and on a regular schedule thereafter. The frequency is usually specified in the sealed source and device registration certificate and/or license.

Legacy site. An existing decommissioning site that is complex and difficult to decommission for a variety of financial, technical, or programmatic reasons.

License Termination Plan (LTP). A detailed description of the activities a reactor licensee intends to use to assess the radiological status of its facility, to remove radioactivity attributable to licensed operations at its facility to levels that permit release of the site in accordance with NRC's regulations and termination of the license, and to demonstrate that the facility meets NRC's requirements for release. An LTP consists of several interrelated components including: (1) a site characterization; (2) identification of remaining dismantlement activities; (3) plans for site remediation; (4) detailed plans for the final radiation survey; (5) a description of the end use of the facility, if restricted; (6) an updated site-specific estimate of remaining decommissioning costs; and (7) a supplement to the environmental report, pursuant to 10 CFR 51.33, describing any new information or significant environmental change associated with the licensee's proposed termination activities (see 10 CFR 50.82).

License Termination Rule (LTR). The License Termination Rule refers to the final rule on "Radiological Criteria for License Termination," published by NRC as Subpart E to 10 CFR 20 on July 21, 1997 (62 FR 39058).

Licensee. A person who possesses a license, or a person who possesses licensable material, who NRC could require to obtain a license.

MARSSIM. The *Multi-Agency Radiation Site Survey and Investigation Manual (NUREG–1575)* is a multi-agency consensus manual that provides information on planning, conducting, evaluating, and documenting building surface and surface soil final status radiological surveys for demonstrating compliance with dose- or risk-based regulations or standards.

Model. A simplified representation of an object or natural phenomenon. The model can be in many possible forms, such as a set of equations or a physical, miniature version of an object or system constructed to allow estimates of the behavior of the actual object or phenomenon when the values of certain variables are changed. Important environmental models include those estimating the transport, dispersion, and fate of chemicals in the environment.

Monitoring. Monitoring (radiation monitoring, radiation protection monitoring) is the measurement of radiation levels, concentrations, surface area concentrations, or quantities of radioactive material and the use of the results of these measurements to evaluate potential exposures and doses (see 10 CFR 20.1003).

mrem/y (millirem per year). One one-thousandth (0.001) of a rem per year. (See also *sievert.*)

National Environmental Policy Act (NEPA). The National Environmental Policy Act of 1969, which requires Federal agencies, as part of their decision-making process, to consider the environmental impacts of actions under their jurisdiction. Both the Council on Environmental Quality (CEQ) and NRC have promulgated regulations to implement NEPA requirements. CEQ regulations are contained in 40 CFR Parts 1500 to 1508, and NRC requirements are provided in 10 CFR Part 51.

Naturally Occurring Radioactive Material (NORM). The natural radioactivity in rocks, soils, air and water. NORM generally refers to materials in which the radionuclide concentrations have not been enhanced by or as a result of human practices. NORM does not include uranium or thorium in source material.

Non-impacted Areas. The areas with no reasonable potential for residual radioactivity in excess of natural background or fallout levels (see 10 CFR 50.2).

Pathway. See *exposure pathway.*

Performance-Based Approach. Regulatory decisionmaking that relies upon measurable or calculable outcomes (i.e., performance results) to be met, but provides more flexibility to the licensee as to the means of meeting those outcomes.

Permeability. The ability of a material to transmit fluid through its pores when subjected to a difference in head (pressure gradient). Permeability depends on the substance transmitted (oil, air, water, and so forth) and on the size and shape of the pores, joints, and fractures in the medium and the manner in which they are interconnected.

Porosity. The ratio of openings, or voids, to the total volume of a soil or rock expressed as a decimal fraction or as a percentage.

Potentiometric Surface. The two-dimensional surface that describes the elevation of the water table. In an unconfined aquifer, the potentiometric surface is at the top of the water level. In a confined aquifer, the potentiometric surface is above the top of the water level because the water is under confining pressure.

Prescribed Amount of Financial Assurance. An amount of financial assurance based on the authorized possession limits of the NRC license, as specified in 10 CFR 30.35(d), 40.36(b), or 70.25(d).

GLOSSARY

Principal Activities. Activities authorized by the license which are essential to achieving the purpose(s) for which the license was issued or amended. Storage during which no licensed material is accessed for use or disposal and activities incidental to decontamination or decommissioning are not principal activities (see 10 CFR 30.4).

Probabilistic. Refers to computer codes or analyses that use a random sampling method to select parameter values from a distribution. Results of the calculations are also in the form of a distribution of values. The results of the calculation do not typically include the probability of the scenario occurring.

Reasonable Alternatives. Those alternatives that are practical or feasible from a technical and economic standpoint.

Reasonably foreseeable land use. Land use scenarios that are likely within 100 years, considering advice from land use planners and stakeholders on land use plans and trends.

rem. The special unit of any of the quantities expressed as dose equivalent. The dose equivalent in rems is equal to the absorbed dose in rads multiplied by the quality factor (1 rem = 0.01 sievert) (see 10 CFR 20.1004).

Remedial Action. See *decontamination*.

Remediation. See *decontamination*.

Residual Radioactivity. Radioactivity in structures, materials, soils, ground water, and other media at a site resulting from activities under the licensee's control. This includes radioactivity from all licensed and unlicensed sources used by the licensee, but excludes background radiation. It also includes radioactive materials remaining at the site as a result of routine or accidental releases of radioactive material at the site and previous burials at the site, even if those burials were made in accordance with the provisions of 10 CFR Part 20 (see 10 CFR 20.1003).

RESRAD Code. A computer code developed by the U.S. Department of Energy and designed to estimate radiation doses and risks from RESidual RADioactive materials in soils.

RESRAD-BUILD Code. A computer code developed by the U.S. Department of Energy and designed to estimate radiation doses and risks from RESidual RADioactive materials in BUILDings.

Restricted Area. Any area to which access is limited by a licensee for the purpose of protecting individuals against undue risks from exposure to radiation and radioactive materials (see 10 CFR 20.1003).

Risk. Defined by the "risk triplet" of a scenario (a combination of events and/or conditions that could occur) or set of scenarios, the probability that the scenario could occur, and the consequence (e.g., dose to an individual) if the scenario were to occur.

Risk-Based Approach. Regulatory decision making that is based solely on the numerical results of a risk assessment. (Note that the Commission does not endorse a risk-based regulatory approach.)

Risk-Informed Approach. Regulatory decision making that represents a philosophy whereby risk insights are considered together with other factors to establish requirements that better focus licensee and regulatory attention on design and operational issues commensurate with their importance to public health and safety.

Risk Insights. Results and findings that come from risk assessments.

Robust engineered barrier. A man-made structure that is designed to mitigate the effect of natural processes or human uses that may initiate or accelerate release of residual radioactivity through environmental pathways. The structure is designed so that the radiological criteria for license termination (10 CFR 20, Subpart E) can be met. Robust engineered barriers are designed to be more substantial, reliable, and sustainable for the time period needed without reliance on active ongoing maintenance.

Safety Evaluation Report. NRC staff's evaluation of the radiological consequences of a licensee's proposed action to determine if that action can be accomplished safely.

Saturated Zone. That part of the earth's crust beneath the regional water table in which all voids, large and small, are ideally filled with water under pressure greater than atmospheric.

Scoping Survey. A type of survey that is conducted to identify (1) radionuclide contaminants, (2) relative radionuclide ratios, and (3) general levels and extent of residual radioactivity.

Screening Approach/Methodology/Process. The use of (1) predetermined building surface concentration and surface soil concentration values, or (2) a predetermined methodology (e.g., use of the DandD code) that meets the radiological decommissioning criteria without further analysis, to simplify decommissioning in cases where low levels of residual radioactivity are achievable.

Sealed Source. Any special nuclear material or byproduct material encased in a capsule designed to prevent leakage or escape of the material.

sievert (Sv). The SI unit of any of the quantities expressed as dose equivalent. The dose equivalent in sieverts is equal to the absorbed dose in grays multiplied by the quality factor (1 sievert = 100 rem) (see 10 CFR 20.1004).

Site. The area of land, along with structures and other facilities, as described in the original NRC license application, plus any property outside the originally licensed boundary added for the purpose of receiving, possessing, or using radioactive material at any time during the term of the license, as well as any property where radioactive material was used or possessed that has been released prior to license termination

Site Characterization. Studies that enable the licensee to sufficiently describe the conditions of the site, separate building, or outdoor area to evaluate the acceptability of the decommissioning plan.

Site Characterization Survey. See *characterization survey*.

Site Decommissioning Management Plan (SDMP). The program established by NRC in March 1990 to help ensure the timely cleanup of sites with limited progress in completing the remediation of the site and the termination of the facility license. SDMP sites typically have buildings, former waste disposal areas, large volumes of tailings, ground-water contamination, and soil contaminated with low levels of uranium or thorium or other radionuclides.

Site-Specific Dose Analysis. Any dose analysis that is done other than by using the default screening tools.

Smear. A radiation survey technique which is used to determine levels of removable surface contamination. A medium (typically filter paper) is rubbed over a surface (typically of area 100 cm^2), followed by a quantification of the activity on the medium. Also known as a swipe.

Source Material. Uranium or thorium, or any combination of uranium and thorium, in any physical or chemical form, or ores that contain by weight one-twentieth of one percent (0.05 percent) or more of uranium, thorium, or any combination of uranium and thorium. Source material does not include special nuclear material (see 10 CFR 20.1003).

Source Term. A conceptual representation of the residual radioactivity at a site or facility.

Special Nuclear Material. (1) Plutonium, uranium-233 (U-233), uranium enriched in the isotope 233 or in the isotope 235, and any other material that the Commission, pursuant to the provisions of Section 51 of the Atomic Energy Act, determines to be special nuclear material, but does not include source material; or (2) any material artificially enriched by any of the foregoing but does not include source material (see 10 CFR 20.1003).

Specific Licenses. Licenses issued to a named person who has filed an application for the license under the provisions of 10 CFR Parts 30, 32 through 36, 39, 40, 61, 70 and 72. Examples of specific licenses are industrial radiography, medical use, irradiators, and well logging.

Survey. An evaluation of the radiological conditions and potential hazards incident to the production, use, transfer, release, disposal, or presence of radioactive material or other sources of radiation. When appropriate, such an evaluation includes a physical survey of the location of radioactive material and measurements or calculations of levels of radiation, or concentrations or quantities of radioactive material present (see 10 CFR 20.1003).

Survey Unit. A geographical area consisting of structures or land areas of specified size and shape at a site for which a separate decision will be made as to whether or not the unit attains the site-specific reference-based cleanup standard for the designated pollution parameter. Survey units are generally formed by grouping contiguous site areas with similar use histories and having the same contamination potential (classification). Survey units are established to facilitate the survey process and the statistical analysis of survey data.

Technologically Enhanced Naturally Occurring Radioactive Material (TENORM). Naturally occurring radioactive material with radionuclide concentrations increased by or as a result of past or present human practices. TENORM does not include background radioactive material or the natural radioactivity of rocks and soils. TENORM does not include uranium or thorium in source material.

Timeliness. Specific time periods stated in NRC regulations for decommissioning unused portions of operating nuclear materials facilities and for decommissioning the entire site upon termination of operations.

Total Effective Dose Equivalent (TEDE). The sum of the deep-dose equivalent (for external exposures) and the committed effective dose equivalent (CEDE) (for internal exposures) (see 10 CFR 20.1003).

Transmissivity. The rate of flow of water through a vertical strip of aquifer which is one unit wide and which extends the full saturated depth of the aquifer.

Unrestricted Area. An area, access to which is neither limited nor controlled by the licensee (see 10 CFR 20.1003).

Unsaturated Zone. The subsurface zone in which the geological material contains both water and air in pore spaces. The top of the unsaturated zone typically is at the land surface, otherwise known as the vadose zone.

Vadose Zone. See *unsaturated zone.*

PART I: DECOMMISSIONING PROCESS AND DECOMMISSIONING GROUPS

1 PURPOSE OF REPORT AND DOCUMENT ROADMAP

1.1 PURPOSE OF THIS REPORT

The purpose of this volume is to:

- illustrate to licensees, and the general public, NMSS's decommissioning process;

- provide guidance to NRC licensees for terminating an NRC nuclear materials license and to make available methods, acceptable to NRC staff, for implementing specific parts of the Commission's decommissioning regulations;

- delineate techniques and criteria used by NRC staff in evaluating decommissioning actions;

- provide guidance to NRC staff overseeing NMSS decommissioning programs to evaluate a licensee's decommissioning actions; and

- maintain a risk-informed, performance-based, and flexible decommissioning approach.

This NUREG provides guidance regarding decommissioning leading to termination of a license. Licensees[4] decommissioning their facilities are required to demonstrate to NRC that their proposed methods will ensure that the decommissioning can be conducted safely and that the facility, at the completion of decommissioning activities, will comply with NRC requirements for license termination. The policies and procedures discussed in this NUREG should be used by NRC staff overseeing the decommissioning program at licensed fuel cycle, fuel storage, and materials sites to evaluate a licensee's decommissioning actions. The Foreword of this volume discusses the applicability of this NUREG to reactor and uranium recovery facilities. This NUREG is also intended to be used in conjunction with NRC Inspection Manual Chapter 2605, "Decommissioning Inspection Program for Fuel Cycle and Materials Licensees."

This NUREG is not a substitute for regulations, and compliance with it is not required. Methods and solutions different than those in this NUREG will be acceptable, if they provide a basis for concluding that the decommissioning actions are in compliance with the Commission's regulations.

1.2 APPROACH

A brief discussion of recent decommissioning regulatory history, an overview of the decommissioning process, and the License Termination Rule (LTR) are discussed in Chapters 4, 5, and 6, respectively.

NRC staff reviewed the numbers and types of licenses issued by the Commission and determined that the majority of licensees were those that used and possessed sealed sources or relatively limited amounts of unsealed radioactive material. Due to the amounts, forms, and types of

---

[4] For purposes of this document, the term "licensee" includes persons in possession of licensable material whom NRC could require to become a licensee.

radioactive material used by these licensees, it did not appear that most licensees would need to submit decommissioning plans (DPs) or perform complex remedial activities to decommission their facilities in accordance with NRC criteria.

However, certain licensees need to submit information regarding either (1) the status of their facilities when they request license termination or (2) the activities that they intend to use to remediate their facilities. The types of information required could range from very simple descriptions of the radiological status of the facilities and the disposition of radioactive material possessed by the licensees to, in the case of licensees who proposed license termination under restricted conditions, very detailed descriptions of institutional controls, dose estimates to potential future critical groups, and arrangements to ensure that adequate financial assurance mechanisms are in place at license termination in the form of a detailed DP.

Based on the above, NRC staff determined that the best approach would be to develop detailed descriptions of the types of information needed to evaluate proposed decommissioning activities and then tailor the information needed from the licensees based on the complexity and safety significance of the decommissioning project. As described in Section 1.3, this approach is implemented through several interactions between NRC staff and licensees.

1.3 DECOMMISSIONING ROADMAP

To implement the risk-informed, iterative approach, the staff developed decommissioning "groups," based on the complexity of the decommissioning and the decommissioning alternatives in the LTR. A roadmap to these groups is provided in Figure 1.1.

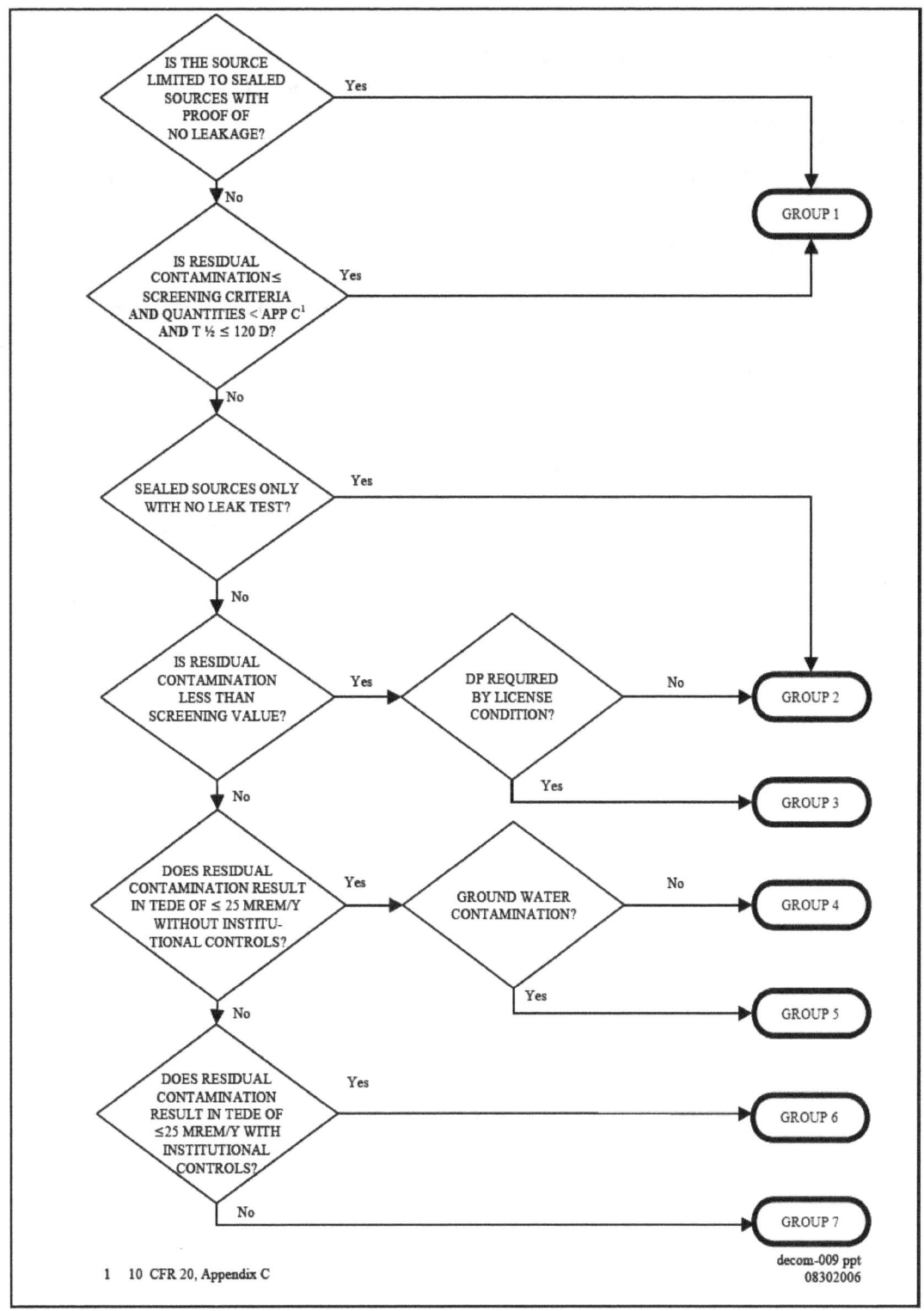

Figure 1.1 Determining the Appropriate Decommissioning Group.

The decommissioning process begins when the licensee determines that decommissioning of all or a portion of a site is necessary or desirable. Decommissioning of a site, or portion of a site, is necessary when certain site use conditions are met. These conditions, related to decommissioning timing, are explained in Chapter 5.

In the past, NRC staff classified facilities undergoing decommissioning by either the activities performed during the operation of the facilities or the types of licensed material possessed by the licensee. However, for purposes of this NUREG, the staff classified facilities undergoing decommissioning into seven (7) groups, based on the amount of residual radioactivity, the location of that material, and the complexity of the activities needed to decommission the site. Group 1 is typically a sealed source facility that has not experienced any leakage; Group 7 would be a large facility with contamination that would result in the license being terminated with restrictions on future site use and require an environmental impact statement (EIS) to support the action. Defined in Chapter 7, these groups have been created for convenience of analysis only and are not based on any specific regulatory requirements. A more detailed description of each group and the action necessary to decommission it are given in Chapters 8–14. For each decommissioning group, Table 1.1 offers an abridged description, illustrates examples, and identifies reference chapters in this document.

Table 1.1   Description, Examples, and Reference Chapters in this Document for Each Decommissioning Group

Group	Brief Description	Examples	Reference Chapter
1	See Section 8.1 for a complete description. Licensed material was not released into the environment, did not cause the activation of adjacent materials, and did not contaminate work areas.	Licensees who used only sealed sources such as radiographers and irradiators	8
2	See Section 9.1 for a complete description. Licensed material was used in a way that resulted in residual radioactivity on building surfaces and/or soils. The licensee is able to demonstrate that the site meets the screening criteria for unrestricted use.	Licensees who used only quantities of loose radioactive material that they routinely cleaned up (e.g., R&D facilities)	9
3	See Section 10.1 for a complete description. Licensed material was used in a way that could meet the screening criteria, but the license needs to be amended to modify or add procedures to remediate buildings or sites.	Licensees who may have occasionally released radioactivity within NRC limits (e.g., broad scope)	10
4	See Section 11.1 for a complete description. Licensed material was used in a way that resulted in residual radiological contamination of building surfaces or soils, or a combination of both (but not ground water). The licensee demonstrates that the site meets unrestricted use levels derived from site-specific dose modeling.	Licensees whose sites released loose or dissolved radioactive material within NRC limits and may have had some operational occurrences that resulted in releases above NRC limits (e.g., waste processors)	11
5	See Section 12.1 for a complete description. Licensed material was used in a way that resulted in residual radiological contamination of building surfaces, soils, or ground water, or a combination of all three. The licensee demonstrates that the site meets unrestricted use levels derived from site-specific dose modeling.	Licensees whose sites released, stored, or disposed of large amounts of loose or dissolved radioactive material onsite (e.g., fuel cycle facilities)	12
6	See Section 13.1 for a complete description. Licensed material was used in a way that resulted in residual radiological contamination of building surfaces, and/or soils, and possibly ground water. The licensee demonstrates that the site meets restricted use levels derived from site-specific dose modeling.	Licensees whose sites would cause more health and safety or environmental impact than could be justified when cleaning up to the unrestricted release limit (e.g., facilities where large inadvertent release(s) occurred)	13
7	See Section 14.1 for a complete description. Licensed material was used in a way that resulted in residual radiological contamination of building surfaces, and/or soils, and possibly ground water. The licensee demonstrates that the site meets alternate restricted use levels derived from site-specific dose modeling.	Licensees whose sites would cause more health and safety or environmental impact than could be justified when cleaning up to the restricted release limit (e.g., facilities where large inadvertent release(s) occurred)	14

Once the decision has been made to decommission, the next step is to determine what information the licensee needs to provide to demonstrate site conditions successfully. When NRC staff is informed that a licensee has decided to permanently cease licensed operations, or has not conducted licensed activities for a period greater than 24 months, and must decommission all or part of its facility, NRC staff should contact the licensee and determine if the licensee will need to submit a DP to support its request for license termination. If the licensee does not need to submit a DP, NRC staff should follow the guidance in this NUREG for that decommissioning group. Licensees needing to submit a DP should follow the requirements of the regulations briefly discussed in Chapters 10, 11, 12, 13, or 14, depending on the decommissioning group. Detailed descriptions of applicable portions of DP contents are found in Chapters 16–18 of this volume. Table 1.2 summarizes information needed for the staff to conduct its technical review for each of the decommissioning groups.

Table 1.2   Principal Regulatory Features of Decommissioning Groups

Description	GROUP 1	GROUP 2	GROUP 3	GROUP 4	GROUP 5	GROUP 6	GROUP 7
Description	Sealed source, screening criteria	Screening criteria, no DP	Screening criteria, DP	Site specific, no ground water contamination	Site specific, ground water contamination	Restricted release	Alternate criteria
NEPA Compliance[a]	Categorical Exclusion	EA	EA	EA	EA	EIS	EIS
Licensee Requests Release for Restricted or Unrestricted Use	Unrestricted use	Unrestricted use	Unrestricted use	Unrestricted use	Unrestricted use	Restricted Use	Restricted use
Decommissioning Plan Required	No	No	Yes	Yes	Yes	Yes	Yes
Decommissioning Plan Review Documentation	N/A	N/A	Letter to the licensee or Safety Evaluation Report	Safety Evaluation Report	Safety Evaluation Report	Safety Evaluation Report	Safety Evaluation Report
Radioactive Material Disposition Documentation	NRC Form 314 or equivalent	NRC Form 314 or equivalent	NRC Form 314 or equivalent	NRC Form 314 or equivalent	NRC Form 314 or equivalent	NRC Form 314 or equivalent	NRC Form 314 or equivalent
Method for Demonstrating Site is Suitable for Release	Survey or demonstration	Survey or demonstration	Survey or demonstration	Site specific	Site specific	Site specific	Site specific
Confirmatory or Side-by Side Survey	Not Customary	Depends on licensee's survey and radioactive material use at facility	Depends on licensee's survey and radioactive material use at facility	Yes	Yes	Yes	Yes
Closeout Inspection	No	As appropriate	As appropriate	Yes	Yes	Yes	Yes
Federal Register Notices used to Inform the Public of Staff Actions	No	Yes—(1) announce FONSI	Yes—(1) announce DP receipt and NRC's intended actions[b] and (2) announce FONSI	Yes—(1) announce DP receipt and NRC's intended actions[b] and (2) announce FONSI	Yes—(1) announce DP receipt and NRC's intended actions[b] and (2) announce FONSI	Yes—(1) announce DP receipt and NRC's intended actions[b] and (2) announce EIS	Yes—(1) announce DP receipt and NRC's intended actions[b] and (2) announce EIS
Documentation Used to Support License Termination	License Amendment	License Amendment	License Amendment	License Amendment	License Amendment	License Amendment	License Amendment

Notes

This table generally describes the major regulatory features of the different decommissioning groups. It does not describe all of the requirements, NRC staff actions, and licensee actions for each group, nor should it be used to determine the appropriate group. Licensees and NRC staff should refer to the detailed descriptions in each of the chapters of this NUREG report.

a   See NUREG–1748 for detailed guidance.

b   The *Federal Register* notice of license amendment for DP receipt provides opportunity for a hearing and opportunity for comment.

1.4   FURTHER INFORMATION FOR LICENSEES AND REVIEWERS

NRC staff should refer to this volume to identify the information to be submitted by the licensee for the staff to conduct its technical review and for what review actions the staff takes for each decommissioning group. A licensing review conducted using this volume of the NUREG is not intended to be a detailed evaluation of all aspects of facility decommissioning. NRC staff should use the approach outlined in this volume in a manner that allows for flexibility. The objectives of the review are to confirm that the decommissioning of the site will be accomplished in a manner consistent with applicable regulatory requirements. In conducting the evaluation, the staff should determine if the proposal submitted by the licensee is acceptable. In most cases, this involves assessing whether the methods and data used by the licensee in support of its proposal are acceptable, and if the results meet the requirements in 10 CFR Part 20, Subpart E.

NRC regulations indicate when a DP is required. Groups 1–2 do not generally require a DP, but Groups 3–7 do. For those that require a DP, the content of the DP is shown in Chapters 16–18, the Checklist in Appendix D, and in Volume 2 of this NUREG. Compliance with the environmental requirements of the National Environmental Policy Act (NEPA) and NRC environmental regulations (10 CFR Part 51) is explained in Section 15.7.

Details of requirements for dose modeling inspections and surveys are presented in Volume 2 of this NUREG. Details of "Financial Assurance, Recordkeeping, and Timeliness" are in Volume 3 of this NUREG. For a complete listing of documents used in the compilation of this work and the status of each at the time of publication, see Chapter 4.

2 REGULATORY AUTHORITY: NRC AND AGREEMENT STATES

Certain States, called Agreement States (see Figure 2.1), have entered into agreements with NRC that give them the authority to license and inspect byproduct, source, or limited quantities of special nuclear materials used or possessed within their borders. Any applicant other than a Federal Agency who wishes to possess or use licensed material in one of these Agreement States needs to contact the responsible officials in that State for guidance on preparing a license application.

In the special situation of work at Federally controlled sites in Agreement States, it is necessary to understand the jurisdictional status of the land in order to determine whether NRC or the Agreement State has regulatory authority. NRC has regulatory authority in areas determined to be of "exclusive Federal jurisdiction," while the Agreement State has jurisdiction in land areas of non-exclusive Federal jurisdiction. Licensees are responsible for determining the jurisdictional status of the specific areas where they plan to conduct licensed operations. NRC recommends that licensees ask their local contact for the Federal Agency controlling the site (e.g., contract officer, base environmental health officer, district office staff) to help determine the jurisdictional status of the land and to provide the information in writing, so that licensees can comply with NRC or Agreement State regulatory requirements, as appropriate. Additional guidance on determining jurisdictional status is found in All Agreement States Letter, SP–96–022, dated February 16, 1996, which is available from NRC upon request.

The Commission shall not discontinue regulatory authority of and shall retain regulatory responsibility for production, utilization, or enrichment facilities, and formula quantities (see 10 CFR 150.11) of special nuclear material.

Table 2.1 provides a quick way to check on which Agency has regulatory authority.

Table 2.1 Who Regulates the Activity?

Applicant and Proposed Location of Work	Regulatory Agency
Federal Agency regardless of location (except that Department of Energy [DOE] and, under most circumstances, its prime contractors are exempt from licensing)	NRC
Non-Federal entity in non-Agreement State, US territory, or possession	NRC
Non-Federal entity in Agreement State at non-Federally controlled site	Agreement State
Non-Federal entity in Agreement State at Federally controlled site not subject to exclusive Federal jurisdiction	Agreement State
Non-Federal entity in Agreement State at Federally controlled site subject to exclusive Federal jurisdiction	NRC

Figure 2.1 shows NRC's four Regional Offices and their respective geographical areas of responsibility for licensing purposes and identifies the Agreement States. Fuel cycle facility inspection activities have been consolidated into Region II, and Region II's nuclear materials licensing and inspection activities have been consolidated into Region I.

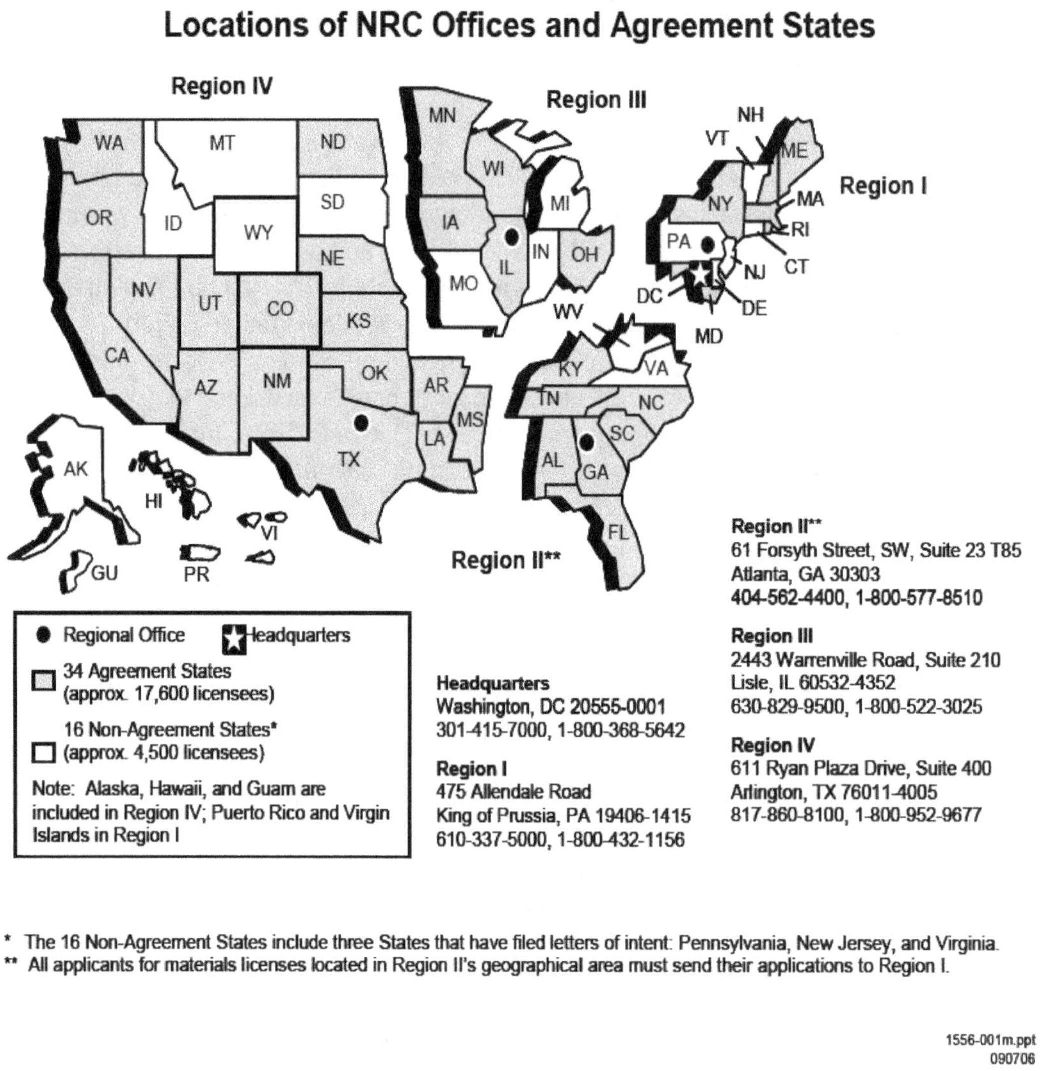
Locations of NRC Offices and Agreement States
Region IV
Region III
Region I
Region II**
WA
MT
ND
OR
ID
SD
WY
NV
UT
NE
CO
KS
CA
AZ
NM
OK
TX
AK
HI
GU
PR
VI
MN
WI
MI
IA
IL
IN
OH
MO
WV
AR
LA
MS
KY
TN
AL
GA
FL
SC
NC
VA
MD
DC
DE
NJ
PA
NY
VT
NH
ME
MA
RI
CT
Regional Office
Headquarters
34 Agreement States (approx. 17,600 licensees)
16 Non-Agreement States* (approx. 4,500 licensees)
Note: Alaska, Hawaii, and Guam are included in Region IV; Puerto Rico and Virgin Islands in Region I
Headquarters
Washington, DC 20555-0001
301-415-7000, 1-800-368-5642
Region I
475 Allendale Road
King of Prussia, PA 19406-1415
610-337-5000, 1-800-432-1156
Region II**
61 Forsyth Street, SW, Suite 23 T85
Atlanta, GA 30303
404-562-4400, 1-800-577-8510
Region III
2443 Warrenville Road, Suite 210
Lisle, IL 60532-4352
630-829-9500, 1-800-522-3025
Region IV
611 Ryan Plaza Drive, Suite 400
Arlington, TX 76011-4005
817-860-8100, 1-800-952-9677
* The 16 Non-Agreement States include three States that have filed letters of intent: Pennsylvania, New Jersey, and Virginia.
** All applicants for materials licenses located in Region II's geographical area must send their applications to Region I.
1556-001m.ppt
090706

Figure 2.1 Locations of NRC Offices and Agreement States in the United States.

Reference: A current list of Agreement States (including names, addresses, and telephone numbers of responsible officials) is available by choosing "Directories" on the NRC Office of State and Tribal Programs' (STP's) Home Page, http://www.hsrd.ornl.gov/nrc. As an alternative, request the list from NRC Regional Offices.

3 LICENSEE'S MANAGEMENT RESPONSIBILITY

NRC recognizes that effective radiation safety program management is vital to achieving safe and compliant operations. NRC believes that consistent compliance with its regulations provides reasonable assurance that licensed activities will be conducted safely.

> "Management" refers to the processes for conducting and controlling a radiation safety program and the individuals who are both responsible for those processes and authorized to provide the necessary resources to achieve regulatory compliance.

NRC and its licensees share a common responsibility to protect public health and safety. Federal regulations and the NRC regulatory program are important elements in the protection of the public. NRC licensees, however, are primarily responsible for safety using nuclear materials and have the primary responsibility for compliance with the LTR. To ensure adequate management involvement, a management representative must sign the license application, acknowledging management's commitment and responsibility for the following:

- radiation safety, security, control of radioactive materials, and compliance with regulations;
- completeness and accuracy of the radiation safety records and all information provided to NRC;
- knowledge about the contents of the license and application;
- meticulous compliance with current NRC and Department of Transportation (DOT) regulations and the licensee's operating and emergency procedures;
- provision of adequate resources (including space, equipment, personnel, time, and, if needed, contractors) to the radiation protection program to ensure that the public and workers are protected from radiation hazards and that meticulous compliance with regulations is maintained;
- selection and assignment of a qualified individual to serve as the Radiation Safety Officer (RSO) for licensed activities;
- prohibiting against discrimination of employees engaged in protected activities;
- provision of information to employees regarding the employee protection and deliberate misconduct provisions;
- obtaining NRC's prior written consent before transferring control of the license; and
- notifying the appropriate NRC Regional Administrator in writing, immediately following the filing of a petition for voluntary or involuntary bankruptcy.

For information on NRC inspection, investigation, enforcement, and other compliance programs, see the current version of "General Statement of Policy and Procedures for NRC Enforcement Actions," NUREG–1600. For hard copies of NUREG–1600, see the Notice of Availability (on the inside front cover of this report).

4 APPLICABLE REGULATIONS, GUIDANCE, AND REFERENCES

> It is the licensee's responsibility to obtain, understand, and abide by each applicable regulation and existing license condition.

4.1 DECOMMISSIONING REGULATORY HISTORY

On June 27, 1988, NRC amended its regulations in 10 CFR Parts 30, 40, 50, 70, and 72 to set forth the technical and financial criteria for decommissioning licensed nuclear facilities (53 *Federal Register* (FR) 24018). These regulations were further amended on July 26, 1993, to establish additional recordkeeping requirements for decommissioning (58 FR 39628); on July 15, 1994, to establish timeframes and schedules for the decommissioning of licensed nuclear facilities (59 FR 36026); and on July 26, 1995, to clarify that financial assurance requirements must be in place during operations and updated when licensed operations cease. NRC promulgated these amendments to ensure that the decommissioning of all licensed nuclear facilities is performed in a safe and timely manner and that adequate funds are available to ensure that the decommissioning of licensed facilities can be accomplished.

On July 21, 1997, NRC published the final rule on "Radiological Criteria for License Termination" (the License Termination Rule (LTR)) as Subpart E to 10 CFR Part 20 (62 FR 39058). The LTR establishes criteria for license termination. The criterion for termination with unrestricted release is residual radioactivity, which is distinguishable from background, results in a total effective dose equivalent (TEDE) to an average member of a critical group that does not exceed 0.25 millisievert per year (mSv/y) (25 mrem/y). In addition, the residual radioactivity has been reduced to levels that are as low as is reasonably achievable (ALARA). For license termination with restrictions on future land use, the LTR establishes criteria of 1.0 mSv/y (100 mrem/y) or 5.0 mSv/y (500 mrem/y) under certain conditions.

Supplemental information regarding implementation of the LTR was published by NRC in the *Federal Register* on November 18, 1998 (63 FR 64132), December 7, 1999 (64 FR 68395), and June 13, 2000 (65 FR 37186). This supplemental information established screening values for building surface contamination for beta/gamma radiation emitters, screening values for surface soil contamination, and clarifying information on the use of the screening values. These screening values correspond to levels of radionuclide contamination that would be deemed in compliance with the unrestricted use dose limit in 10 CFR 20.1402 (i.e., 0.25 mSv/y (25 mrem/y)).

4.2 STATUTES

NRC's decommissioning and environmental protection regulations derive their authority from the following statutes:

- Atomic Energy Act (AEA) of 1954, as amended;

- Energy Reorganization Act of 1974, as amended; and

- National Environmental Policy Act of 1969, as amended.

4.3 DECOMMISSIONING REGULATIONS

The following Parts of 10 CFR contain regulations applicable to decommissioning materials licenses:

- 10 CFR Part 2, "Rules of Practice for Domestic Licensing Proceedings and Issuance of Orders." Section 10 CFR 2.1205 discusses the public's opportunities to request hearings on licensing actions.

- 10 CFR Part 19, "Notices, Instructions and Reports to Workers: Inspection and Investigations."

- 10 CFR Part 20, "Standards for Protection Against Radiation," especially Subpart E – Radiological Criteria for License Termination. The requirements for release criteria are contained in 10 CFR 20.1402, 20.1403, and 20.1404. The requirements for final status surveys are contained in 10 CFR 20.1501(a).

- 10 CFR Part 30, "Rules of General Applicability to Domestic Licensing of Byproduct Material." Termination of licenses and decommissioning are discussed in 10 CFR 30.36. Financial assurance requirements are found in 10 CFR 30.35 and 30.36. Completeness and accuracy of the radiation safety records and information provided to NRC is addressed in 10 CFR 30.9.

- 10 CFR Part 40, "Domestic Licensing of Source Material." Termination of licenses and decommissioning are discussed in 10 CFR 40.42. Financial assurance requirements are found in 10 CFR 40.36 and 40.42. Completeness and accuracy of the radiation safety records and information provided to NRC is addressed in 10 CFR 40.9. Note that this NUREG does not apply to uranium recovery facilities.

- 10 CFR Part 51, "Environmental Protection Regulations for Domestic Licensing and Related Regulatory Functions."

- 10 CFR Part 70, "Domestic Licensing of Special Nuclear Material." Termination of licenses and decommissioning are discussed in 10 CFR 70.38. Financial assurance requirements are found in 10 CFR 70.25 and 70.38. Completeness and accuracy of the radiation safety records and information provided to NRC are addressed in 10 CFR 70.9.

- 10 CFR Part 71, "Packaging and Transportation of Radioactive Material." Part 71 requires that licensees or applicants who transport licensed material, or who may offer such material to a carrier for transport, must comply with the applicable requirements of the DOT that are found in 49 CFR Parts 170 through 189. Copies of DOT regulations can be ordered from the Government Printing Office (GPO), whose address and telephone number are listed in Section 4.11.

- 10 CFR Part 72, "Licensing Requirements for the Independent Storage of Spent Nuclear Fuel and High-Level Radioactive Waste." Termination of licenses and decommissioning are discussed in 10 CFR 72.54. Financial assurance requirements are found in 10 CFR 72.30 and 72.54. Criteria for decommissioning are found in 10 CFR 72.130. Completeness and accuracy of the radiation safety records and information provided to NRC are addressed in 10 CFR 72.11.

4.4   DECOMMISSIONING INSPECTION MANUAL CHAPTERS

- Nuclear Regulatory Commission (U.S.) (NRC). Inspection Manual Chapter 2605, "Decommissioning Procedures for Fuel Cycle and Materials Licensees." NRC: Washington, DC. November 1996.

- —————. Inspection Manual Chapter 2602, "Decommissioning Inspection Program for Fuel Cycle Facilities and Materials Licensees." NRC: Washington, DC. June 1997.

4.5   DECOMMISSIONING INSPECTION PROCEDURES

- Nuclear Regulatory Commission (U.S.) (NRC). Inspection Procedure 87104, "Decommissioning Inspection Procedure for Materials Licensees." NRC: Washington, DC. June 1997.

- —————. Inspection Procedure 88104, "Decommissioning Inspection Procedure for Fuel Cycle Facilities." NRC: Washington, DC. June 1997.

- —————. Inspection Procedure 83890, "Closeout Inspection and Survey." NRC: Washington, DC. March 1994.

- —————. Temporary Instruction 2800/026, "Follow-up Inspection of Formerly Licensed Sites Identified as Potentially Contaminated." NRC: Washington, DC. July 2000.

4.6   OTHER NRC DOCUMENTS REFERENCED IN THIS NUREG WITH APPLICATION OUTSIDE OF DECOMMISSIONING

- Nuclear Regulatory Commission (U.S.) (NRC). Branch Technical Position, "License Condition for Leak Testing Sealed Byproduct Material Sources." NRC: Washington, DC. April 1993.

- —————. Branch Technical Position, "License Condition for Leak Testing Sealed Plutonium Sources." NRC: Washington, DC. April 1993.

- —————. Branch Technical Position, "License Condition for Leak Testing Sealed Source Which Contains Alpha and/or Beta-Gamma Emitters." NRC: Washington, DC. April 1993.

- —————. Branch Technical Position, "License Condition for Leak Testing Sealed Uranium Sources." NRC: Washington, DC. April 1993.

- —————. "Guidelines for Decontamination of Facilities and Equipment Prior to Release for Unrestricted Use or Termination of Licenses for Byproduct, Source, or Special Nuclear Material." NRC, Division of Fuel Cycle Safety and Safeguards: Washington, DC. April 1993.

- —————. IE Circular No. 81–07, "Control of Radioactively Contaminated Material." NRC: Washington, DC. May 1981.

- —————. Information Notice 85–92, "Surveys of Wastes Before Disposal from Nuclear Reactor Facilities." NRC: Washington, DC. December 1985.

- —————. Information Notice 88–22, "Disposal of Sludge from Onsite Sewage Treatment Facilities at Nuclear Power Stations." NRC: Washington, DC. May 1988.

- —————. Information Notice 94–07, "Solubility Criteria for Liquid Effluent Releases to Sanitary Sewerage Under the Revised 10 CFR Part 20." NRC: Washington, DC. January 28, 1994.

- —————. Information Notice 94–23, "Guidance to Hazardous, Radioactive and Mixed Waste Generators on the Elements of a Waste Minimization Program." NRC: Washington, DC. March 25, 1994.

- —————. Information Notice 96–28, "Suggested Guidance Relating to Development and Implementation of Corrective Action." NRC: Washington, DC. May 1, 1996.

- —————. Information Notice 97–55, "Calculation of Surface Activity for Contaminated Equipment and Materials." NRC: Washington, DC. July 23, 1997.

- —————. Memorandum from William F. Kane, Director, Office of Nuclear Material Safety and Safeguards, and Samuel J. Collins, Director, Office of Nuclear Reactor Regulation, "Case Specific Licensing Decisions on Release of Solid Materials from Licensed Facilities." NRC: Washington, DC. August 7, 2000.

- —————. Memorandum from Donald A. Cool, Director, Division of Industrial and Medical Nuclear Safety, Office of Nuclear Material Safety and Safeguards, "Update of the August 7, 2000 Memo from William Kane, NMSS and Samuel Collins, NRR – Case-Specific Licensing Decisions on Release of Soils from Licensed Facilities." NRC: Washington, DC. July 27, 2001.

- —————. Memorandum from Donald A. Cool, Director, Division of Industrial and Medical Nuclear Safety, Office of Nuclear Material Safety and Safeguards, "Update on Case-Specific Licensing Decisions on Controlled Release of Concrete from Licensed Facilities." NRC: Washington, DC. December 27, 2002.

- —————. NUREG–0041, Rev. 1, "Manual of Respiratory Protection Against Airborne Radioactive Material." NRC: Washington, DC. October 1976.

- —————. NUREG–1460, Rev. 1, "Guide to NRC Reporting and Recordkeeping Requirements." NRC: Washington, DC. July 1994.

- —————. NUREG–1556. Vol. 15, "Guidance About Changes of Control and About Bankruptcy Involving Byproduct, Source or Special Nuclear Material Licenses." NRC: Washington, DC. November 2000.

- —————. NUREG/CR–5569, Rev. 1, "Health Physics Positions Data Base." NRC: Washington, DC. February 1994.

- —————. NUREG–1600, "General Statement of Policy and Procedures for NRC Enforcement Actions." NRC: Washington, DC. May 1, 2000.

- —————. NUREG–1660, "Specific Schedules of Requirements for Transport of Specified Types of Radioactive Material Consignments." NRC: Washington, DC. November 1998.

- —————. NUREG–1748, "Environmental Review Guidance for Licensing Actions Associated with NMSS Programs." NRC: Washington, DC. August 2003.

- —————. Regulatory Guide 1.23, "Onsite Meteorological Programs." NRC: Washington, DC. February 1972.

- —————. Regulatory Guide 1.86, "Termination of Operating Licenses for Nuclear Reactors." NRC: Washington, DC. June 1974.

- —————. Regulatory Guide 3.71, "Nuclear Criticality Safety Standards for Fuels and Materials Facilities." NRC: Washington, DC. August 1998.

- —————. Regulatory Guide 4.15, "Quality Assurance for Radiological Monitoring Programs (Normal Operations) – Effluent Streams and the Environment." NRC: Washington, DC. February 1979.

- —————. Regulatory Guide 4.16, "Monitoring and Reporting Radioactivity in Releases of Radioactive Materials in Liquid and Gaseous Effluents from Nuclear Fuel Processing and Fabrication Plants and Uranium Hexafluoride Production Plants." NRC: Washington, DC. December 1985.

- —————. Regulatory Guide 4.20, "Constraint on Releases of Airborne Radioactive Materials to the Environment for Licensees Other Than Power Reactors." NRC: Washington, DC. December 1996.

- —————. Regulatory Guide 8.4, "Direct-reading and Indirect-reading Pocket Dosimeters." NRC: Washington, DC. February 1973.

- —————. Regulatory Guide 8.7, "Instructions for Recording and Reporting Occupational Radiation Exposure Data." NRC: Washington, DC. June 1992.

- —————. Regulatory Guide 8.9, "Acceptable Concepts, Models Equations, and Assumptions for a Bioassay Program." NRC: Washington, DC. July 1993.

- —————. Regulatory Guide 8.15, "Acceptable Programs for Respiratory Protection." NRC: Washington, DC. October 1999.

- —————. Regulatory Guide 8.21, "Health Physics Surveys for Byproduct Material at NRC–Licensed Processing and Manufacturing Plants." NRC: Washington, DC. October 1979.

- —————. Regulatory Guide 8.23, "Radiation Surveys at Medical Institutions." NRC: Washington, DC. January 1981.

- —————. Regulatory Guide 8.24, "Health Physics Surveys During Enriched Uranium-235 Processing and Fuel Fabrication." NRC: Washington, DC. October 1979.

- —————. Regulatory Guide 8.25, "Air Sampling in the Workplace." NRC: Washington, DC. June 1992.

- —————. Regulatory Guide 8.28, "Audible-Alarm Dosimeters." NRC: Washington, DC. August 1981.

- —————. Regulatory Guide 8.34, "Monitoring Criteria and Methods to Calculate Occupational Radiation Doses." NRC: Washington, DC. July 1992.

- —————. Regulatory Guide 8.36, "Radiation Dose to the Embryo/Fetus." NRC: Washington, DC. July 1992.

- —————. Regulatory Guide 8.37, "ALARA Levels for Effluents from Materials Facilities." NRC: Washington, DC. July 1993.

4.7 DECOMMISSIONING DOCUMENTS REFERENCED IN THIS NUREG

- Nuclear Regulatory Commission (U.S.) (NRC), Washington, D.C. "Action Plan to Ensure Timely Cleanup of Site Decommissioning Management Plan Sites." *Federal Register*: Vol. 57, No. 74, pp. 13389–13392. April 1992.

- —————. Information Notice 96–47, "Recordkeeping, Decommissioning Notification for Disposals of Radioactive Waste by Land Burial Authorized Under Former 10 CFR 20.304, 20.302 and 20.2002." NRC: Washington, DC. August 16, 1996.

- —————. NRC 2003. SECY-03-0069, "Results of the License Termination Rule Analysis," May 2, 2003.

- —————. NRC 2003. SRM-SECY-03-0069, "Staff Requirements - SECY-03-0069 - Results of the License Termination Rule Analysis," November 17, 2003.

- —————. NRC 2004. SECY-04-0035, "Results of the License Termination Rule Anlysis of the Use of Intentional Mixing of Contaminated Soil," March 1, 2004.

- —————. NRC 2004. SRM-SECY-04-0035, "Staff Requirements - SECY-04-0035 - Results of the License Termination Rule Anlysis of the Use of Intentional Mixing of Contaminated Soil," May 11, 2004.

- —————. NRC 2004. Regulatory Issue Summary 2004-08, "Results of the License Termination Rule Analysis," May 28, 2004.

- —————.. NRC 2006. SECY-06-0143, "Stakeholder comments and Path Forward on Decommissioning Guidance to Address License Termination Rule Analysis Issues," July 5, 2006.

- —————. NRC 2006. SRM-SECY-06-0143, "Staff Requirements - SECY-06-0143 - Stakeholder Comments and Path Forward on Decommissioning Guidance to Address License Termination Rule Analysis Issues," September 19, 2006.

- —————. NUREG/BR–0241, "NMSS Handbook for Decommissioning Fuel Cycle and Materials Licensees." NRC: Washington, DC. March 1997.

- —————. NUREG–0586, "Final Generic Environmental Impact Statement on Decommissioning of Nuclear Facilities." NRC: Washington, DC. August 1988.

- —————. NUREG–1496, "Generic Environmental Impact Statement in Support of Rulemaking on Radiological Criteria for License Termination of NRC–Licensed Nuclear Facilities." NRC: Washington, DC. July 1997.

- —————. NUREG–1506, "Measurement Methods for Radiological Surveys in Support of New Decommissioning Criteria." NRC: Washington, DC. August 1995.

- —————. NUREG–1507, "Minimum Detectable Concentrations with Typical Radiation Survey Instruments for Various Contaminants and Field Conditions." NRC: Washington, DC. August 1995.

- —————. NUREG–1549, "Decision Methods for Dose Assessment to Comply with Radiological Criteria for License Termination, Draft." NRC: Washington, DC. July 1998.

- —————. NUREG–1575, "Multi-Agency Radiation Survey and Site Investigation Manual (MARSSIM), Rev. 1." NRC: Washington, DC. August 2000.

- —————. NUREG–1727, "NMSS Decommissioning Standard Review Plan." NRC: Washington, DC. September 2000.

- —————. NUREG/CR–5512, Volume 1, "Residual Radioactive Contamination from Decommissioning: Technical Basis for Translating Contamination Levels to Annual Total Effective Dose Equivalent." NRC: Washington, DC. October 1992.

- —————. Draft NUREG/CR–5512, Volume 2, "Residual Radioactive Contamination from Decommissioning: User's Manual DandD Version 2.1." NRC: Washington, DC. April 2001.

- —————. Draft NUREG/CR–5512, Volume 3, "Residual Radioactive Contamination From Decommissioning: Parameter Analysis." NRC: Washington, DC. October 1999.

- — — — — —. Draft NUREG/CR–5512, Volume 4, "Comparison of the Models and Assumptions used in the DandD 1.0, RESRAD 5.61, and RESRAD–Build Computer Codes with Respect to the Residential Farmer and Industrial Occupant Scenarios Provided in NUREG/CR–5512." NRC: Washington, DC. October 1999.

- — — — — —. NUREG/CR–5621, "Groundwater Models in Support of NUREG/CR–5512." NRC: Washington, DC. December 1998.

- — — — — —. NUREG/CR–5849, "Manual for Conducting Radiological Surveys in Support of License Termination." NRC: Washington, DC. Draft for Comment, June 1992.

- — — — — —. NUREG/CR–6692, "Probabilistic Modules for the RESRAD and RESRAD–Build Computer Codes." NRC: Washington, DC. November 2000.

- — — — — —. Policy and Guidance Directive Fuel Cycle 83–23, "Termination of Byproduct, Source and Special Nuclear Material Licenses." NRC: Washington, DC. November 4, 1983.

- — — — — —. Regulatory Issue Summary 2000–09, "Standard Review Plan for Licensee Requests to Extend the Time Periods Established for Initiation of Decommissioning Activities." NRC: Washington, DC. June 26, 2000.

4.8 PUBLIC INTERACTION DOCUMENTS REFERENCED IN THIS NUREG

- Nuclear Regulatory Commission (U.S.) (NRC). NUREG/BR–0199, "Responsiveness to the Public." NRC: Washington, DC. January 1996.

- — — — — —. NUREG/BR–0224, "Guidelines for Conducting Public Meetings." NRC: Washington, DC. February 1996.

- — — — — —. "Management Directive 3.4, Release of Information to the Public." NRC: Washington, DC. December 1, 1999.

- — — — — —. "Management Directive 3.5, Public Attendance at Certain Meetings Involving NRC Staff." NRC: Washington, DC. May 24, 1996.

- — — — — —. "Policy Statement on Staff Meetings Open to the Public." NRC: Washington, DC. 65 FR 56964, September 20, 2000.

- — — — — —. "Public Outreach Handbook." NRC: Washington, DC. March 1995.

- — — — — —. "Regulation of Decommissioning Communications Plan." NRC: Washington, DC. March 26, 2001.

- — — — — —. "Enhancing Public Participation in NRC Meetings: Policy Statement." NRC: Washington, DC. 67 FR 36920, May 28, 2002.

4.9 OTHER DOCUMENTS REFERENCED IN THIS NUREG

- American National Standards Institute (ANSI)–publications available at http://www.ansi.org.

- International Atomic Energy Agency (IAEA). IAEA No. 16, "Manual on Environmental Monitoring in Normal Operations." IAEA: Vienna, Austria. 1996.

- —————. IAEA Series No. 18, "Environmental Monitoring in Emergency Situations." IAEA: Vienna, Austria. 1966.

- —————. IAEA Safety Series No. 41, "Objectives and Design of Environmental Monitoring Programs for Radioactive Contaminants." IAEA: Vienna, Austria. 1975.

- International Commission on Radiological Protection (ICRP). ICRP 30, "Limits for Intakes of Radionuclides by Workers." ICRP: Stockholm, Sweden. 1978.

- National Fire Protection Association (NFPA) Standard 232, "Standards for the Protection of Records." NFPA: Quincy, MA. 1986.

- National Council on Radiation Protection and Measurements (NCRP) Report 50, "Environmental Radiation Measurements." NCRP: Bethesda, MD. December 1976.

- —————. NCRP Report 123, "Screening Models for Releases of Radionuclides to Atmosphere, Surface Water, and Ground." NCRP: Bethesda, MD. January 1996.

- —————. NCRP Report 127, "Operational Radiation Safety Program." NCRP: Bethesda, MD. 1998.

- Slade, D. (ed.), "Meteorology and Atomic Energy – 1968." TID–24190, July 1968 (available from the National Technical Information Service, Springfield, Virginia).

- Thom, H.C.S., "New Distribution of Extreme Winds in the United States." *Journal of the Structural Division, Proceedings of the American Society of Civil Engineers*, pp. 1787–1801. July 1968.

- U.S. Department of Commerce (USDC), "Climatic Atlas of the United States." USDC: Washington, DC. Environmental Data Service, Environmental Science Service Administration, 1968.

- U.S. Environmental Protection Agency (EPA). "Limiting Values of Radionuclide Intake and Air Concentration and Dose Conversion Factors for Inhalation, Submersion, and Ingestion." EPA: Washington, DC. Federal Guidance Report No.11, September 1988.

- U.S. Geological Survey (USGS) and U.S. Bureau of Mines (USBM), Circular 831, "Principles of a Resource/Reserve Classification for Minerals." USGS: Reston, VA. 1980.

4.10 DOCUMENTS SUPERSEDED BY THIS REPORT

This volume supersedes the Regulatory Guides (RG) and Policy and Guidance Directives (P&GD) listed in Table 4.1, and they should no longer be used.

Table 4.1   List of Documents Superseded by this Report

Document Identification	Title	Date
RG 3.65	Standard Format and Content Decommissioning Plans for Licensees Under 10 CFR Parts 30, 40, and 70	6/1989
RG 3.66	Standard Format and Content of Financial Assurance Mechanisms Required for Decommissioning Under 10 CFR Parts 30, 40, 70, and 72	6/1990
P&GD FC 90–2	Standard Review Plan for Evaluating Compliance with Decommissioning Requirements for Source, Byproduct, and Special Nuclear Material License Applications	4/1991
P&GD FC 91–2	Standard Review Plan: Evaluating Decommissioning Plans for Licensees Under 10 CFR Parts 30, 40, 70	8/1991
P&GD FC 83–3	Standard Review Plan for Termination of Special Nuclear Material Licenses of Fuel Cycle Facilities	3/1983

In addition, this volume supersedes most of NUREG/BR–0241, "NMSS Handbook for Decommissioning Fuel Cycle and Materials Facilities," except for those portions of the handbook covering decommissioning financial assurance and recordkeeping (Chapter 5 and Appendices D and P), which are addressed in Volume 3.

Volume 1 of this NUREG also incorporates and updates numerous portions of NUREG–1727, "NMSS Decommissioning Standard Review Plan," specifically, Chapters 1–4, 8–13, portions of 14 and 15, 16, and Appendix A, portions of Appendix C dealing with screening, I, and J. This three-volume NUREG series supersedes NUREG/BR–0241 and NUREG–1727 in their entirety and should be used as guidance for decommissioning.

4.11   TO REQUEST COPIES

To request copies of the regulations cited in Section 4.3, call the GPO order desk in Washington, DC, at (202) 512–1800. Order the two-volume bound version of Title 10, Code of Federal Regulations, Parts 0–50 and 51–199 from the GPO, Superintendent of Documents, Post Office Box 371954, Pittsburgh, Pennsylvania 15250–7954. You may also contact the GPO electronically at http://www.gpo.gov. Request single copies of NRC documents from NRC's Regional Offices (see Figure 2.1 for addresses and telephone numbers). Note that NRC publishes amendments to its regulations in the *Federal Register*.

Appendix I explains how to use the Internet to obtain copies of NRC documents and other information.

5   THE DECOMMISSIONING PROCESS

> **NOTE: In addition to the guidance in this chapter,**
> **licensees are encouraged to contact NRC, or the appropriate Agreement State authority,**
> **to assure an understanding of what actions should be taken**
> **to initiate and complete the license termination process**
> **on a license- or facility-specific basis.**
> **Cases where licensees abandon a site or refuse to decommission a site would be**
> **considered for civil or criminal action, as warranted.**

Decommissioning means to safely remove a facility or site from service and reduce residual radioactivity to a level that permits release of the property and termination of the license (see 10 CFR 20.1003). The following sections discuss timing and activities associated with decommissioning.

The regulations in 10 CFR 20.1406 establish requirements on minimizing contamination during operations. While the requirements apply only to applications filed after August 20, 1997, all licensees are strongly encouraged to remediate any contamination immediately after it occurs. If license amendments to authorize specific activities are necessary to remediate the results of unplanned events, these actions should be initiated promptly. If contamination is reduced to acceptable release levels during the operational phase of the facility, it will significantly reduce the regulatory burden during decommissioning. For example, if any remaining contamination (after operations cease) can be remediated without new procedures or activities, a DP may not be required.

5.1   TIMING OF DECOMMISSIONING

Decommissioning normally occurs after a licensee decides to stop operating. However, there are other requirements to decommission parts of a facility prior to complete shutdown (see 10 CFR 30.36(d), 40.42(d), 70.38(d), and 72.54(d)). Collectively, these are known as the Timeliness Rule. In short, any separate building or area that has not been used for two years must be promptly remediated if the remediation activities are allowed by the existing license (see Section 15.5 for an additional discussion of partial site decommissioning). If the remediation activities are not currently allowed under an existing license, the licensee must develop a DP and submit a request for a license amendment within one year. The decommissioning process is to be completed within two years, unless an alternative schedule is approved. Figure 5.1 shows how to determine if decommissioning is needed and the actions necessary to achieve it.

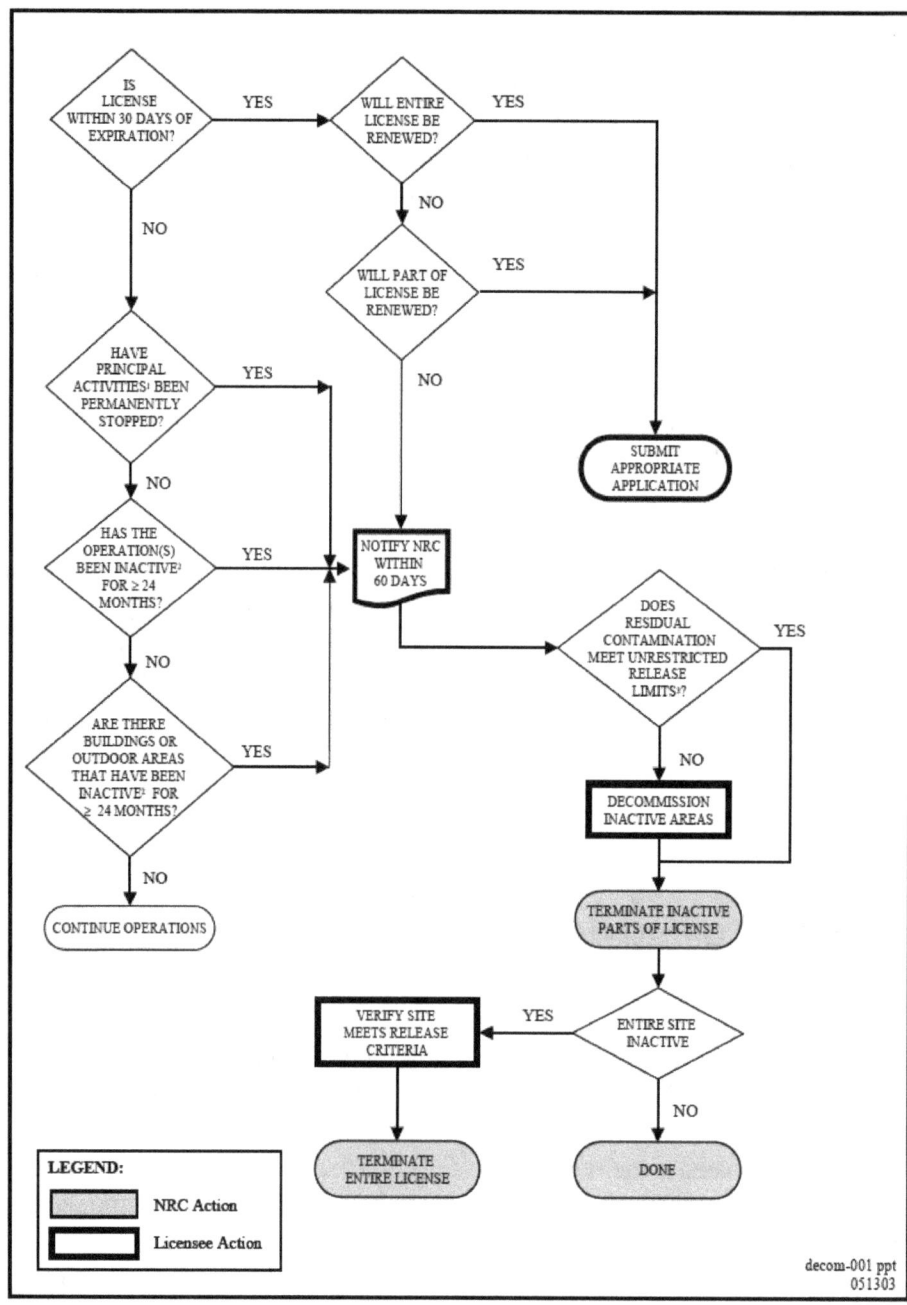

Figure 5.1   Do I Need to Decommission?

Notes:

1   Principal activities are defined as those identified in the license and necessary supporting functions.

2   Inactive means not used for principal activities for a period of >24 months.

3   10 CFR 20 Subpart E defines limits for residual radioactivity based on calculated dose; 10 CFR 20.1402 defines unrestricted release limits ≤25 mrem/y plus ALARA to an average member of the critical group for the approved land use scenario.

Licensed facilities, areas, and buildings convert from "active" status to "decommissioning" status when one of the following occurs:

- The license expires or is revoked by the Commission.

- The licensee decides to permanently cease operations with licensed material at the entire site or in any separate building or outdoor area that contains residual radioactivity, such that the area is unsuitable for release in accordance with NRC requirements.[5]

- Twenty-four (24) months have elapsed since principal activities have been conducted under the license, or

- No principal activities have been conducted in a separate building or outdoor area for a period of 24 months, and residual radioactivity is present that would preclude its release in accordance with NRC requirements.

Within 60 days of the occurrence of any of the above, the licensee is required to inform NRC of the occurrence in writing. In addition, the licensee is required to (a) begin decommissioning the facility, (b) submit within 12 months a DP[6] to NRC, and (c) begin decommissioning in accordance with the plan when it is approved by NRC. Unless otherwise approved by NRC, licensees are required to complete decommissioning their facilities within 24 months of initiating decommissioning operations.

NRC staff has also determined that the Timeliness Rule and the LTR on decommissioning materials facilities applies to previous onsite burial of radioactive material, if the former disposal site met the definition of an inactive outdoor area. NRC regulations require licensees to notify NRC if they have burial sites that may require decommissioning and to maintain records of these burials. Disposals made pursuant to former 10 CFR 20.304, 20.302 and current 20.2002 at facilities licensed under 10 CFR Parts 30, 40, 70, and 72, and that have been unused for NRC licensed operations for a period of 24 months, are subject to the requirements of the Timeliness Rule and the dose standards of the LTR (i.e., dose maintained ALARA and within LTR limit). The requirements for recordkeeping and application of the timeliness rule to former onsite disposals are discussed in Information Notice 96–47, "Recordkeeping, Decommissioning Notification for Disposals of Radioactive Waste by Land Burial Authorized under Former 10 CFR 20.304, 20.302 and 20.2002," August 16, 1996 (see Volume 3 of this NUREG series).

Pursuant to 10 CFR 30.36(f), 40.42(f), 70.38(f), and 72.54(f)(1), the Commission may grant a request to extend the time periods outlined above, if the Commission determines that the relief is not detrimental to the public health and safety and is otherwise in the public interest. In order for a licensee's request for an alternative schedule to be considered, the licensee must submit the request to the Commission not later than 30 days before notification is required. The schedule

---

[5]   Outdoor areas where radioactive materials were used that currently meet NRC criteria for unrestricted use are not subject to the timeliness rule's notification requirements.

[6]   A DP is not required if no new procedures and activities are necessary (see 10 CFR 30.36(g), 40.42(g), and 70.38(g)).

for decommissioning the site will be held in suspension until a decision on the licensee's request is made by the Commission. To review a licensee's request for an alternate schedule, the staff will use the criteria presented in Section 2.6 of NUREG-1757, Vol. 3.

The Timeliness Rule provides for two alternative schedules: (1) an alternative schedule for submitting a DP; and (2) an alternative schedule for completion of site decommissioning. The Commission may approve an alternate DP *submission date* after considering all of the following:

- if the Commission determines the alternative schedule is necessary for effective conduct of decommissioning operations;

- if the delay presents no undue risk from radiation to the public health and safety; and

- if the alternative DP submission schedule is otherwise in the public interest.

A request for an alternative *schedule for completion* of decommissioning may be approved, if warranted, after considering all of the following:

- whether it is technically feasible to complete the decommissioning within the 24-month period;

- whether sufficient waste disposal capacity is available to allow the completion of the decommissioning within the 24-month period;

- whether a significant volume reduction in waste requiring disposal will be achieved by allowing short-lived radionuclides to decay;

- whether a significant reduction in radiation exposure to workers can be achieved by allowing short-lived radionuclides to decay; and

- other site-specific factors such as the regulatory requirements of other agencies, lawsuits, ground water treatment activities, monitored natural groundwater restoration, actions that could result in more environmental harm than deferred cleanup, and other factors beyond the control of the licensee.

In addition, approval of the request must also be in the "public interest." NRC has determined that it is normally in the public's interest to have radiologically contaminated areas remediated soon after permanent cessation of operations. When decommissioning is delayed for long periods following cessation of operations, there is a risk that safety practices may become lax as key personnel relocate and management interest wanes. In addition, bankruptcy, corporate takeover, or other unforeseen changes in a company's financial status may complicate and perhaps further delay decommissioning." Further, waste disposal costs have, in the past, increased at rates significantly higher than the rate of inflation and therefore delaying remediation will result in higher costs to the public if the government eventually assumes responsibility for the decommissioning. Therefore, in evaluating a licensee's request for an alternative completion schedule, NRC staff should consider whether the licensee has adequately addressed how postponing decommissioning would be in the public's interest. For example, the licensee might demonstrate that delaying remediation reduces or eliminates overall health risk to the public and/or impact to the human environment and is thus in the "public interest."

5.2   DECOMMISSIONING PROCESS

The decommissioning process consists of a series of integrated activities, beginning with the licensee notifying NRC and changing the licensee's program from "active" to "decommissioning" status, and concluding with the termination of the license and release of the site pursuant to 10 CFR 30.36(k), 40.42(k), 70.38(k) or 72.54. Depending on several factors, including the type of license, the use of radioactive material at the facility, or past management of radioactive material at the facility, the decommissioning may be either relatively simple and straightforward or complex.

While the steps may vary for different sites, the basic process is the same. Figure 5.2 illustrates the steps in a flow chart format, showing licensee and NRC actions. The steps in the process are as follows:

- Stop operations, either in a specific area or building (see Section 15.5 for a discussion on partial site decommissioning) or for the entire facility.

- Notify NRC of the decision within 60 days.

- Determine locations and concentrations of remaining radiological contamination.

- If necessary, develop a DP (see Figure 5.3) that includes all of the following:

  — the current radiological contamination at the site;

  — the criteria for the final condition of the site;

  — the activities to remediate existing contamination that are not currently authorized by the license;

  — procedures to protect workers;

  — decommissioning cost estimates;

  — the final survey method to demonstrate compliance with NRC criteria; and

  — provides the schedule for remediation activities and license termination.

- If necessary, provide environmental information on NEPA Compliance as described in Section 15.7.

- Clean up contamination, as needed.

- Conduct Final Status Survey to show compliance with dose limits for license termination.

- Request that NRC terminate the license.

Note that it is important for licensees to notify NRC promptly when operations cease. It is also important that the staff meet with the licensee to discuss the decommissioning requirements early in the process.

In 2002, NRC and EPA entered into a Memorandum of Understanding (MOU) entitled, "Consultation and Finality on Decommissioning and Decontamination of Contaminated Sites." The MOU continues the 1983 EPA policy that EPA will defer Comprehensive Environmental Response, Compensation, and Liability Act (CERCLA) authority at NRC-licensed sites that are decommissioned, unless otherwise requested by NRC. The MOU identifies the criteria under which NRC will consult with EPA on sites undergoing decommissioning under NRC authority and outlines the process under which NRC will consult with EPA. The intent of the process established under the MOU is to minimize the occurrence of so called "dual regulation," where EPA is required to respond under CERCLA to conditions at a site cleaned up to the radiological criteria for license termination in 10 CFR Part 20, Subpart E. The MOU is included as Appendix H in this Volume.

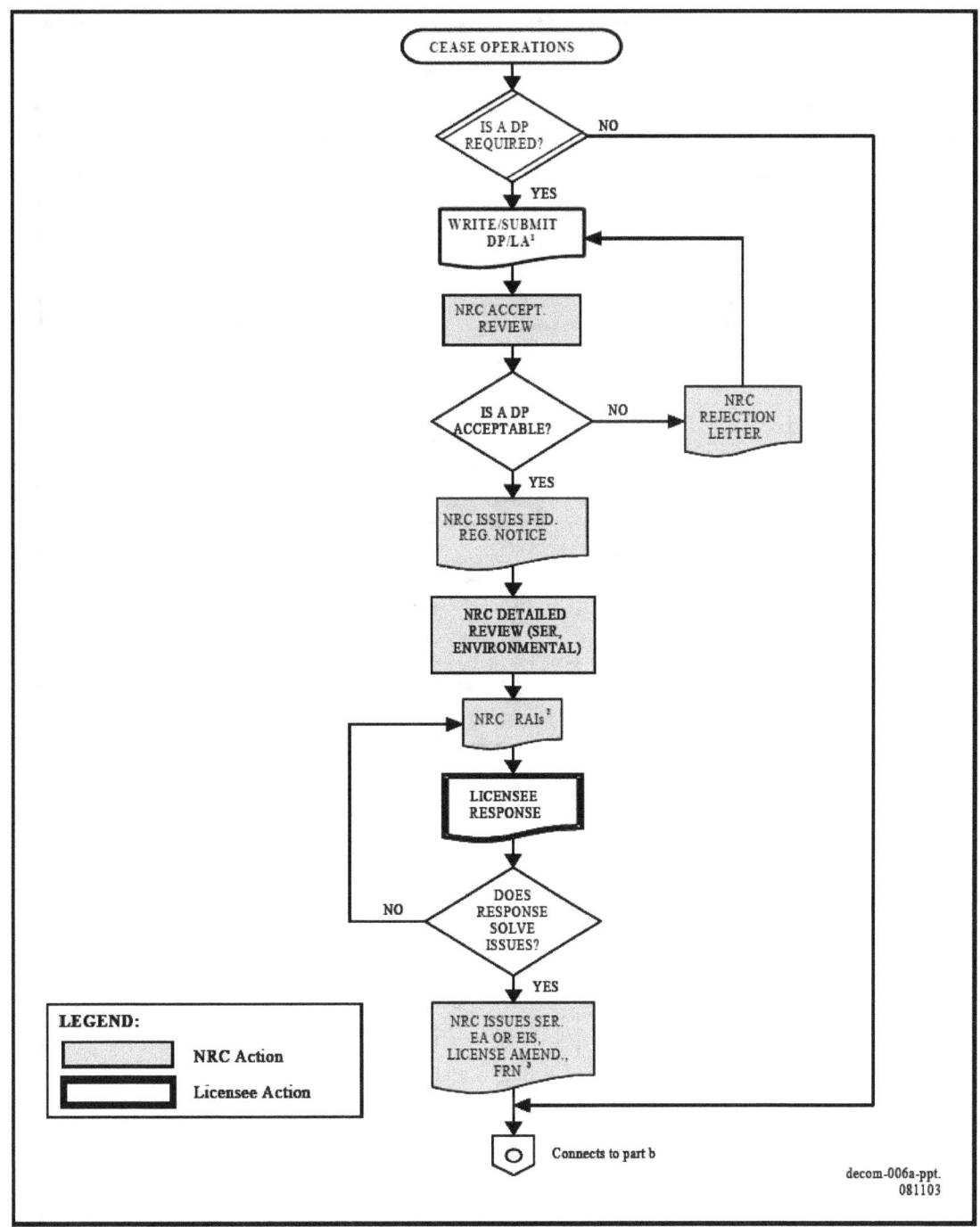

Figure 5.2a   The Decommissioning Process (1 of 2).

Notes:

1   LA is a request for License Amendment.

2   RAI is a Request for Additional Information.

3   FRN is a *Federal Register* notice.

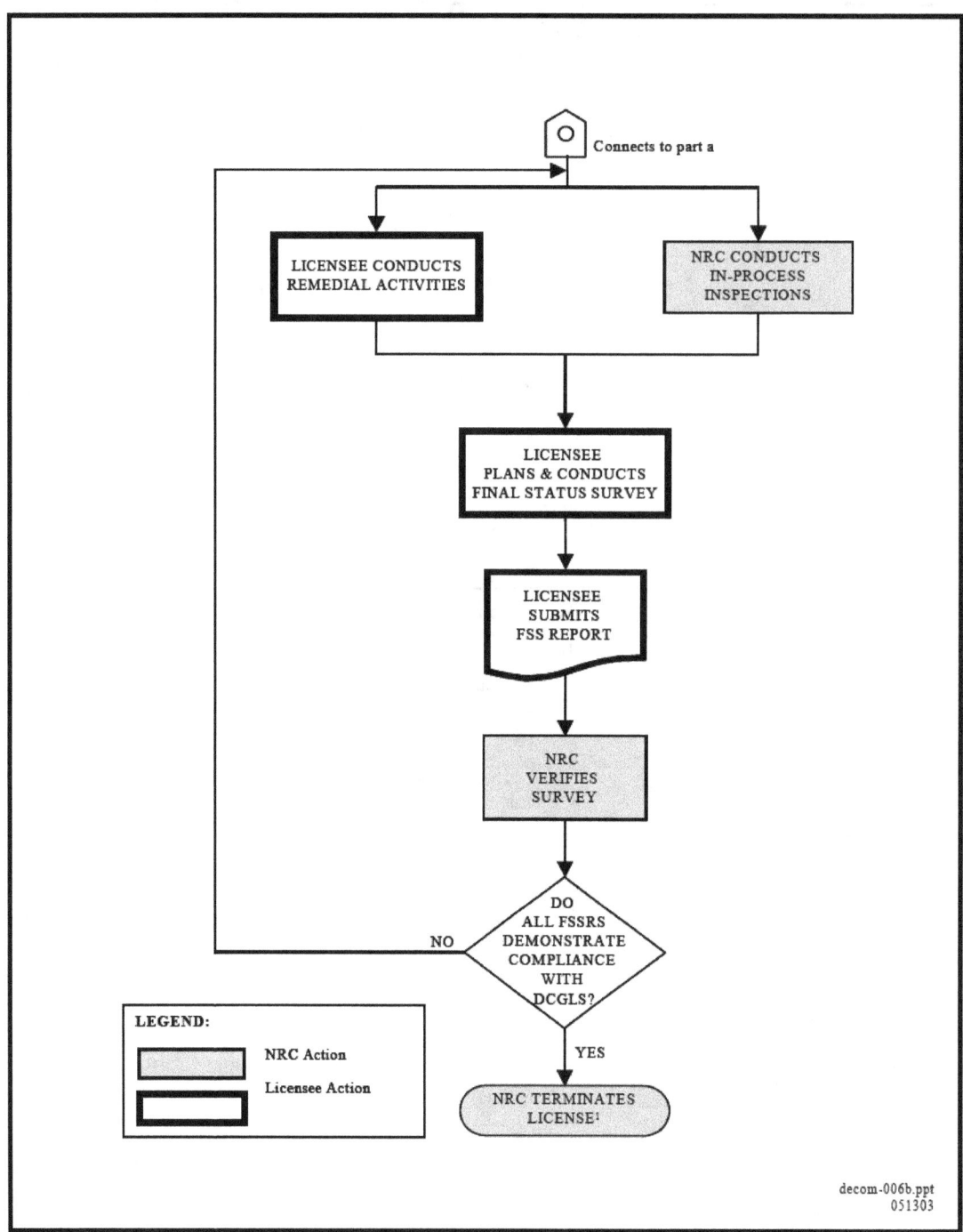

Figure 5.2b   The Decommissioning Process (2 of 2).

Notes:

1   *Federal Register* notice issued.

5.3 DECOMMISSIONING PLAN REVIEW

Acceptance Review

The staff should review the DP to ensure that, at a minimum, the DP contains the information from the Appendix D checklist that NRC staff and the licensee have previously agreed upon. NRC staff should conduct a limited technical review of the DP. The technical accuracy and completeness of the information should be assessed during the detailed technical review. This NRC staff review of the DP will determine that enough information is included and that the level of detail appears to be adequate for the staff to perform a detailed technical review, or the plan will be rejected.

If a DP is required, licensees are strongly encouraged to meet with NRC prior to the submittal of their DP and at any stage in this process. The conditions requiring a DP are specified in 10 CFR 30.36(g), 40.42(g), 70.38(g) or 72.54(g). In short, a DP is required if one is specified in the existing license or if new activities or procedures–those not currently authorized in the license–are needed to conduct remediation. Figure 5.3 illustrates these conditions in the form of a flowchart; if any of the specified conditions exist, a DP is required. The DP is processed as follows (see Figure 5.2):

- NRC meets with licensee to determine which items in the DP Evaluation Checklist in Appendix D are applicable.

- Licensee submits DP for all or part of the facility.

- NRC conducts an acceptance review to decide if the plan is complete:

  — NRC determines if all of the items identified in Chapters 16 and 17 and the DP Evaluation Checklist are present.

  — NRC determines if there is sufficient information in each section for NRC to evaluate the proposed decommissioning alternative:

    - current condition of site;

    - release criteria and important values (e.g., residual concentrations);

    - land use scenario and critical group(s) (See Consolidated Decommissioning Guidance, Volume 2); and

    - final survey plan.

- If the DP is not complete, NRC rejects it, and the licensee is informed in writing.

- After acceptance for technical review, NRC conducts a detailed evaluation of the plan from environmental (NEPA) and safety perspectives.

- NRC solicits comments from stakeholders in accordance with 10 CFR 20.1405.

- If the information in the plan is not sufficient for NRC to complete the environment and safety reviews, NRC requests additional information (RAI).

- Upon receipt of the RAI, the licensee revises the plan; the revised plan is reviewed, as above.

- NRC issues license amendment approving the DP.

- Once the plan is approved, the licensee implements the plan. NRC should conduct in-process inspections to verify compliance (see Section 15.3, "Decommissioning Inspections").

- At the completion of remediation, the licensee conducts a final status survey to demonstrate compliance with license termination criteria.

- NRC verifies the survey by one or more of the following:

   — QA/QC reviews;

   — side-by-side or split sampling; and

   — independent, confirmatory surveys.

- If the survey does not demonstrate compliance, additional remediation and/or surveys are required.

- When the survey demonstrates compliance with release criteria, NRC terminates, or modifies the license for partial site release.

Safety Evaluation

The staff should review the technical content of the information provided by the licensee to ensure that the licensee used defensible assumptions and models to calculate the potential dose to the average member of the critical group. The staff should also verify that the licensee provided enough information to allow an independent evaluation of the potential dose resulting from the residual radioactivity after license termination and provided reasonable assurance that the decommissioning option will comply with regulations.

For sites that require a DP, NRC publishes notices in the *Federal Register*. Once NRC staff find the DP acceptable for review, NRC issues a *Federal Register* notice to announce (1) staff consideration of a license amendment, (2) receipt of the DP, (3) opportunity for a hearing, (4) public comment solicitations, and (5) any public meetings. Following this, there may be a public meeting to discuss the proposed actions with interested and affected parties. Following approval of the DP, NRC issues one or more *Federal Register* notices to announce (1) the approval of the DP by license amendment and (2) the results of staff's environmental review. If a site is on the SDMP list, NRC also issues a *Federal Register* notice announcing removal of the site from the SDMP list at the completion of remediation.

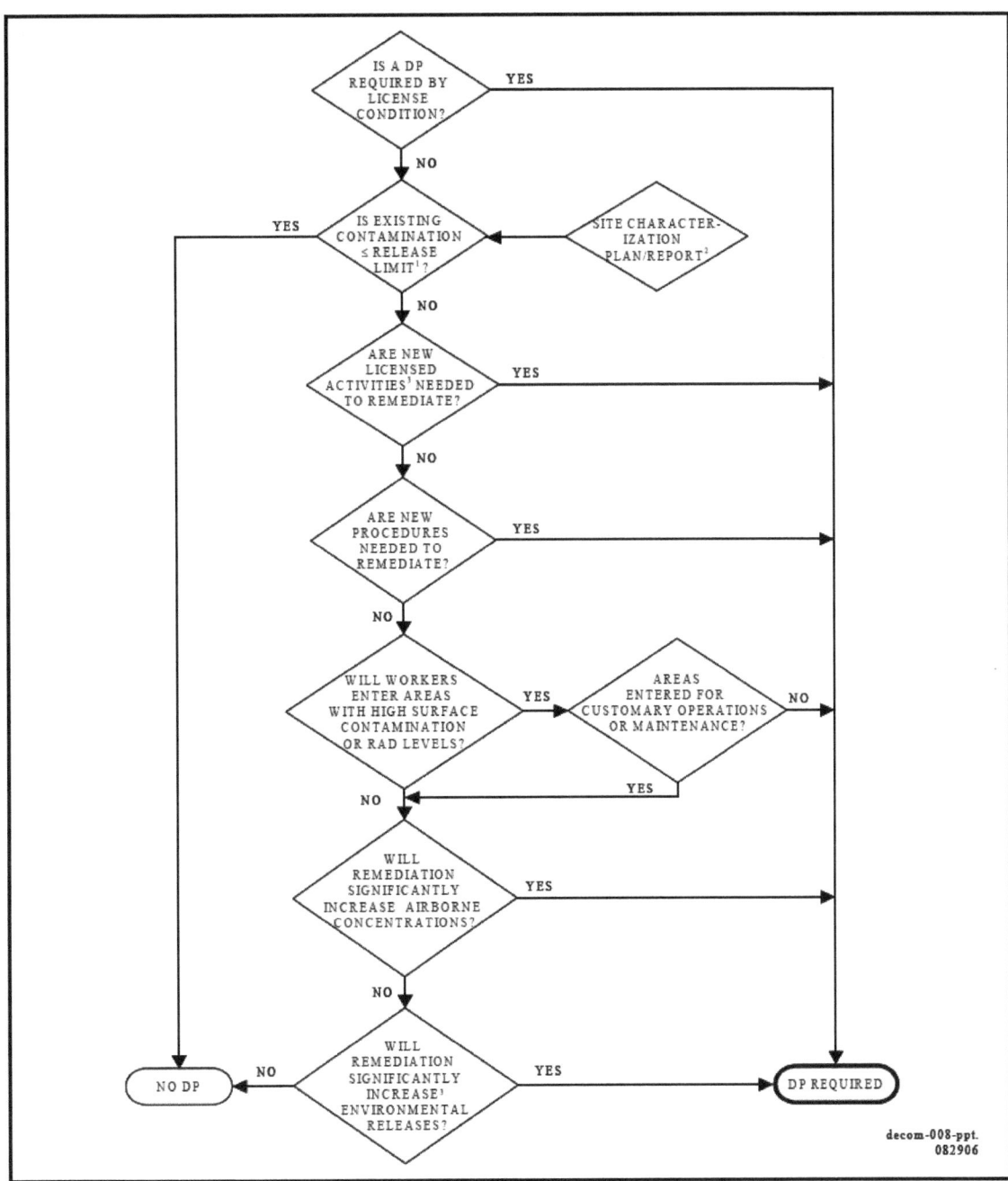

Figure 5.3 Is a Decommissioning Plan Required?

Notes:

1   "Release limits" are defined in 10 CFR 20, Subpart E

2   Site characterization plan and report are not required to be submitted to the NRC, unless by license condition (see Section15 4 1)

3   "New licensed activity" means any activity at the facility involving radioactive materials that is not authorized in the license prior to decommissioning Examples of activities not typically authorized include building demolition and exhumation of burial areas

4   "Significantly increase" means any increase that initiates or changes any report to NRC

6   RADIOLOGICAL CRITERIA FOR DECOMMISSIONING

Dose-based requirements for licensees seeking license termination are found in 10 CFR 20, Subpart E. These regulations establish two final states for licensee termination: unrestricted use and restricted use. In addition to the specific limits for each state, specified in Sections 6.1 and 6.2, NRC requires licensees to maintain ALARA doses. This means the licensee must make every reasonable effort to reduce the dose as far below the specified limits as is practical, taking into account the state of technology and economics (see 10 CFR 20.1003).

The use of a dose limit allows both the licensee and the regulator to take site-specific information into account in determining acceptable concentrations of residual radioactivity at the site using dose models and exposure scenarios that are as realistic as necessary. Section 6.6 describes the NRC Technical Basis for Dose Modeling Evaluations (Screening). Chapters 11–13 in this document discuss procedures, acceptance criteria, and evaluation findings acceptable to NRC staff for limited dose analyses. Dose analyses for more complicated decommissioning projects, that is, projects requiring collection of site-specific parameters and the submission of a DP, are to be discussed in the Consolidated Decommissioning Guidance, Volume 2.

6.1   UNRESTRICTED USE

Residual radioactivity, distinguishable from background, results in a calculated dose from all pathways to the average member of the critical group that is not in excess of 0.25 mSv/y (25 mrem/y).

6.2   RESTRICTED CONDITIONS

The basic requirement for license termination under restricted conditions is that the licensee provide institutional controls that limit the calculated dose to 0.25 mSv/y (25 mrem/y). Further, the licensee must reduce residual radioactivity so that if these controls fail, the calculated dose would not exceed 1 mSv/y (100 mrem/y). In rare instances, the calculated dose may exceed 1 mSv/y (100 mrem/y), but it may not exceed 5 mSv/y (500 mrem/y). Additional institutional controls would be established to meet regulatory requirements (see Chapter 13 for a discussion of institutional controls).

To qualify for license termination under restricted conditions, the licensee must meet all of the criteria:

- Demonstrate that further reductions in residual radioactivity would either cause net environmental harm or are technically or economically not feasible.

- Demonstrate provisions for legally enforceable controls to limit dose to 0.25 mSv/y (25 mrem/y).

- Provide financial assurance to allow a third party to control and maintain the site.

- Demonstrate that advice from affected parties on the adequacy of the proposed institutional controls and financial assurance has been obtained and used in developing the DP. See Sections 17.7.5 and M.6 for guidance on obtaining advice from the public.

6.3 ALTERNATE CRITERIA

In the unlikely event that a licensee is not able to reduce residual radioactivity to a level that limits the calculated dose such that it is not in excess of 0.25 mSv/y (25 mrem/y) with restrictions in place, the licensee may request permission from the Commission to use alternate criteria. In doing so, the licensee must demonstrate all of the following:

- The calculated dose from all man-made sources is unlikely to exceed 100 mrem (1 mSv) per year by identifying these sources and the expected dose from each.

- Institutional controls will minimize the dose from the site.

- The licensee has obtained public advice on the proposed institutional controls and financial assurance. See Sections 17.7.5 and M.6 for guidance on obtaining advice from the public.

NRC staff will review the application, publish a notice in the *Federal Register*, solicit comments from State and local governments and from potentially affected parties, Indian Nations, and from the Environmental Protection Agency (EPA). NRC staff will then make a recommendation. The Commission will consider the comments from the public, the EPA, and NRC staff and make the final decision on the acceptability of the proposed criteria, which will be published in the *Federal Register* (see 10 CFR 2.105(e), 20.1404(b), and 20.1405).

6.4 RELEASE CRITERIA

NRC staff reviews the release criteria to verify that the licensee has developed appropriate release criteria, referred to as the derived concentration guideline levels (DCGLs). Volume 2 of this NUREG discusses the information to be submitted by the licensee and provides details of the staff's review.

6.5 GRANDFATHERED SITES

Sites being decommissioned under approved DPs, submitted before August 20, 1998, are grandfathered from the provisions of 10 CFR 20 Subpart E. Specifically, the criteria in the Site Decommissioning Management Plan (SDMP) Action Plan (57 FR 13389) are reasonably consistent with the dose-based criteria and are within the range of measurable values that could be derived through the site-specific screening and modeling approaches used in dose-based site analysis. See Section 15.6 for a discussion of SDMP sites. In the event a licensee makes significant changes to the DP–those requiring formal NRC approval–or cannot demonstrate compliance with approved residual concentrations, the grandfathering provisions of the LTR will not continue. The revised DP will be subject to 10 CFR 20, Subpart E.

Furthermore, the grandfathering provision does not generally extend to all pre–LTR decommissioning actions, because they were not all done under the criteria of the SDMP Action Plan and therefore would not provide assurance that such actions were adequate to protect the public. NRC has conducted a systematic review of terminated licenses and identified any sites warranting further NRC attention under the requirements of 10 CFR 20, Subpart E.

6.6 DECOMMISSIONING SCREENING CRITERIA

Both the decommissioning roadmap (Figure 1.1) and the regulatory features of decommissioning groups in Table 1.2 describe the use of decommissioning screening criteria for Groups 1–3. The technical basis, scope, criteria, qualification, and recommended approaches and tools for the use of screening criteria are presented in this section.

6.6.1 TECHNICAL BASIS FOR SCREENING

On July 21, 1997, NRC published a final rule on "Radiological Criteria for License Termination," in the *Federal Register* (62 FR 39058), which was incorporated as Subpart E to 10 CFR Part 20. On July 8, 1998, the Commission directed staff to develop a standard review plan (SRP) for decommissioning. The staff completed development of the SRP in September 2000, in part, as a technical information support document for performing the staff's evaluations of the licensee's dose modeling. It presented detailed technical approaches, methodologies, criteria, and guidance for staff reviewing dose modeling to demonstrate compliance with the dose criteria in 10 CFR Part 20, Subpart E, and has been incorporated into this NUREG series.

NRC staff developed building surface concentration screening values and surface soil concentration values to support implementation of the LTR and to simplify decommissioning in cases where low levels of contamination exist. These values were published in the *Federal Register* on November 18, 1998, December 7, 1999, and June 13, 2000 (see Section 4.1), and their use is discussed in this section. The use of the screening values provides reasonable assurance that the dose criterion in 10 CFR 20.1402 will be met. This section explains the staff's review when the licensee proposes to use these screening values. In addition to these screening criteria, NRC has developed a screening code "DandD" for demonstrating compliance with the dose criteria in Part 20, Subpart E and to simplify decommissioning in cases where low levels of contamination exist. A full discussion of the use of screening criteria to evaluate site conditions can be found in Appendix B of this volume and Volume 2 of this NUREG.

6.6.2 BRIEF DESCRIPTION AND SCOPE

The screening process, discussed here in Sections 6.6.3–6.6.4, is fully described in Volume 2 and should be used by licensees for demonstrating compliance with the unrestricted release dose criteria in 10 CFR 20, Subpart E. The sections of Volume 2 specific to screening are summarized below.

Sections 6.6.1–6.6.4 of this volume summarize the Volume 2 discussions of the following:

- acceptable approaches, look-up tables, and screening models for evaluating a licensee's demonstration of compliance with the dose criteria, using a screening methodology;

- the attributes of screening and site-specific analysis, to evaluate the merits of both approaches; and

- the criteria for qualification of the site for this screening approach.

6.6.3   CRITERIA FOR CONDUCTING SCREENING

This section pertains to the licensee's demonstration of compliance with the dose criteria in Part 20, Subpart E, using a screening approach instead of dose analysis. The licensee's use of the screening analysis should be performed using one or more of the currently available screening tools:

- a look-up table for common beta-/gamma-emitting radionuclides for building surface contamination (63 FR 64132, November 18, 1998);

- a look-up table for common radionuclides for soil surface contamination (64 FR 68395, December 7, 1999) (tabulated in Appendix B); and

- screening levels derived using DandD, Version 2.0, for the specific radionuclide(s), using the code's default parameters.

A full discussion of the use of screening criteria to evaluate site conditions can be found in Volume 2 of this NUREG.

A screening analysis is usually conducted for simple sites with building surface (e.g., non-volumetric) contamination and/or with surficial soil contamination (considered to be within approximately the first 15 cm (6 in) of soil).

The licensee should demonstrate qualification of the site for screening in terms of site physical conditions and compatibility with the modeling code's assumptions and default parameters and the acceptable screening tools (e.g., code, look-up tables), approaches, and parameters that staff can use to translate the dose into equivalent screening concentration levels. When using the screening approach for demonstrating compliance with the dose criteria in Part 20, Subpart E, licensees need to demonstrate that the particular site conditions (e.g., physical and source-term conditions) are compatible and consistent with the DandD model assumptions (NUREG/CR–5512, Volume 1).

6.6.4 QUALIFICATION OF THE SITE FOR SCREENING

Sites exhibiting any of the conditions found in Figure 6.1 (excluding those caused by sources of background radiation) would probably not be a Group 1–3 candidate, but would probably be a Group 4–7 candidate.

Criteria that Exclude Sites from Groups 1–3

Criteria that would preclude a site from being classified in Groups 1–3 include any one of the following:

- soil contamination greater than 15 cm (6 inches) below the ground surface;

- radionuclide residual radioactivity present in an aquifer;

- buildings with volumetrically contaminated material;

- radionuclide concentrations in surface water sediments; and

- sites that have an infiltration rate greater than the vertical saturated hydraulic conductivity (i.e., resulting in the water running off the surface rather than purely seeping into the ground).

These are limitations caused by the conceptual models used in developing the screening analysis. In other words, the conceptual model, parameters, and scenarios in the DandD computer code are generally incompatible with such conditions. However, situations do exist where you can still use the analyses using scenario assumptions to modify the source term. For example, by assuming buried radioactive material is excavated and spread across the surface, the screening criteria may be applicable for use at the site.

Figure 6.1 Groups 1–3 Exclusion Criteria.

Licensees should be aware that a screening analysis, for demonstrating compliance with the dose criteria in Part 20, Subpart E, may not be applicable for certain sites because of the status of contaminants (e.g., location and distribution of radionuclides) or because of site-specific physical conditions. Therefore, licensees should assess the site source-term (e.g., radionuclide distribution) characteristics to ensure consistency with the source-term assumptions in the screening model/code used (e.g., DandD). See Figure 6.1 for a description of these limitations. In addition, licensees should determine if specific physical conditions at the site would invalidate the model and code assumptions associated with the screening code/model. Licensees should review the selected screening parameters and pathways to ensure that they are conservative and consistent with the parameters and pathways of the DandD code. Further, licensees may determine that there could be conditions at their site that cannot be handled by the simple screening model, either because of the complex nature of the site or because of the simple conceptual model in the DandD screening code. Recommended approaches to address and resolve these screening issues are presented in Appendix B.

6.6.5 SCREENING DEFINITION AND APPROACHES FOR THE TRANSITION FROM SCREENING TO SITE-SPECIFIC ANALYSIS

Licensees may also consider the use of other screening tools on a case-by-case basis (e.g., other look-up tables or other conservative codes/models) after evaluation and comparison of the level of conservatism, compatibilities, and consistencies of these tools with the DandD code default conditions and with site-specific conditions. Scenario descriptions used in generic screening are developed and discussed in NUREG/CR–5512, Volumes 1–3. This NUREG and NUREG–1549 provide both the rationale for applicability of the generic scenarios, criteria, rationale, and assumptions and the associated parameter values or ranges. In general, licensees should recognize that when they select other approaches or models for the dose analysis, or modify the DandD code default parameters, scenarios, and/or pathways, they will be performing a site-specific analysis.

6.6.6 SITE SCREENING: QUALIFICATION OF ASSUMPTIONS

When using the screening approach for demonstrating compliance with the unrestricted release dose criteria in Part 20, Subpart E, licensees need to demonstrate that the particular site conditions (e.g., physical and source-term conditions) are compatible and consistent with the DandD model assumptions (NUREG/CR–5512, Volume 1). In addition, the default parameters and default scenarios/pathways must be used in the screening dose analysis. Therefore, reviewers should examine the site conceptual model, the generic source-term characteristics, and other attributes of the sites to ensure that the site is qualified for screening.

NRC staff should verify that all of the following site conditions exist:

- Building Surface Contamination

  — The contamination on building surfaces (e.g., walls, floors, ceilings) should be surficial and non-volumetric (e.g., < 10 mm (0.4 in)).

  — Contamination on surfaces is mostly fixed (not loose), with the fraction of loose contamination not to exceed 10 percent of the total surface activity. For exceptions, see footnote a in Table B.1 in Appendix B.

  — The screening criteria may not be applied to surfaces such as buried structures (e.g., drainage or sewer pipes) or mobile equipment within the building; such structures and buried surfaces will be treated on a case-by-case basis.

- Surface Soil Contamination

  — The initial residual radioactivity (after decommissioning) is contained in the top layer of the surface soil (e.g., approximately 15 cm (6 in)).

  — The unsaturated zone and the ground water are initially free of contamination.

— The vertical saturated hydraulic conductivity at the specific site is greater than the infiltration rate.

After verifying that a site qualifies for screening, staff may compare the actual level of contamination at the site with the screening levels published in NRC's look-up tables or may use the latest version of the DandD code.

It should be noted that NRC staff should also evaluate complex site conditions that may disqualify the site for screening. Examples of such complex site conditions may include: highly fractured formation, karst conditions, extensive surface-water contamination, and/or a highly non-homogeneous distribution of contamination. Therefore, reviewers should ensure that the site meets the definition of a "simple site" to qualify for screening.

6.6.7   ACCEPTABLE SCREENING TOOLS

The currently available screening tools that NRC will accept for a screening analysis are the following:

- a look-up table (Table B.1 in Appendix B) for common beta-/gamma-emitting radionuclides for building-surface contamination;

- a look-up table (Table B.2 in Appendix B) for common radionuclides for soil-surface contamination;

- the screening values in Tables B.1 and B.2 are intended for single radionuclides (for radionuclides in mixtures, the "sum of fractions" rule should be used); and

- screening levels derived using DandD, latest version, for the specific radionuclide and using code default parameters. (The DandD code may be accessed at the Web site: http://techconf.llnl.gov.)

A comprehensive discussion of the screening methodology for dose calculations can be found in Volume 2 of this NUREG.

6.7   CHANGING DERIVED CONCENTRATION GUIDELINE LEVELS

Modifying the DCGLs, or model assumptions, after the DP is approved, typically will involve a license amendment and *Federal Register* notices announcing and issuing the amendment (10 CFR 2.105 and 2.106). Licensees can derive "operational" DCGLs for purpose of remediation activities without NRC review required. The common situation for these would involve sites with multiple radionuclides or sources; this is discussed in Section 2.7 of Volume 2. As discussed in Section 4.3 of Volume 2, remedial action support surveys are a very good method to reduce the risk of failing the compliance measure with the FSS results.

It is important to note that there may be situations that arise, that require the licensee to submit a license amendment to address dose modeling. Two of the situations that may arise involve new information gathered during remediation activities that was not identified during a desultory site characterization. One situation arises when new sources (e.g., contaminated ground water) or new radionuclides are discovered during remediation. Another situation arises when new information invalidates the assumptions used in the dose modeling. Examples of important assumptions can include, but are not limited to, extent and depth of contamination, area of influence for waterborne pathways, and physical characteristics such as K_d or porosity.

If the new information were either (a) to decrease the single radionuclide or single source DCGLs or (b) to require new DCGLs to be approved, the licensee would need to submit a license amendment, in most cases. The licensee should contact NRC staff to discuss the situation and to scope out the extent of the license amendment.

7 DECOMMISSIONING GROUPS

> **NOTE:** In addition to the guidance in this chapter,
> licensees are encouraged to contact NRC, or the appropriate Agreement State authority,
> to assure an understanding of what actions should be taken
> to initiate and complete the license termination process
> on a license- or facility-specific basis.
> Cases where licensees abandon a site or refuse to decommission a site would be
> considered for civil or criminal action, as warranted.

7.1 INTRODUCTION

Activities to decommission a site depend on the type of operations conducted by the licensee and the residual radioactivity. The various site conditions have been divided into seven decommissioning groups. These groups are defined in Sections 7.2–7.9.

The decommissioning actions that are typically applicable to each decommissioning group are summarized in the following chapters. It is important to recognize that every applicable NRC action cannot be fully addressed. NRC staff and licensees should use the decommissioning groups described in this document as a general guide to the actions and scope of the decommissioning process, while remaining flexible with respect to the appropriate actions that they will be required to undertake.

Although it is anticipated that most licensees will fall under one of the decommissioning groups as described, it is not expected that all actions will be appropriate for each licensee. The intent is to present the general information needed by NRC and the actions to be taken by the licensee, recognizing that the unique nature of some facilities may require site-specific modifications to the procedures.

NRC should review the information supplied by the licensee to determine if the description of the current radiological status of the facility is adequate to allow NRC to fully understand the types, levels, and extent of radioactive material contamination at the facility. This information should include summaries of the types and extent of radionuclide contamination in all media at the facility, including buildings, systems and equipment, surface and subsurface soil, and surface and ground water.

7.2 CRITERIA

Generally, the staff will evaluate the decommissioning of nuclear facilities using one of seven reviews (referred to as "Groups"), summarized below and described in the following sections. Typically, Groups 1 and 2 will not require a DP and will be able to demonstrate compliance with 10 CFR Part 20.1402. Group 3 sites will require an abbreviated DP, without a site-specific dose modeling analysis. Group 4 through 7 sites and all Part 72 licensees are required to submit a DP with site-specific dose modeling in accordance with NRC regulations in 10 CFR 30.36(g)(1), 40.42(g)(1), 70.38(g)(1), or 72.54(d).

Figure 7.1 provides a decision tree for determining the appropriate decommissioning group.

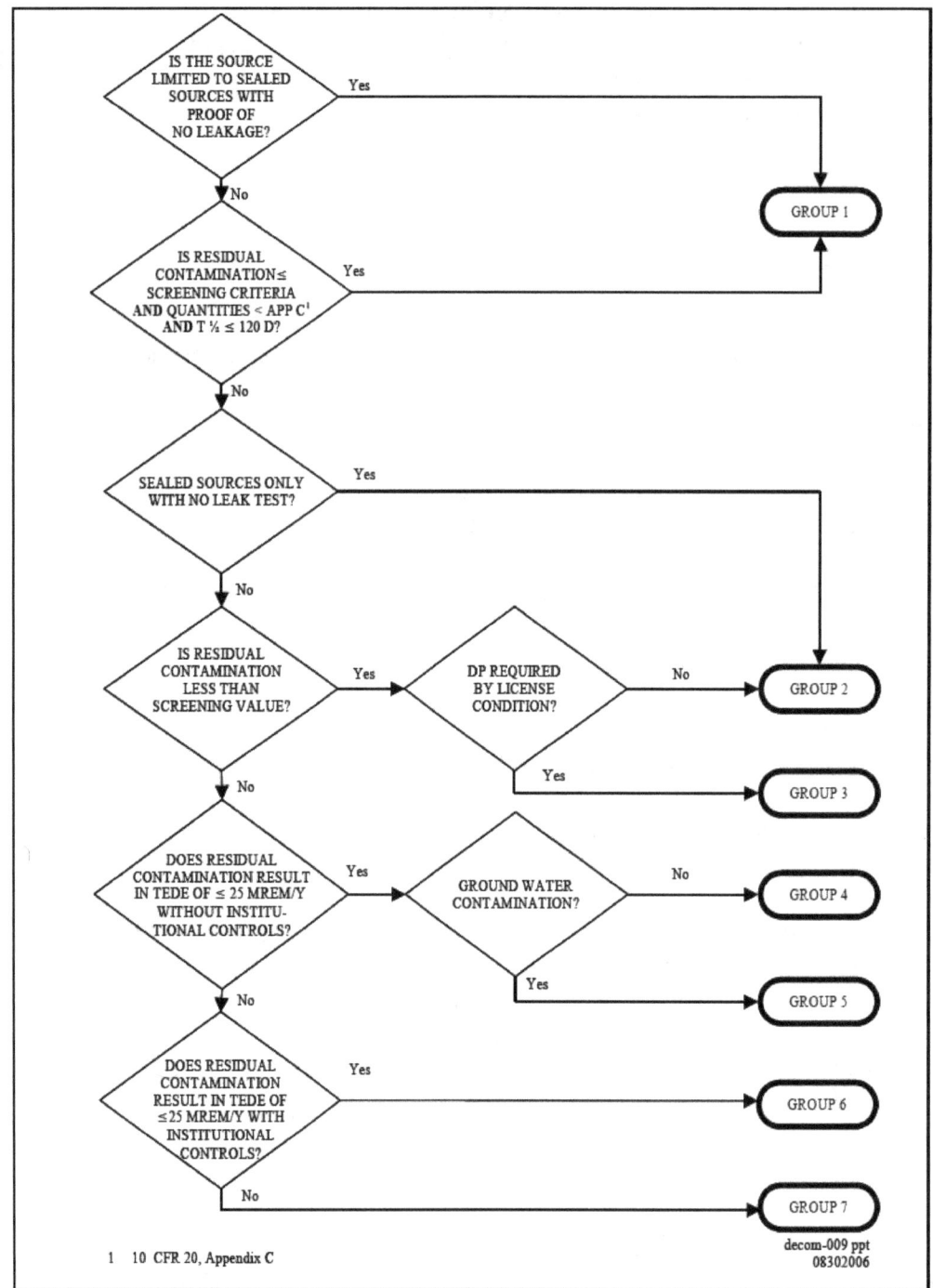

Figure 7.1   Determining the Appropriate Decommissioning Group.

7.3   GROUP 1: UNRESTRICTED RELEASE; NO DECOMMISSIONING PLAN REQUIRED

Group 1 facilities typically involve licensed material used in a way that would preclude its release into the environment, would not cause the activation of adjacent materials, or would not have contaminated work areas above the levels of the decommissioning screening criteria (see Appendix B). Activities that may fall into the Group 1 category are:

- Licensees who possessed and used only sealed sources, and whose most recent leak test results are current and demonstrate that the source(s) did not leak while in the licensee's possession; or

- Licensees who possessed and used relatively short-lived radioactive material (i.e., $T_{1/2}$ less that or equal to 120 days) in an unsealed form and, within timeliness constraints, the maximum activity authorized under the license has decayed to less than the quantity specified in 10 CFR Part 20, Appendix C, **and** the licensee's survey, performed in accordance with 10 CFR Part 30.36, does not identify any residual levels of radiological contamination greater than decommissioning screening criteria.

7.4   GROUP 2: UNRESTRICTED RELEASE USING SCREENING CRITERIA; NO DECOMMISSIONING PLAN REQUIRED

Group 2 facilities may have residual radiological contamination present in building surfaces and soils. However, licensees are able to demonstrate that their facilities meet the provisions of 10 CFR 20.1402 ("Radiological Criteria for Unrestricted Use") by applying the screening approach dose analysis described in Chapter 6.

Additionally, licensees in Group 2 typically possess historical records of material receipt, use, and disposal, such that quantifying past radiological material possession and use may be developed with a high degree of confidence. Furthermore, these licensees have radiological survey records that characterize the residual radiological contamination levels present within the facilities and at their sites. That is, they are able to demonstrate residual radiological contamination levels without more sophisticated survey procedures (greater than those used for operational surveys) or dose modeling. These licensees do not need to use site-specific parameters or establish site-specific DCGLs in order to demonstrate acceptability for release of their sites.

For Group 2 facilities, a DP is not required, but licensees will have to demonstrate that the site meets the screening criteria assumptions described in Chapter 6. A DP is not required because worker cleanup activities and procedures are consistent with those approved for routine operations, and no dose analysis is required.

Activities that may fall into the Group 2 category are:

• The licensee possessed and used only sealed sources, but the most recent leak tests indicate that the sources leaked.

• The licensee used unsealed radioactive material, and the licensee's survey demonstrated that levels of radiological contamination on building surfaces or surface soils are less than decommissioning screening criteria.

7.5 GROUP 3: UNRESTRICTED RELEASE USING SCREENING CRITERIA; DECOMMISSIONING PLAN REQUIRED

NRC recognizes that circumstances exist where licensees possess prerequisite expertise, equipment and facilities to remediate their facilities, but have not incorporated remediation procedures into their license prior to license termination. A license amendment is necessary to authorize the activities for decommissioning. Group 3 facilities could meet the screening criteria, but they need to submit a DP, and their license needs to be amended to modify or add to existing procedures, in order to remediate buildings or sites.

For Group 3 facilities, licensees will also have to demonstrate that the site meets the screening criteria assumptions described in Chapter 6. A site-specific dose analysis is not required.

7.6 GROUP 4: UNRESTRICTED RELEASE WITH SITE-SPECIFIC DOSE ANALYSIS AND NO GROUND WATER CONTAMINATION; DECOMMISSIONING PLAN REQUIRED

Group 4 facilities have residual radiological contamination present in building surfaces and soils, but the licensee cannot meet, or chooses not to use, screening criteria, and the ground water is demonstrably not contaminated. The licensees are able to demonstrate that residual radioactive material may remain at their site but within the levels specified in NRC criteria for unrestricted use (10 CFR 20.1402, "Radiological Criteria for Unrestricted Use") by applying site-specific criteria in a comprehensive dose analysis.

A site DP is required and should characterize the location and extent of radiological contamination. The DP should also identify the land use, exposure pathways, and critical group for the dose analysis.

7.7 GROUP 5: UNRESTRICTED RELEASE WITH GROUND WATER CONTAMINATION; DECOMMISSIONING PLAN REQUIRED

Group 5 facilities have residual radiological contamination present in building surfaces, soils, and the ground water. The licensees are able to demonstrate that residual radioactive material may remain at their site but within the levels specified in NRC criteria for unrestricted use (10 CFR 20.1402, "Radiological Criteria for Unrestricted Use") by applying site-specific criteria in a comprehensive dose analysis.

A site DP is required and should characterize the location and extent of radiological contamination. The DP should also identify the land use, exposure pathways, and critical group for the dose analysis.

7.8 GROUP 6: RESTRICTED RELEASE; DECOMMISSIONING PLAN REQUIRED

Group 6 facilities have residual radiological contamination present in building surfaces, soils, and possibly the ground water. The licensees are able to demonstrate that proposed residual radioactivity at the facility is in excess of the levels specified in NRC criteria for unrestricted use but within the levels specified for restricted use (10 CFR 20.1403) by applying site-specific criteria in a comprehensive dose analysis.

A site DP is required and should characterize the location and extent of radiological contamination. The DP should also identify the land use, exposure pathways, institutional controls, and critical group for the dose analysis.

These sites require extensive NRC review and are typically handled on a case-by-case basis.

7.9 GROUP 7: RESTRICTED RELEASE USING ALTERNATE CRITERIA; DECOMMISSIONING PLAN REQUIRED

Group 7 facilities have residual radiological contamination present in building surfaces, soils, and possibly ground water. These licensees intend to decommission their facilities such that residual radioactive material remaining at their site is in excess of the levels specified in NRC criteria for unrestricted use. The licensees will apply site-specific criteria in a comprehensive dose analysis in accordance with alternate criteria for license termination (10 CFR 20.1404). A site DP that identifies the land use, exposure pathways, institutional controls, and critical group for the dose analysis is required. These sites require extensive NRC review and are handled on a case-by-case basis. License termination must be approved by the NRC Commissioners.

8 GROUP 1 DECOMMISSIONING

> **NOTE: In addition to the guidance in this chapter,**
> **licensees are encouraged to contact NRC, or the appropriate Agreement State authority,**
> **to assure an understanding of what actions should be taken**
> **to initiate and complete the license termination process**
> **on a license- or facility-specific basis.**
> **Cases where licensees abandon a site or refuse to decommission a site would be**
> **considered for civil or criminal action, as warranted.**

8.1 INTRODUCTION

Group 1 decommissioning activities involve licensees only using licensed material in a manner that would preclude release of the licensed material to the environment, would not cause the activation of adjacent materials, or would not contaminate work areas above the levels of the decommissioning screening criteria (Appendix B). Termination of these licenses would not require the licensee to submit a DP.

Group 1 includes the following licensees:

- Licensees who possessed and used only sealed sources and whose most recent leak tests are current and demonstrate that the sealed sources did not leak while in the licensee's possession.

- Licensees who possessed and used relatively short-lived radioactive material (i.e., $T_{1/2}$ less than or equal to 120 days) in an unsealed form, the maximum activity authorized under the license has decayed to less than the quantity specified in 10 CFR Part 20, Appendix C, and the licensee's survey performed in accordance with 10 CFR Part 30.36 does not identify any residual levels of radiological contamination greater than decommissioning screening criteria.

8.2 LICENSEE ACTIONS

For Group 1 decommissioning, the following licensee actions are required:

- Notify NRC as required by 10 CFR 30.36(d), 40.42(d), 70.38(d) and 72.54.

- Dispose of the licensed material in accordance with NRC requirements, usually by returning the material to the manufacturer.

- For other than sealed sources, perform a radiation survey and submit the results in accordance with 10 CFR 30.36(j), 40.42(j), 70.38(j), or 72.54(l), or demonstrate that the facility, or portion of the facility, meets NRC criteria for unrestricted use by using the dose screening methodology described in Section 6.6.

- Guidance on surveys is found in Figure 8.1, Section 15.4 of this volume, and Consolidated Decommissioning Guidance, Volume 2.

- For all sealed sources, including those no longer in licensee's possession, provide to NRC results from the most recent leak tests demonstrating there has been no leakage.

- Transfer the decommissioning records discussed in 10 CFR 30.35, 30.36, and 30.51; 40.36, 40.42, and 40.61; 70.25, 70.38, and 70.51; or 72.80, as appropriate, or affirm that they are not required to retain or transfer these records.

- Submit NRC Form 314, "Certificate of Disposition of Materials," or equivalent information to NRC. Written confirmation from the recipient listed on NRC Form 314 that the material has been transferred to them should be attached to the Form 314, shown in Appendix A.

Simplified Survey Procedures

In preparing for the FSS, the licensee should establish a method to identify individual measurement/sampling points on each surface in the indoor area that was involved in licensed material use. At a minimum, the licensee's termination survey should consist of the following:

- One hundred percent scanning of all surfaces in the area of the facility where licensed material was used or stored, using an appropriate radiation detection instrument (including scan sensitivity);

- Evaluations for total and removable radioactive material at each area exhibiting elevated radiation levels, or at a frequency of one wipe comprising 100 cm^2 per 300 ft^2; and

- Evaluations of radiation levels at one meter above surfaces.

Particular attention should be afforded any drains, air vents, or other fixtures or equipment that may have become contaminated during licensed material use. This is especially significant in situations where renovations have occurred and potentially contaminated areas may be inaccessible under current conditions.

Figure 8.1 Simplified Survey Procedures.

8.3 NRC ACTIONS

For Group 1 decommissioning, NRC staff:

- Should determine that the facility meets the Group 1 criteria.

- Should initiate initial processing of the decommissioning action using the Appendix C checklist.

- Should review, after verifying the disposition of the licensed material, the information submitted by the licensee to demonstrate that its facility is suitable for unrestricted use.

- Should review leak test results, verify that the type and number of sources on the license and NRC Form 314 are in agreement and the most recent leak test results are current and indicate that the sources did not leak.

- Should review licensee's survey, paying particular attention to anomalies such as the use of inappropriate radiation survey and analytical instrumentation, incomplete evaluation of radioactive material use/storage areas, and spurious survey results.

- Should contact the licensee if the licensee's FSS does not appear valid or if the leak test results are inconclusive with respect to the condition of the sealed sources.

- Does not need to prepare an environmental assessment for termination of the license, since this action is categorically excluded under 10 CFR 51.22(c)20.

- Will notify the licensee by license amendment after NRC has verified the suitability of its facility for unrestricted use. This amendment will be placed in the license docket file, and the license will be terminated. The completed license amendment and transmittal letter will be included in the official docket file for the license, and a copy shall be placed into NRC's Agencywide Documents Access and Management System (ADAMS).[7]

- Will retire the records in accordance with current records management guidance (e.g., RMG 92–01 and 93–03, or see Volume 3). Retired records will be included in the official docket file for the license, and a copy shall be placed into ADAMS.

---

[7] For certain types of facilities, NRC will notice the amendment issue in the *Federal Register* (see 10 CFR 2.106).

9 GROUP 2 DECOMMISSIONING

> **NOTE: In addition to the guidance in this chapter,**
> **licensees are encouraged to contact NRC, or the appropriate Agreement State authority,**
> **to assure an understanding of what actions should be taken**
> **to initiate and complete the license termination process**
> **on a license- or facility-specific basis.**
> **Cases where licensees abandon a site or refuse to decommission a site would be**
> **considered for civil or criminal action, as warranted.**

9.1 INTRODUCTION

Licensees who decommission under Group 2 did not have releases into the environment in excess of 10 CFR 20 limits and did not activate adjacent materials. These licensees would not be able to decommission under Group 1 because levels of persistent contamination of work areas, building surfaces, and limited surface soil contamination may exist.

Group 2 includes the following licensees:

- licensees who can demonstrate compliance with 10 CFR Part 20.1402 ("Radiological Criteria for Unrestricted Use") using the screening methodology discussed in Section 6.6;

- licensees who possess and use only sealed sources that cannot demonstrate current leak-tight integrity; and

Licensees decommissioning under Group 2 would not be required to develop a DP (10 CFR 30.36(f)(1), 40.42(f)(1), and 70.38(f)(1)) for the following reasons:

- Decommissioning workers would not be entering areas normally occupied where surface contamination and radiation levels are significantly higher than routinely encountered during operation.

- Procedures would involve techniques applied routinely during cleanup or maintenance operations.

- Procedures would not result in significantly greater airborne concentrations of radioactive materials than are present during operation.

- Procedures would not result in significantly greater releases of radioactive material to the environment than those associated with operation.

Licensees using small quantities of C-14 or H-3 may be decommissioned under Group 2 depending on the total activity of C-14 or H-3 possessed under the license and the authorized use of the radioactive material.

9.2   LICENSEE ACTIONS

Although submission of a DP is not required for decommissioning under Group 2, these licensees are required to determine the radiological status of their facility and demonstrate that their facility meets NRC requirements for unrestricted use. This is accomplished by remediating the site as necessary, performing a radiation survey, and conducting dose evaluations using the screening methodology described in Section 6.6.

For Group 2 decommissioning, the following licensee actions are required:

- Notify NRC as required by 10 CFR 30.36(d), 40.42(d), and 70.38(d).

- Dispose of the licensed material in accordance with NRC requirements, usually by returning sealed sources to the manufacturer or disposing of licensed material as outlined in NRC regulations.

- For all sealed sources, including sources no longer in the licensee's possession, provide to NRC results from the most recent leak tests.

- Transfer the decommissioning records discussed in 10 CFR 30.35, 30.36, and 30.51; 40.36, 40.42, and 40.61; or 70.25, 70.38, 70.51, 72.30, and 72.80, as appropriate, or affirm that they are not required to retain or transfer these records.

- Determine the radiological status of the facility and perform further remediation, if necessary, to meet NRC screening criteria for unrestricted use (10 CFR 20.1402).

- Submit an FSSR, or demonstrate that the facility, or portion of the facility, meets NRC criteria for unrestricted use by using the dose screening methodology described in Section 6.6. Guidance on surveys is found in Figure 8.1 and Section 15.4 of this volume and Volume 2 of this NUREG.

- Submit NRC Form 314, "Certificate of Disposition of Materials," or equivalent information to NRC. Written confirmation from the recipient listed on NRC Form 314 that the material has been transferred to them should be attached to the Form 314, shown in Appendix A.

In performing the decommissioning of its facility, the licensee should first identify any areas in the facility that were involved in licensed material use by reviewing facility records and conducting a survey of the licensed material use area. This survey should be similar to the routine contamination surveys conducted under the licensee's radiological safety plan. The licensee should then remediate all surfaces in the areas at the facility that were involved in licensed material use or storage and dispose of all radioactive material and waste as discussed in NRC regulations in 10 CFR Part 20, Subpart K.

If an FSS is required to demonstrate that its facility is suitable for unrestricted use, the licensee should design the survey so it is of sufficient scope and quality to make this demonstration. More information on surveys is contained in Figure 8.1 and Section 15.4 of this volume and Volume 2 of this NUREG.

9.3   NRC ACTIONS

For Group 2 decommissioning, the following are NRC actions:

- Determine whether the decommissioning meets the Group 2 criteria summarized above.

- Initiate initial processing of the decommissioning action using the Appendix C checklist.

- Determine whether a Technical Assistance Control number for the decommissioning action should be assigned and, if so, arrange for one to be assigned to the decommissioning.

- Ensure that the notification of cessation of operations is placed in the licensee's docket file.

- If an EA is needed, consider using the license termination rule Generic Environmental Impact Statement (GEIS), as described in Section 15.7 of this guidance.

- Acknowledge, in writing, the receipt of the notification and inform the licensee of any additional information required to support the licensee's request to terminate the license.

- Contact the licensee by telephone to determine the licensee's estimated decommissioning schedule and confirm that the schedule conforms with NRC requirements. This information will be useful in scheduling any confirmatory surveys or closeout inspections that NRC may undertake as part of the decommissioning of the facility and ensure that the licensee will conduct the decommissioning of its facility in accordance with the schedules discussed in 10 CFR Parts 30.36, 40.42, 70.38, and 72.54.

- Upon receipt of the radiation survey from the licensee, perform a "completeness" review to determine whether the radiation survey contains sufficient type and quality of information to begin the indepth technical review. Inform the licensee of the results of the acceptance review.

- Review the radiation survey to ensure that it adequately demonstrates that the facility is suitable for unrestricted use. See Section 15.4 for a list of radiation survey requirements and contents, and Chapter 4 in Volume 2 of this NUREG for additional information on surveys.

- If there is an issue related to the EPA/NRC MOU (e.g., residual radioactivity levels exceed the concentrations specified in the EPA/NRC MOU), the NRC reviewer should coordinate with DWMEP for the appropriate EPA notification. The EPA/NRC MOU is provided in Appendix H of this volume.

- Ensure that the licensee has submitted NRC Form 314 at the completion of the decommissioning operations.

- Ensure that the licensee has transferred the decommissioning records discussed in 10 CFR Parts 30.35, 30.36, and 30.51; 40.36, 40.42, and 40.61; or 70.25, 70.38, and 70.51, as appropriate, or has affirmed that they are not required to retain or transfer these records.

- As the final step in terminating the license, notify the licensee by license amendment after NRC has verified the suitability of its facility for unrestricted use. This amendment shall be placed in the license docket file, and the license shall be terminated. The completed license amendment and transmittal letter shall be included in the official docket file for the license, and a copy shall be placed into ADAMS.[8]

• Retire the records in accordance with current management directives (e.g., RMG 92–01 and 93–03 or see Volume 3). Retired records shall be included in the official docket file for the license, and a copy shall be placed into ADAMS.

---

[8] Certain types of facilities require an additional *Federal Register* notice at issuance of license amendment (see 10 CFR 2.106).

10 GROUP 3 DECOMMISSIONING

> **NOTE:** In addition to the guidance in this chapter,
> licensees are encouraged to contact NRC, or the appropriate Agreement State authority,
> to assure an understanding of what actions should be taken
> to initiate and complete the license termination process
> on a license- or facility-specific basis.
> Cases where licensees abandon a site or refuse to decommission a site would be
> considered for civil or criminal action, as warranted.

10.1 INTRODUCTION

Group 3 is similar to Group 2 in site conditions, i.e., levels of persistent contamination of work areas, building surfaces, and limited surface soil contamination may exist. Additionally, the types of licensees to which Group 3 applies are similar to those described in Group 2. However, licensees decommissioning under Group 3, and not Group 2, pursuant to 10 CFR 30.36(f)(1), 40.42(f)(1), 70.38(f)(1), and 72.54(d), must develop a DP. (See Chapter 5 for a description of when a DP is required.) This is because they have not incorporated the necessary activities and procedures into their license prior to ceasing operations.

10.2 LICENSEE ACTIONS

Licensee actions are the same as those for Group 2 (see Section 9.2), except that Group 3 licensees need to submit a DP. Even though the submission of a DP is required for decommissioning under Group 3, these licensees, in most cases, will not be expected to submit the same level of detail as required for Groups 4–7. Chapters 16–18 of this volume describe information necessary for the preparation of a DP. Licensees decommissioning under the provisions of Group 3 may find that most of the information below may be excerpted from their current license and that they may only need to develop the limited information not contained in their license.

NRC regulations at 10 CFR 30.36(g)(4)(ii) and (iii), 40.42(g)(4)(ii) and (iii), and 70.38(g)(4)(ii) and (iii) require that DPs contain "a description of the planned decommissioning activities" and "a description of the methods used to ensure protection of workers and the environment against radiation hazards during decommissioning." NRC regulations 10 CFR 72.54(g)(2), (3), and (6) require that DPs contain "the choice of the alternative for decommissioning with a description of the activities involved," "a description of the controls and limits on procedures and equipment to protect occupational and public health and safety," and "a description of technical specifications and quality assurance provisions in place during decommissioning." Licensees decommissioning under Group 3 are required to demonstrate that their facility meets NRC requirements for unrestricted use. Generally, this information is developed by the licensee after determining the radiological status of the facility and is presented to NRC for review and approval in the form of a license amendment request to authorize decommissioning in accordance with the DP.

GROUP 3 DECOMMISSIONING

The general outline of a Group 3 decommissioning plan should include information in Chapters 16–18 as follows:

- Executive Summary
- Facility Operating History
  - License Number/Status/Authorized Activities
  - License History
  - Previous Decommissioning Activities
  - Spills
- Facility Description
  - Site Location and Description
  - Radiological Status of Facility
  - Contaminated Structures
  - Contaminated Systems and Equipment
  - Surface Soil Contamination
- Unrestricted Release using Screening Criteria
  - Building Surfaces
  - Surface Soil
  - Planned Decommissioning Activities
  - Contaminated Structures
  - Contaminated Systems and Equipment
  - Soil
  - Schedules
- Project Management and Organization
  - Radiation Safety Officer
  - Training
  - Contractor Support
- Radiation Safety and Health Program
  - Radiation Safety Controls and Monitoring for Workers
    - Workplace Air Sampling Program
    - Respiratory Protection Program

- ■ Internal Exposure Determination

- ■ Contamination Control Program

- ■ Instrumentation Program

— Health Physics Audits and Recordkeeping Program

- Environmental Monitoring Program

— Effluent Monitoring Program

— Effluent Control Program

- Radioactive Waste Management Program

— Solid Radioactive Waste

— Liquid Radioactive Waste

- Facility Radiation Survey

- Financial Assurance

10.3   NRC ACTIONS

10.3.1   UPON RECEIPT OF THE REQUIRED NOTIFICATION

NRC actions are the same as for Group 2 (see Section 9.3) except that the staff should also review the DP using the process explained in Chapter 5 and the criteria given in Chapters 16–18.

Upon receipt of the required notification from the licensee, NRC staff should:

- Verify the decommissioning group and review required, acknowledge receipt of the notification, and file the notification as discussed in Chapter 5 of this volume.

- Ensure that the licensee has submitted the decommissioning records discussed in 10 CFR Parts 30.35, 30.36, and 30.51; 40.36, 40.42, and 40.61; 70.25, 70.38, and 70.51; or 72.30 and 72.54, as appropriate, or has affirmed that they are not required to retain or submit these records.

- Request that the Regional Office and Headquarters determine whether the lead office for the decommissioning will be the NRC Regional Office or NRC Headquarters.[9]

- Contact the licensee to discuss the decommissioning process and NRC criteria for releasing licensed sites (Appendix D contains a checklist that should be used during NRC staff's discussion with the licensee).

---

[9]   Group 3 sites can be managed by either the NRC Regional office or by NRC Headquarters. Regional staff and management should discuss the decommissioning with NRC Headquarters to determine which office will assume the lead for management of the decommissioning.

- Coordinate with any other groups that may have regulatory authority at the site. This may include State radiation and hazardous materials control authorities, regional radioactive waste compacts, the EPA, or the Occupational Safety and Health Administration (OSHA).

- Determine whether local citizen or environmental groups have an interest in the site, as well as the appropriate individuals to be included on the external distribution list for documents pertaining to the decommissioning. In the past, NRC staff has found that local (State, county, town) regulatory, land use, or public works authorities, and State representatives or county executive offices can be useful in contacting these groups. See Section 15.10 for information on the actions to be taken if local citizen or environmental groups have an interest in the site.

- If warranted by local citizen interest, determine that local libraries have Internet access and provide instructions to the interested local citizens on how to access documents through ADAMS.

10.3.2   UPON RECEIPT OF THE DECOMMISSIONING PLAN

Upon receipt of the DP from the licensee, NRC staff should perform the following:

- Perform an "acceptance" or "completeness" review of the DP to determine if it contains sufficient type and quality of information to begin the indepth technical review of the DP using the Appendix D checklist; it should be completed within 90 days (30 days for most DP amendments).

- Begin initial processing of the decommissioning action using the Appendix C checklist.

- Inform the licensee of the results of the acceptance review. If the DP is not acceptable, inform the licensee of the deficiencies;

- Once the DP is acceptable for review, prepare and publish in the *Federal Register*, and in local media, a notice announcing the receipt of the DP, offering the opportunity for a hearing, and soliciting public comments. Also contact local and State governments and any Indian Nation, seeking their comments (see 10 CFR 20.1405).

- Review the DP as described in Chapter 16 of this volume, using the checklist in Appendix D.

- If the technical review indicates that the DP cannot be approved as submitted, inform the licensee of the need for supplementary information. Coordinate the resolution of the deficiencies with the licensee and any other appropriate organizations exercising regulatory authority at the facility;

- Document the review of the DP in a letter or an SER, using the Appendix G SER outline, as appropriate.

- If an EA is required, consider using the license termination rule GEIS, NUREG–1496, as described in Section 15.7.3 of this guidance; publish the FONSI in the *Federal Register*.

- During the review of the DP, hold a public meeting, if warranted based on discussions with NRC management, the licensee, other regulatory authorities, or interested members of the public. See Section 15.10 for a discussion on decommissioning communications planning and guidance on planning public meetings. Chapter 4 discusses applicable regulations and contains a list of guidance on public meetings.

- Upon approval of the DP, incorporate it into the license as a license amendment.[10]

- Upon completion of decommissioning, terminate the license as described in Section 9.3.

10.4   DOSE MODELING INFORMATION TO BE SUBMITTED

The licensee's dose information for unrestricted release using screening criteria is explained in Section 6.6 of this NUREG.

---

[10]   Certain types of facilities require an additional *Federal Register* notice at issuance of license amendment (see 10 CFR 2.106). See Section 5.3 of this volume.

11 GROUP 4 DECOMMISSIONING

> **NOTE:** In addition to the guidance in this chapter,
> licensees are encouraged to contact NRC, or the appropriate Agreement State authority,
> to assure an understanding of what actions should be taken
> to initiate and complete the license termination process
> on a license- or facility-specific basis.
> Cases where licensees abandon a site or refuse to decommission a site would be
> considered for civil or criminal action, as warranted.

11.1 INTRODUCTION

Facilities that decommission under Group 4 have used licensed material in a manner that resulted in its release into the environment, activated adjacent materials, or resulted in persistent contamination of work areas, but did not result in contamination of ground water. While these facilities have residual radiological contamination present in building surfaces and soils, the licensee cannot meet or chooses not to use screening criteria. The licensees are able to demonstrate that residual radioactive material may remain at their site but within the levels specified in NRC criteria for unrestricted use (10 CFR 20.1402, "Radiological Criteria for Unrestricted Use") by applying site-specific criteria in a comprehensive dose analysis.

11.2 LICENSEE ACTIONS

These licensees should:

- Submit the notification required under 10 CFR 30.36(d), 40.42(d), 70.38(d), and 72.54(d).

- Transfer the decommissioning records discussed in 10 CFR Parts 30.35, 30.36, 30.51; 40.36, 40.42, 40.61; 70.25, 70.38, 70.51; 72.30, and 72.54, as appropriate, or affirm that they are not required to retain or transfer these records.

- Perform a preliminary assessment of the facility, including a document review and a scoping survey.

- Perform site characterization in sufficient detail to support the planned activities and demonstrate that there is no existing ground water contamination.[11]

- Submit a DP in accordance with 10 CFR 30.36(g), 40.42(g), 70.38(g), and 72.54(g) to NRC for review and approval as a license amendment request. Chapters 16–18 describe information necessary for the preparation of a DP. Appendix D contains a DP checklist.

---

[11] Note that ground water must be monitored throughout remediation because these activities could cause ground water contamination.

- Submit an ER. Section 15.7 (NEPA Compliance) describes information necessary for the preparation of an ER.

- Perform the remediation using the approved DP and financial assurance mechanism (FA) (see Volume 3 of this guidance).

- Transfer or dispose of all radioactive material and waste resulting from the decommissioning in accordance with the approved DP and 10 CFR Part 20, Subpart K.

- Perform an FSS in accordance with the procedures approved in the DP.

- Submit the FSSR to NRC for review and approval.

- Submit NRC Form 314 to NRC. A sample Form 314 is shown in Appendix A.

11.3   NRC ACTIONS

11.3.1   UPON RECEIPT OF THE REQUIRED NOTIFICATION

Upon receipt of the required notification from the licensee, NRC staff should:

- Verify the decommissioning group and review required, acknowledge receipt of the notification, and file the notification as discussed in Chapter 5 of this volume.

- Ensure that the licensee has submitted the decommissioning records discussed in 10 CFR Parts 30.35, 30.36, and 30.51; 40.36, 40.42, and 40.61; 70.25, 70.38, and 70.51; or 72.30 and 72.54, as appropriate, or has affirmed that they are not required to retain or submit these records.

- Request that the Regional Office and Headquarters determine whether the lead office for the decommissioning will be the NRC Regional Office or NRC Headquarters.[12]

- Contact the licensee to discuss the decommissioning process and NRC criteria for releasing licensed sites (Appendix D contains a checklist that may be used during NRC staff's discussion with the licensee).

- Coordinate with any other groups that may have regulatory authority at the site. This may include State radiation and hazardous materials control authorities, regional radioactive waste compacts, the EPA, or the Occupational Safety and Health Administration (OSHA).

- Determine whether local citizen or environmental groups have an interest in the site, as well as the appropriate individuals to be included on the external distribution list for documents pertaining to the decommissioning. In the past, NRC staff has found that local (State, county, town) regulatory, land use, or public works authorities, and State representatives or county executive offices can be useful in contacting these groups. See Section 15.10 for information

---

[12] In general, NRC Headquarters will have responsibility for managing decommissioning projects for material sites in Groups 4-7, since they require site-specific dose modeling evaluations, have contaminated groundwater, or are requesting release in accordance with 10 CFR 20.1403 or 10 CFR 20.1404.

on the actions to be taken if local citizen or environmental groups have an interest in the site. Staff should also develop a specific site communication plan.

- If warranted by local citizen interest, arrange to establish a Local Public Document Room (LPDR) or, in lieu of establishing a formal LPDR, arrange with a local library to act as an informal LPDR.

11.3.2 UPON RECEIPT OF THE DECOMMISSIONING PLAN

Upon receipt of the DP from the licensee, NRC staff should:

- Perform an "acceptance" or "completeness" review of the DP to determine if it contains sufficient type and quality of information to begin the indepth technical review of the DP using the Appendix D2 checklist. The acceptance review will be done by a team that includes the requisite disciplines led by the PM; it should be completed within 90 days.

- Initiate processing of the decommissioning action using the Appendix D checklist.

- Inform the licensee of the results of the acceptance review. If the DP is not acceptable, inform the licensee of the deficiencies.

- Once the DP is acceptable for review, prepare and publish in the *Federal Register*, and in local media, a notice announcing the receipt of the DP, offering the opportunity for a hearing, and soliciting public comments. Also contact local and State governments and any Indian Nation, seeking their comments (see 10 CFR 20.1405).

- Review the DP as described in Chapter 16 of this volume, using the checklist in Appendix D.

- If the technical review indicates that the DP cannot be approved as submitted, inform the licensee of the need for supplementary information. Coordinate the resolution of the deficiencies with the licensee and any other appropriate organizations exercising regulatory authority at the facility.

- Evaluate the licensee's ALARA analysis. Volume 2 of this guidance provides more details on completing ALARA analyses.

- Evaluate the licensee's dose analysis (see Volume 2 for details of the technical review).

- Document the review of the DP in an SER, using the Appendix G SER outline.

- Prepare an EA, considering the use of the license termination rule GEIS, NUREG–1496, as described in Section 15.7.3 of this guidance; publish the FONSI in the *Federal Register*.

- During the review of the DP, hold a public meeting, if warranted based on discussions with NRC management, the licensee, other regulatory authorities, or interested members of the public. See Chapter 4, Applicable Regulations, for a list of guidance on public meetings.

- Upon approval of the DP, incorporate it into the license as a license amendment.[13]

---

[13] Certain types of facilities require an additional *Federal Register* notice at issuance of license amendment (see 10 CFR 2.106). See Section 5.3 of this volume.

11.3.3 DURING REMEDIAL ACTIVITIES

NRC staff:

- Should plan to visit the facility during each significant phase of the decommissioning (i.e., characterization, cleanup, final status survey). The number of inspections and surveys will be determined by the Inspection staff and project manager. The facility type and need for survey will be considered. This visit may be coordinated with a scheduled inspection of the facility by qualified inspectors. See Section 15.3 for a discussion of inspections of facilities undergoing decommissioning.

- Should maintain contact with other interested parties, such as other regulatory authorities and members of the public, and make every effort to keep these individuals or groups informed of the progress of the decommissioning.

- Will ensure that all documents relating to the decommissioning are entered into ADAMS.

- As appropriate, should coordinate the review and approval of modifications to the DP with any other groups exercising regulatory authority at the facility.

11.3.4 AFTER COMPLETION OF REMEDIATION

Upon receipt of the FSSR from the licensee, NRC staff should:

- Perform an "acceptance" or "completeness" review of the FSSR, if necessary, to determine whether it contains sufficient type and quality of information to begin the indepth technical review of the FSSR.

- Inform the licensee of the results of the acceptance review. If the FSSR is not acceptable, inform the licensee of the deficiencies.

- If the acceptance review indicates that the FSSR is acceptable, perform the technical review of the FSSR in accordance with Section 15.4.4 of this volume and with Volume 2 of this NUREG.

- If the technical review indicates that the FSSR is unacceptable, inform the licensee of the deficiencies. Coordinate the resolution of the deficiencies with the licensee and any other appropriate organizations exercising regulatory authority at the facility.

Upon completion of the review and acceptance of the FSSR, NRC staff should:

- Conduct a confirmatory survey (if necessary) at the facility following the procedures discussed in Section 15.4.5.

- Upon approval of the confirmatory survey report (if required), NRC staff should verify that the licensed material has been disposed of in accordance with NRC requirements. This may be accomplished by having the licensee provide written confirmation from the recipient listed on NRC Form 314 that the material has been transferred to them, or by NRC staff contacting the recipient listed on NRC Form 314 directly.

- NRC staff should also perform or arrange to have a closeout inspection performed at the facility as discussed in Section 15.3.

- If there is an issue related to the EPA/NRC MOU (e.g., residual radioactivity levels exceed the concentrations specified in the EPA/NRC MOU), the NRC reviewer should coordinate with DWMEP for the appropriate EPA notification. The EPA/NRC MOU is provided in Appendix H of this volume.

- After verifying the disposition of the licensed material and ensuring that a satisfactory closeout inspection was performed, NRC staff will prepare a license amendment and inform the licensee that the license has been terminated.[14]

11.4   DOSE MODELING INFORMATION TO BE SUBMITTED

The licensee's dose modeling for unrestricted release using site-specific information should include the information listed below. For a complete discussion of dose modeling, see Volume 2 of this NUREG.

- Source term information, including nuclides of interest, configuration of the source, and areal variability of the source. Three key areas of review for the source term assumptions are (a) the configuration; (b) the residual radioactivity spatial variability; and (c) the chemical form(s).

- A description of the exposure scenario, including a description of the critical group.

- A description of the conceptual model of the site including the source term, physical features important to modeling the transport pathways, and the critical group.

- Identification, description, and justification of the mathematical model used (e.g., hand calculations, DandD Screen v1.0, and RESRAD v6.0).

- A description of the parameters used in the analysis.

- A discussion about the effect of uncertainty on the results.

- Input and output files or printouts, if a computer program was used.

---

[14] Certain types of facilities require an additional *Federal Register* notice at issuance of license amendment (see 10 CFR 2.106). See Section 5.3 of this volume.

12 GROUP 5 DECOMMISSIONING

> **NOTE:** In addition to the guidance in this chapter,
> licensees are encouraged to contact NRC, or the appropriate Agreement State authority,
> to assure an understanding of what actions should be taken
> to initiate and complete the license termination process
> on a license- or facility-specific basis.
> Cases where licensees abandon a site or refuse to decommission a site would be
> considered for civil or criminal action, as warranted.

12.1 INTRODUCTION

Facilities that decommission under Group 5 have used licensed material in a manner that resulted in its release into the environment, activated adjacent materials or resulted in persistent contamination of work areas, and resulted in contamination of ground water. Group 5 decommissioning includes licensees who intend to decommission their facilities in accordance with the NRC criteria for unrestricted use as described in 10 CFR 20.1402.

Sites with both ground water contamination and any of the following characteristics are in Group 5:

- The near surface ground water is either potable or allowed to be used for irrigation, and provides sufficient yields for those purposes.

- Aquifer volume is sufficient to provide the necessary yields.

- Current and informed consideration of future land use patterns do not preclude ground water use (i.e., material either has a long half-life, with peak exposures occurring later than 1000 years, or the site is in non-industrial areas).

Descriptions of water quality and quantity in the saturated zone should be based on the classification systems used by EPA or the State, as appropriate. For cases where the aquifer is classified as not being a source of drinking water and is adequate for stock watering and irrigation, the licensee does not need to consider the drinking water pathway (and generally, the fish pathway, depending on the model) but should still maintain the irrigation and meat/milk pathways.

12.2 LICENSEE ACTIONS

Group 5 decommissioning requires all the information specified in Chapter 11 plus a description of the extent of ground water contamination and proposed activities to remediate the ground water to meet criteria for unrestricted release.

The information supplied by the licensee should be sufficient to allow NRC staff to determine how the ground water characteristics of the site affect the doses to onsite or offsite individuals during or at the completion of decommissioning. The following information should be included in the ground water hydrology section of the DP (see Section 16.3 for details):

- a description of the saturated and unsaturated zones, including all potentially affected aquifers, the lateral extent, thickness, water-transmitting properties, recharge and discharge zones, and ground water flow directions and velocities;

- descriptions for monitor wells, including location, elevation, screened intervals, depths, construction and completion details, and hydrogeologic units monitored;

- physical parameters such as storage coefficients, transmissivities, hydraulic conductivities, porosities, and intrinsic permeabilities;

- a description of the unsaturated zone, including descriptions of the lateral extent and thickness of permeable and impermeable zones, potential conduits of anomalously high flux, and the direction and velocity of unsaturated flow;

- information on all monitor stations, including location and depth;

- a description of physical parameters, including the spatial and stratigraphic distribution of the total and effective porosity; water content variations with time; saturated hydraulic conductivity; characteristic relationships between water content, pressure head, and hydraulic conductivity; and hysteretic behavior during wetting and drying cycles, especially during extreme conditions;

- a description of the numerical analyses techniques used to characterize the unsaturated and saturated zones, including the model type, justification, documentation, verification, calibration and other associated information. In addition, the description should include the input data, data generation or reduction techniques, and any modifications to these data; and

- the distribution coefficients of the radionuclides of interest at the site.

12.3   NRC ACTIONS

NRC actions[15] are the same as in Section 11.3 with the following additions as described in this section.

12.3.1   UPON RECEIPT OF THE REQUIRED NOTIFICATION

Same as Section 11.3.1.

---

[15]   In general, NRC Headquarters will have responsibility for managing decommissioning projects for material sites in Groups 4-7, since they require site-specific dose modeling evaluations, have contaminated groundwater, or are requesting release in accordance with 10 CFR 20.1403 or 10 CFR 20.1404.

12.3.2 UPON RECEIPT OF THE DECOMMISSIONING PLAN

In addition to those actions in Section 11.3.2, perform the following:

- If the EA does not conclude with a FONSI, NRC staff will conduct an EIS.

- NRC staff should evaluate all of the following in the licensee's description of the ground water hydrology:

  — testing and monitoring program and sample collection procedure;

  — rationale for choosing particular sampling locations;

  — adequacy of non-licensee-constructed monitoring devices used in the characterization;

  — aquifer tests and results derived from testing;

  — potential interactions of ground water with the residual radioactive material; and

  — major hydrologic parameters, aerial extent of aquifers, recharge-discharge zones, flow rates and directions, and travel times, including seasonal fluctuations and long-term trends.

- NRC staff should evaluate all of the following in the licensee's conceptual model:

  — hydrogeologic processes and features, areas of anomalous physical parameters affecting regional processes, extent of aquifers and confining layers, and interactions between aquifers;

  — movement of ground water in the saturated and unsaturated zones;

  — numerical analyses of ground water data collected by the licensee for the site and vicinity (this will normally involve analytical or numerical modeling);

  — model type chosen for analysis is properly documented, verified, and calibrated and adequately simulates the physical system of the site and vicinity;

  — modeling strategy used by the licensee to assure that it is logical and defensible; and

  — adequacy of the model input data generation and reduction techniques. Modifications of input data required for calibration should be reviewed to ensure that the new values are realistic and defensible.

12.3.3 DURING REMEDIAL ACTIVITIES

Same as Section 11.3.3.

12.3.4   AFTER COMPLETION OF REMEDIATION

Same as Section 11.3.4.

Following its review of this information, NRC staff will determine whether the licensee's conclusions are adequate. Alternatively, NRC staff may decide to conduct an independent analysis. If NRC staff were to conduct an independent analysis, they would compare the results with those derived by the licensee to determine if the licensee's results were adequate.

12.4   DOSE MODELING INFORMATION TO BE SUBMITTED

The licensee's dose modeling for unrestricted release using site-specific information should include the information listed in Section 11.4 and should address the following considerations discussed in this section. For a complete discussion of dose modeling, see Volume 2 of this NUREG.

Dose modeling involving ground water contamination presents particular problems in all of the following areas:

- configuration of the source and areal variability of the source;
- exposure scenario, including a description of the critical group;
- conceptual model of the site; and
- mathematical model used.

13 GROUP 6 DECOMMISSIONING

> **NOTE: In addition to the guidance in this chapter,
> licensees are encouraged to contact NRC, or the appropriate Agreement State authority,
> to assure an understanding of what actions should be taken
> to initiate and complete the license termination process
> on a license- or facility-specific basis.
> Cases where licensees abandon a site or refuse to decommission a site would be
> considered for civil or criminal action, as warranted.**

13.1 INTRODUCTION

Facilities that decommission under Group 6 have used licensed material in a manner that resulted in releases to the environment, activated adjacent materials, or resulted in persistent contamination of work areas or ground water. Group 6 decommissioning includes licensees who intend to decommission its facility in accordance with the NRC criteria for restricted use as described in 10 CFR 20.1403. These sites require extensive NRC review and are typically handled on a case-by-case basis.

13.2 LICENSEE ACTIONS

Licensees should include all the relevant information required in Chapter 12, describing the extent of the residual radioactivity and proposed activities to remediate. Additionally, licensees must demonstrate that the site is acceptable for license termination under restricted conditions. This information should be sufficient to allow NRC staff to determine that:

- Further reductions in residual radioactivity would:

  — result in net environmental harm (a demonstration that the benefits of dose reduction are less than the cost of doses, injuries and fatalities) or

  — not be necessary because the proposed levels are ALARA. This analysis should include a complete cost-benefit calculation, because the potential dose exceeds 0.25 mSv/y (25 mrem/y) and is beyond the scope of the GEIS for the LTR. Licensees should use estimates from their decommissioning funding plan as a baseline for ALARA calculations (see Section 17.7.6).

- There are adequate institutional controls to limit TEDE to the public to less than 0.25 mSv/y (25 mrem/year).

- There is sufficient financial assurance for an independent, third party to assume control of the site and perform necessary maintenance at no cost.

- There is agreement by a competent party to assume control of and responsibility for maintenance of the site (see Sections 17.7.3 and 17.7.4).

13.3   DOSE MODELING INFORMATION TO BE SUBMITTED

Dose modeling is required for two conditions: (1) when institutional controls (ICs) are in place, and (2) when ICs fail.

13.3.1   DOSE MODELS

The models should include the following information (see Volume 2 of this NUREG for a complete discussion):

- configuration and areal variability of the source;
- conceptual model of the site;
- mathematical model used; and
- exposure scenarios, including a description of the critical group(s), for the two separate conditions.

The results must demonstrate that with the ICs in place, the TEDE to the critical group is less than or equal to 0.25 mSv/y (25 mrem/y), and if the ICs fail, the TEDE to the critical group is ALARA and may not exceed either:

- 1.0 mSv/y (100 mrem/y); or,
- 5.0 mSv/y (500 mrem/y), provided that the licensee does all of the following:
  — Demonstrates that further reductions in residual radioactivity to meet the 1.0 mSv/y (100 mrem/y) limit (see also Section 17.7.6):
    - are not technically feasible;
    - are prohibitively expensive; or
    - would result in net environmental harm.
  — Provides durable institutional controls (see below).
  — Provides financial assurance for an independent third party to:
    - verify institutional controls remain in place;
    - conduct periodic inspections of the site at least once every five years; and
    - assume control of the site and perform necessary maintenance.

13.3.2 INSTITUTIONAL CONTROLS

The licensee must demonstrate the adequacy of the proposed institutional controls and that the public has had opportunity to comment on them. See Appendix M for additional information on restricted use and institutional controls.

13.3.2.1 Control Adequacy

Requirements for demonstrating adequate institutional controls include:

- The proposed controls are adequate to limit the dose to the public under reasonably foreseeable conditions.

  — For sites exceeding the 1.0 mSv/y (100 mrem/y) dose limit but meeting the 5.0 mSv/y (500 mrem/y) limit, and for sites with long-lived radionuclides such as uranium, controls must be durable, meaning they must be expected to last in perpetuity. State and Federal Agencies are examples of such acceptable organizations.

  — For sites meeting the 1.0 mSv/y (100 mrem/y) limit that do not have long-lived radionuclides (e.g., uranium and thorium), the institutional controls may be of the conventional sort, such as deed restrictions that are legally enforceable by an independent party (e.g., County Zoning Board).

- There is adequate money available to the responsible party in a usable form in order to provide for control and maintenance activities for reasonably foreseeable conditions (see Volume 3 for more information).

- There is an agreement from the proposed control party that it is able and willing to assume responsibility for the site.

As part of the detailed evaluation of the DP, NRC staff's review should verify that the following information is included in the description of institutional controls that the licensee plans to use or has provided for the site:

- a description of the legally enforceable institutional control(s) and an explanation of how the institutional control is a legally enforceable mechanism;

- a description of any detriments associated with the maintenance of the institutional control(s);

- a description of the restrictions on present and future landowners;

- a description of the entities enforcing, and their authority to enforce, the institutional control(s);

- a discussion of the durability[16] of the institutional control(s);

- a description of the activities that the entity with the authority to enforce the institutional control(s) may undertake to do so;

- a description of the manner in which the entity with the authority to enforce the institutional control(s) will be replaced if that entity is no longer willing or able to do so (this may not be needed for Federal or State entities);

- a description of the duration of the institutional control(s), the basis for the duration, the conditions that will end the institutional control(s), and the activities that will be undertaken to end the institutional control(s);

- a description of the plans for corrective actions that may be undertaken in the event the institutional control(s) fail; and

- a description of the records pertaining to the institutional controls, how and where they will be maintained, and how the public will have access to the records.

13.3.2.2   Public Interaction

For sites proposing restricted release for license termination, the licensee shall comply with the provisions of 10 CFR 20.1403, which require the licensee to perform the following:

- Seek the advice of individuals and institutions in the community who may be affected by the decommissioning. Licensees shall seek advice from such affected parties regarding the following matters concerning the proposed decommissioning:

  — Whether provisions for ICs proposed by the licensee will

    - provide reasonable assurance that the TEDE from residual radioactivity distinguishable from background to the average member of the critical group will not exceed 0.25 mSv (25 mrem) TEDE per year;

    - be enforceable; and

    - not impose undue burden on the local community or other affected parties.

  — Whether the licensee has provided sufficient financial assurance to enable an independent third party, including a government custodian of a site, to assume and carry out responsibilities for any necessary control and maintenance of the site.

---

[16]   The Commission has stated (see Section B.3.3 of the "Statements of Consideration" for 10 CFR Part 20, Subpart E, "Radiological Criteria for License Termination") that stringent institutional controls would be needed for sites involving large quantities of uranium and thorium contamination. Typically, these would involve legally enforceable deed restrictions and/or controls backed up by State and local government control or ownership, engineered barriers, and as appropriate, Federal ownership.

- To provide for sufficient opportunity for the public to participate, the licensee shall provide for the following:

  — participation by representatives of a broad cross section of community interests who may be affected by the decommissioning;

  — an opportunity for a comprehensive, collective discussion on the issues by the participants represented; and,

  — a publicly available summary of the results of all such discussions, including a description of the individual viewpoints of the participants on the issues and the extent of agreement or disagreement among the participants on the issues.

- Document in the DP how the advice of individuals and institutions in the community who may be affected by the decommissioning has been sought and incorporated, as appropriate, following analysis of that advice.

Additional details on required information for decommissioning with restricted release and how NRC staff evaluates them are contained in Section 17.7 of this volume.

13.4   NRC ACTIONS

NRC actions[17] are the same as in Section 12.3 with the following additions as described in this section.

13.4.1   UPON RECEIPT OF THE REQUIRED NOTIFICATION

Same as Section 12.3.1.

13.4.2   UPON RECEIPT OF THE DECOMMISSIONING PLAN

In addition to those actions in Section 12.3.2, perform the following:

- Prior to the detailed technical review of the DP, NRC staff should determine that the licensee has provided for adequate institutional controls;

- NRC staff should evaluate the licensee's financial assurance;

- Because the licensee plans to limit future land uses at the site, NRC staff should prepare an EIS. NUREG–1748 and Section 15.7 discuss the process of preparing an EIS, the environmental information that should be considered by licensees in their environmental report, and the content of the EIS;

---

[17] In general, NRC Headquarters will have responsibility for managing decommissioning projects for material sites in Groups 4-7, since they require site-specific dose modeling evaluations, have contaminated groundwater, or are requesting release in accordance with 10 CFR 20.1403 or 10 CFR 20.1404.

- NRC staff should evaluate the licensee's interactions with the public (see Sections 17.7.5 and M.6).

13.4.3   DURING REMEDIAL ACTIVITIES

Same as Section 12.3.3.

13.4.4   AFTER COMPLETION OF REMEDIATION

Same as Section 12.3.4.

Following its review of this information, NRC staff will determine whether the licensee's conclusions are adequate. If NRC staff conduct an independent analysis, they would compare staff results with the licensee's to determine if the licensee's results are adequate.

14   GROUP 7 DECOMMISSIONING

> **NOTE: In addition to the guidance in this chapter,**
> **licensees are encouraged to contact NRC, or the appropriate Agreement State authority,**
> **to assure an understanding of what actions should be taken**
> **to initiate and complete the license termination process**
> **on a license- or facility-specific basis.**
> **Cases where licensees abandon a site or refuse to decommission a site would be**
> **considered for civil or criminal action, as warranted.**

14.1   INTRODUCTION

Group 7 facilities have residual radiological contamination present in building surfaces, soils, and possibly ground water. These licensees intend to decommission their facilities such that residual radioactive material remaining at their site is in excess of the levels specified in NRC criteria for unrestricted use. These sites are not in Group 6 because they are not able to demonstrate that residual radioactivity will meet limits for restricted use at license termination and it is not feasible to make further reductions. The licensees will apply site-specific criteria in a comprehensive dose analysis in accordance with alternate criteria for license termination (10 CFR 20.1404). A site DP that identifies the land use, exposure pathways, institutional controls, and critical group for the dose analysis is required. These sites require extensive NRC review and are handled on a case-by-case basis. License termination criteria must be specifically approved by a vote of the NRC Commissioners.

14.2   LICENSEE ACTIONS

Licensees should include all the relevant information required in Chapter 13 describing the extent of residual radioactivity and proposed activities to remediate it, and demonstrating that the site is acceptable for license termination under restricted conditions.

The information supplied by the licensee should be sufficient to allow the staff to determine whether the residual radioactive material at the site will result in a dose that exceeds 0.25 mSv/y (25 mrem/y) but will not exceed 1 mSv/y (100 mrem/y), considering all man-made sources other than medical, when the radionuclide levels are at the DCGL and are ALARA and when institutional controls are in place. The information should also demonstrate that the financial assurance mechanism(s) are adequate for the site. Finally, the information should be adequate to allow the staff to determine if the institutional controls, site maintenance activities, and the manner in which advice from individuals or institutions that could be affected by the decommissioning was sought, obtained, evaluated, and, as appropriate, addressed in accordance with NRC requirements.

NRC staff should verify that the following information is included in the discussion of why the licensee is requesting license termination under the provisions of 10 CFR 20.1404:

- a summary of the dose in TEDE(s) to the average member of the critical group when the radionuclide levels are at the DCGL (considering all man-made sources other than medical);

- a summary of the evaluation performed pursuant to Volume 2 of this guidance demonstrating that these doses are ALARA;

- an analysis of all possible sources of exposure to radiation at the site and a discussion of why it is unlikely that the doses from all man-made sources, other than medical, will be more than 1 mSv/y (100 mrem/y);

- a description of the legally enforceable institutional control(s) and an explanation of how the institutional control is a legally enforceable mechanism;

- a description of any detriments associated with the maintenance of the institutional control(s);

- a description of the restrictions on present and future landowners;

- a description of the entities enforcing and their authority to enforce the institutional control(s);

- a discussion of the durability[18] of the institutional control(s);

- a description of the activities that the party with the authority to enforce the institutional controls will undertake to do so;

- a description of the manner in which the entity with the authority to enforce the institutional control(s) will be replaced if that entity is no longer willing or able to do so;

- a description of the duration of the institutional control(s), the basis for the duration, the conditions that will end the institutional control(s), and the activities that will be undertaken to end them;

- a description of the corrective actions that will be undertaken in the event the institutional control(s) fail;

- a description of the records pertaining to the institutional controls, how and where they will be maintained, and how the public will have access to the records;

- a description of how individuals and institutions that may be affected by the decommissioning were identified and informed of the opportunity to provide advice to the licensee;

- a description of the manner in which the licensee obtained advice from affected individuals, the local community, or institutions;

---

[18] The Commission has stated (see Section B.3.3 of the "Statements of Consideration" for 10 CFR Part 20, Subpart E, "Radiological Criteria for License Termination") that stringent institutional controls would be needed for sites involving large quantities of uranium and thorium contamination. Typically, these would involve legally enforceable deed restrictions and/or controls backed up by State and local government control or ownership, engineered barriers, and as appropriate, Federal ownership.

- a description of how the licensee provided for participation by a broad cross-section of community interests in obtaining the advice;

- a description of how the licensee provided for a comprehensive, collective discussion on the issues by the participants represented;

- a copy of the publicly available summary of the results of discussions, including individual viewpoints of the participants on the issues and the extent of agreement and disagreement among the participants;

- a description of how this summary has been made available to the public; and

- a description of how the licensee evaluated advice from individuals and institutions that could be affected by the decommissioning, and the manner in which the advice was addressed.

14.3   DOSE MODELING INFORMATION TO BE SUBMITTED

Same as Section 13.3.

14.4   NRC ACTIONS

NRC actions[19] are the same as in Section 13.4, except as noted below.

14.4.1   UPON RECEIPT OF THE REQUIRED NOTIFICATION

Same as Section 13.4.1.

14.4.2   UPON RECEIPT OF THE DECOMMISSIONING PLAN

Perform actions in Section 13.4.2. In addition, because the licensee plans to limit future land uses at the site, the staff should prepare an EIS. NUREG–1748 and Section 15.7 discuss the process of preparing an EIS, environmental information that should be considered by licensees in their environmental report, and the content of the EIS.

- In addition to the public and governmental contacts previously identified, the staff shall also contact and solicit comments from the EPA.

- Following its review of this information, the staff will determine whether the licensee's conclusions are adequate. The staff may decide to conduct an independent analysis, which would be compared to the licensee's results.

---

[19] In general, NRC Headquarters will have responsibility for managing decommissioning projects for material sites in Groups 4-7, since they require site-specific dose modeling evaluations, have contaminated groundwater, or are requesting release in accordance with 10 CFR 20.1403 or 10 CFR 20.1404.

- After the DP review is complete, the staff will make a recommendation to the Commission that addresses comments by EPA and the public.

- The Commission will consider the staff recommendation in approval of the proposed limits and direct the staff to prepare the licensing action.

14.4.3 DURING REMEDIAL ACTIVITIES

Same as Section 13.4.3.

14.4.4 AFTER COMPLETION OF REMEDIATION

Same as Section 13.4.4.

15 OTHER DECOMMISSIONING CONSIDERATIONS

15.1 INTRODUCTION

This chapter discusses various policies and procedures related to decommissioning. The topics include the following:

15.2 Financial Assurance

15.3 Decommissioning Inspections

15.4 Decommissioning Surveys

15.5 Partial Site Decommissioning

15.6 Site Decommissioning Management Plan Sites

15.7 National Environmental Policy Act Compliance

15.8 Nuclear Materials Management and Safeguards System

15.9 Decommissioning Contractors

15.10 Decommissioning Communication Planning

15.11 Controlling the Disposition of Solid Materials

15.2   FINANCIAL ASSURANCE

15.2.1   INTRODUCTION

Financial assurance requirements help ensure that adequate funds will be available to pay for certain costs (e.g., decommissioning) in a timely manner. Financial assurance is achieved through the use of financial instruments. Some financial instruments provide a special account into which the licensee may prepay the applicable costs. Other financial instruments guarantee funding by a suitably qualified third party, thereby providing "defense in depth" in the event the licensee is unable or unwilling to pay these costs when they arise. Licensees with assets that substantially exceed the cost of decommissioning may provide a self-guarantee for financial assurance. Financial assurance for decommissioning must be obtained prior to the commencement of licensed activities or receipt of licensed material, and it must be maintained until termination of the license. If the license is being terminated under restricted conditions, then financial assurance for site control and maintenance must be obtained prior to license termination. The amount of financial assurance obtained is often based on a site-specific cost estimate and must be increased if the cost estimate increases. Under NRC regulations, a number of different types of financial instruments may be used to demonstrate financial assurance, including trusts, letters of credit, surety bonds, and guarantees.

At the end of licensed operations, licensees must maintain all financial assurance established pursuant to 10 CFR Parts 30, 40, 70, or 72. NRC licensees must demonstrate financial assurance for decommissioning and, if applicable, for site control and maintenance following license termination. Volume 3 of this guidance establishes a standard format for presenting the information to NRC that will (a) aid the licensee in ensuring that the information is complete; (b) ensure that applicable requirements in 10 CFR Parts 30, 40, 70, and 72 have been met; and (c) achieve the intent of the regulations. This will ensure that the decommissioning of all licensed facilities will be accomplished in a safe and timely manner and that licensees will provide adequate funds to cover all costs associated with decommissioning and, if applicable, with site control and maintenance.

15.2.2   WHEN IS FINANCIAL ASSURANCE REQUIRED?

NRC's financial assurance requirements for decommissioning apply only to licensees authorized to possess or use certain quantities and types of licensed materials. The minimum possession or use thresholds that trigger the requirements vary, depending on the type of license and the types and quantities of materials authorized under the particular license (see Table 15.1). Any license that authorizes the possession or use of types or quantities of materials exceeding these thresholds is subject to NRC's decommissioning financial assurance requirements. Note that the relevant quantities and types of materials are those authorized under a particular license, even if a licensee does not currently or usually possess or use these same quantities and types of materials.

Table 15.1   Minimum License Thresholds to Demonstrate Financial Assurance

Type of License	Minimum License Threshold Requiring Financial Assistance
PART 30	Unsealed byproduct material with a half-life greater than 120 days in amounts greater than 10^3 times the applicable quantities of Appendix B to Part 30 or, for a combination of isotopes, if R divided by 10^3 is greater than 1 when R is defined as the sum of the ratios of the quantity of each isotope to the applicable value in Appendix B to Part 30; **OR** Sealed sources or plated foils with a half-life greater than 120 days in amounts greater than 10^{10} times the applicable quantities of Appendix B to Part 30 or, for a combination of isotopes, if R divided by 10^{10} is greater than 1 when R is defined as the sum of the ratios of the quantity of each isotope to the applicable value in Appendix B to Part 30.
PART 40	Source material in a readily dispersible form exceeding 10 millicuries (mCi).
PART 70	Unsealed special nuclear material in amounts greater than 10^3 times the applicable quantities of Appendix B to Part 30 or, for a combination of isotopes, if R divided by 10^3 is greater than 1 where R is defined as the sum of the ratios of the quantity of each isotope to the applicable value in Appendix B to Part 30.
PART 72	Any amount of spent fuel or high-level radioactive waste.

Licensees who exceed the minimum thresholds outlined above are required to demonstrate financial assurance for decommissioning that is acceptable to NRC until decommissioning has been completed and the license has been terminated. License applicants must have financial assurance in place prior to the receipt of licensed materials.

If the license is being terminated under restricted conditions pursuant to 10 CFR 20.1403, a licensee must provide financial assurance for site control and maintenance following license termination. This assurance must be in place before the license is terminated, and it must be sufficient to enable an independent third party to assume and carry out responsibilities for any necessary control and maintenance of the site. Figure 15.1 lists financial assurance mechanisms.

FINANCIAL ASSURANCE MECHANISMS

Licensees may choose among a number of different mechanisms to comply with the financial assurance requirements for decommissioning (see Volume 3 of this NUREG series for more information). The following financial assurance "methods" are specifically allowed under 10 CFR Parts 30, 40, 70, or 72:

- Prepayment;

- Surety, insurance, or guarantee;

- External sinking fund coupled with a surety method or insurance; and

- Statement of intent by a Federal, State, or local government.

Figure 15.1 Financial Assurance Mechanisms.

15.2.3   DECOMMISSIONING PLAN

At the end of licensed operations, licensees must maintain all decommissioning financial assurance established pursuant to 10 CFR 30.35, 40.36, 70.25, or 72.30. In addition, licensees who submit a DP must demonstrate financial assurance pursuant to 10 CFR 30.36, 40.42, 70.38, or 72.54.

The decommissioning financial assurance demonstration must include:

- an updated, detailed cost estimate for decommissioning and, if the license is being terminated under restricted conditions, for control and maintenance of the site following license termination;

- one or more financial assurance mechanisms (including supporting documentation);

- a comparison of the cost estimate to the level of coverage provided by the financial assurance mechanisms and, if the license is being terminated under restricted conditions, for control and maintenance of the site following license termination; and

- if applicable, a description of the means to be employed for adjusting the cost estimate and associated funding level over any storage or surveillance period.

Table 15.2 shows the financial assurance needs for each decommissioning group. Volume 3 of this NUREG provides guidance to licensees on preparing the financial assurance demonstration that is to be included as part of a DP.

Table 15.2 Financial Assurance Needs by Decommissioning Group

Decommissioning Group(s)	Financial Assurance Needs
1	Not normally required
2	Not likely to be required
3–7	Decommissioning Funding Plan required as part of DP

15.2.4 NRC REVIEW

NRC staff should evaluate the decommissioning financial assurance demonstrations submitted by licensees pursuant to the requirements in 10 CFR Parts 30, 40, 70, and 72. The staff's review ensures that sufficient funds will be available to carry out decommissioning activities and site control and maintenance (if applicable) in a safe and timely manner. Volume 3 provides specific guidelines on the review process. In general, the staff should review:

- the accuracy and appropriateness of the methods used to estimate decommissioning costs and, if the license is being terminated under restricted conditions, the costs of site control and maintenance;

- the acceptability of the financial assurance mechanism(s) for decommissioning and, if the license is being terminated under restricted conditions, for site control and maintenance; and

- the means identified in the DP for adjusting the cost estimate and associated funding level over any storage or surveillance period.

NRC staff should make a quantitative evaluation of the licensee's (a) cost estimate or certification amount; and (b) financial assurance mechanism(s).

NRC maintains control and security of the financial instruments. The staff follows NRC Management Directive 8.12, "Decommissioning Financial Assurance Instrument Security Program," to ensure security and control of the instrument. In the event a licensee defaults before completing the decommissioning, the management directive specifies authority for drawing on the instrument. Policy and Guidance Directive PG 8–11, "NMSS Procedures for Reviewing Declarations of Bankruptcy," specifies bankruptcy procedures. In the event of a bankruptcy, the staff should follow the procedures in the policy and guidance directive to ensure control of the radioactive material and maximum use of any remaining licensee resources for protection of the public.

15.2.5 REFERENCES

- NRC Management Directive 8.12, "Decommissioning Financial Assurance Instrument Security Program."

- Policy and Guidance Directive PG 8–11, "NMSS Procedures for Reviewing Declarations of Bankruptcy."

15.3 DECOMMISSIONING INSPECTIONS

15.3.1 INSPECTION POLICY

Licensees undergoing decommissioning will be periodically inspected. Therefore, it is important for licensees to understand the potential enforcement options available to NRC during the course of these periodic inspections. NUREG–1600, "General Statement of Policy and Procedure for NRC Enforcement Actions," describes the Commission's current Enforcement Policy for materials licensees.

The AEA establishes "adequate protection" as the standard of safety on which NRC regulations are based. In the context of NRC regulations, safety means avoiding undue risk or, stated another way, providing reasonable assurance of adequate protection to workers and the public in connection with the use of source, byproduct, and special nuclear materials.

While safety is the fundamental regulatory objective, compliance with NRC requirements plays an important role in giving NRC confidence that safety is being maintained. NRC requirements, including technical specifications, other license conditions, orders, and regulations, have been designed to ensure adequate protection—which corresponds to "no undue risk to public health and safety"—through acceptable design, construction, operation, maintenance, modification, and quality assurance measures. In the context of risk-informed regulation, compliance plays a very important role in ensuring that key assumptions used in underlying risk and engineering analyses remain valid.

While adequate protection is presumptively assured by compliance with NRC requirements, circumstances may arise where new information reveals that an unforeseen hazard exists or that there is a substantially greater potential for a known hazard to occur. In such situations, NRC has the statutory authority to require licensee action above and beyond existing regulations to maintain the level of protection necessary to avoid undue risk to public health and safety.

Based on NRC evaluation of noncompliance, the appropriate action could include refraining from taking any action, taking specific enforcement action, issuing orders, or providing input to other regulatory actions or assessments, such as increased oversight (e.g., increased inspection). Since some requirements are more important to safety than others, NRC endeavors to use a risk-informed approach when applying NRC resources to the oversight of licensed activities, including enforcement activities.

The primary purpose of NRC's Enforcement Policy is to support NRC's overall safety mission in protecting the public health and safety and the environment. Consistent with that purpose, the policy endeavors to:

- Deter noncompliance by emphasizing the importance of compliance with NRC requirements.

- Encourage prompt identification and prompt, comprehensive correction of violations of NRC requirements.

Therefore, licensees, contractors, and their employees who do not achieve the high standard of compliance that NRC expects will be subject to enforcement sanctions. NRC holds the licensee ultimately responsible, including the performance of any contractor. Each enforcement action is dependent on the circumstances of the case. However, in no case will licensees who cannot achieve and maintain adequate levels of safety be permitted to continue to conduct licensed activities.

15.3.2 INSPECTIONS

At the onset of decommissioning, a site-specific inspection plan and inspection schedule will be developed by NRC inspection staff and PM (or other staff having licensing authority). The Plan and Schedule are based on planned site characterization, remediation, final and confirmatory surveys, and other decommissioning activities to be conducted at the facility. Typically, this site-specific plan (commonly referred to as the Master Inspection Plan (MIP)) is coordinated with the licensee prior to being finalized. The purposes of this coordination are to ensure that inspections are performed at times when significant decommissioning activities are underway and to inform the licensee of the areas of the licensee's program that will be inspected. A discussion of the MIP is in Appendix F. As shown in the MIP, licensee procedures are subject to the NRC inspection process.

The Regions are taking the following actions to increase efficiency in the decommissioning inspection program:

- Linking inspections to the licensee's onsite activities, so that inspectors can make side-by-side observations and measurements during licensee-conducted surveys.

- Interacting with the licensees to ensure complete and appropriate submittals.

- Conducting inprocess inspections only at sites that are actively being remediated, which consequently reduces onsite inspection time and limits the scope and depth of inspections to examining key decommissioning activities.

15.4 DECOMMISSIONING SURVEYS

Following the decision to cease operations, a number of surveys may be needed to determine the site radiological status, monitor progress during remediation, and confirm that the site meets the radiological release criteria.

15.4.1 SITE CHARACTERIZATION SURVEY

Licensees conduct site characterization surveys to determine the type and extent of radiological contamination of structures and environmental media. This information is typically provided as part of the DP. The staff reviews the information in the DP to determine whether or not there is sufficient information to permit planning for site remediation that will be effective and will not endanger the remediation workers, to demonstrate that it is unlikely that significant quantities of residual radioactivity have gone undetected, and to provide information that will be used to design the final status survey. Volume 2 of this guidance discusses the information to be submitted by the licensee and provides details of the staff's review.

Generally, the type and scope of the characterization survey information are less detailed than those required for a final radiological survey. However, licensees may use characterization survey data to support the final radiological survey, as long as they can demonstrate that non-impacted areas at the site have not been adversely impacted by decommissioning operations, and the characterization survey data are of sufficient scope and detail to meet the information needs of a final survey (see Volume 2).

Licensees typically submit site characterization information as part of their DP. However, submission of incomplete site characterization information may result in NRC declining to accept and review the DP until appropriate site characterization information is obtained. The licensee may be requested to submit Site Characterization Plans (SCPs) or other site characterization information prior to submitting the DP or NRC may elect to meet with the licensee prior to, or during, site characterization work. However, it is important to note that, unless required by a license condition, licensees are not required under NRC regulations to submit a separate SCP or Site Characterization Report (SCR), only that site characterization information is required as a component of the DP. So, NRC staff will only request this information when necessary to ensure safety and compliance with NRC regulations.

The regulatory requirements for site characterization surveys are contained in 10 CFR 30.36(g)(4)(i), 40.42(g)(4)(i), 70.38(g)(4)(i), and 72.54(g)(1).

15.4.2 IN-PROCESS SURVEYS

These surveys, conducted during remediation, will assist the licensee in determining when remedial actions have been successful and when the final status survey may commence. In addition, information from these surveys may be used to provide the principal estimate of contaminant variability that will be used to calculate the final status survey sample size in a remediated survey unit. Volume 2 discusses the information to be submitted by the licensee and provides details of NRC staff's review. NRC surveys are conducted in accordance with Inspection Procedure 87104, "Decommissioning Inspection Procedures for Materials Licenses."

The regulatory requirements for in-process surveys are contained in 10 CFR 30.36(g)(4)(ii), 40.42(g)(4)(ii), 70.38(g)(4)(ii), and 72.54(g)(2).

15.4.3   FINAL STATUS SURVEY

Licensees wishing to terminate their licenses must demonstrate to NRC that residual radioactive material at their facility attributable to past licensed operations does not exceed NRC criteria for release of the facility. To the extent that unlicensed sources above background levels of radiation are commingled with licensed material, they are also remediated in decommissioning, and would be included in the source term for dose calculations. The final radiation survey demonstrates that the facility meets NRC criteria for release and termination of the license.

NRC staff will review the final status survey design, as part of the DP review, to determine whether the survey design is adequate for demonstrating compliance with the radiological criteria for license termination. Volume 2 of this guidance discusses the information to be submitted by the licensee and provides details of the staff's review.

NRC regulations require that DPs include a description of the planned final radiological survey. Note that some survey methods, such as MARSSIM, require that certain information needed to develop the final radiological survey be developed as part of the remedial activities at the site and should be submitted in accordance with the instructions in Volume 2.

NRC staff, in conjunction with other Federal Agencies, developed a comprehensive manual for conducting final status surveys (Multi-Agency Radiological Survey and Site Investigation Manual (MARSSIM), NUREG–1575). The purpose of MARSSIM is to describe the procedures for designing and conducting surveys to demonstrate that the residual radioactive material at a facility meets NRC criteria for release of the facility and termination of the facility license.

There are limitations to the applicability of MARSSIM; for example, the methodology currently cannot be applied to volumetric or ground water contamination (see Section 4.6 from Volume 2 of this NUREG series and Table 1.1 in NUREG–1575).

Licensees may submit information on facility radiation surveys using one of the four methods described below.

- Method 1:

    Licensees may submit the information on the release criteria, site characterization survey, and remedial action support surveys, along with a commitment to use the MARSSIM approach in developing the final radiological survey. See Volume 2 of this guidance for further details.

- Method 2:

    For Groups 1–3, a simplified survey may be used, as discussed in Chapters 8–10.

- Method 3:

    Many surveys for Groups 4–7 are addressed by the MARSSIM methodology. However, in some cases, site conditions for Groups 4–7, such as volumetric soil contamination and ground water contamination, are beyond the scope of MARSSIM's statistical applicability. Then a site-specific approach should be developed. (See Appendix I of Volume 2 of this NUREG, Section 2.6 of NUREG–1575 for alternate statistical methodologies, and Sections 4.2.4 and 6.5.5 of NUREG–5849 for general context.) Licensees should coordinate the approach with NRC staff, based on site conditions, and historical site assessment.

- Method 4:

    Licensees may propose other survey methodologies, as appropriate, as part of their DP. Use of an alternate methodology may require an indepth NRC technical review.

The regulatory requirements for FSSes are contained in 10 CFR 20.1501(a), 30.36(g)(4)(iv), 40.42(g)(4)(iv), 70.38(g)(4)(iv), and 72.54(g)(4).

15.4.4 FINAL STATUS SURVEY REPORT

The results of final status surveys are documented in a detailed report that becomes part of the licensee's application to terminate the license. The purpose of the staff's review is to verify that the results of the final status survey demonstrate that the site, area, or building meets the radiological criteria for license termination. Section 4.5 from Volume 2 of this NUREG provides guidance on the acceptable format and content of this report. This section also contains guidance for reviewing FSSRs.

The regulatory requirements for FSSRs are contained in 10 CFR 20.1402, 20.1403, 20.1501, 30.36(j)(2), 40.42(j)(2), 70.38(j)(2), and 72.54(i)(2).

15.4.5 CONFIRMATORY SURVEYS

After acceptance of the licensee's FSSR, NRC may conduct a confirmatory survey. Inspection Procedure 83890, "Closeout Inspection and Survey," discusses the procedures to be followed to determine whether a confirmatory survey is required at a licensed facility and the procedures for performing confirmatory surveys. NRC staff should assign higher priority for conducting confirmatory surveys at sites that may pose a greater potential threat to the public health and safety. The confirmatory survey develops radiological data of the same type as that presented by the licensee, but it is usually limited in scope to spot-checking conditions at selected site locations, comparing findings with those of the licensee, and performing independent statistical evaluations of the data developed by the two surveys. An objective of the confirmatory survey is to verify the accuracy of the licensee's measurement technique. Only limited statistical information is developed to compare with the information submitted by the licensee. NRC uses the report of this survey to support a decision on the licensee's application to terminate a license and release the site. NRC regulations do not include specific requirements for the confirmatory survey.

Any decommissioning facility could undergo a confirmatory survey. NRC has implemented a risk-informed process that assigns higher priority for conducting confirmatory surveys at sites that may pose a greater potential threat to the public health and safety. NRC's approach assumes that inprocess inspections are more efficient than one-time confirmatory surveys. This approach would allow the release of some facilities from regulatory control based solely on past operations and performance, NRC's confidence that the facility was adequately remediated by the licensee, and a satisfactory closeout inspection.

If a confirmatory survey will be performed by an NRC contractor, the staff reviewer should coordinate NRC activities with DWMEP. Volume 2 of this NUREG contains further guidance on confirmatory surveys.

15.5 PARTIAL SITE DECOMMISSIONING

A licensee who has submitted a DP that has not yet been approved or a licensee who has an approved DP may opt to release a portion of its site early. For the case of partial site release, the licensee must submit a request for a license amendment to the extent that the actions are not described in the DP and follow the decommissioning process (characterize contamination and surveys) described in Chapter 5.

A site enters into partial site decommissioning in one of two ways:

- The licensee requests a portion of its facility be removed from the license.
- A licensed facility is required per 10 CFR 30.36(d)(1–4), 40.42(d)(1–4), 70.38(d)(1–4), and 72.54(d)(1–3) to begin decommissioning at a portion of its facility (see below and Figure 5.1).

15.5.1 RECORDS

10 CFR 30.35(g), 40.36(f), 70.25(g), and 72.30(d) describe the requirements for the maintenance of records pertaining to decommissioning licensed facilities that would also apply to decommissioning a portion of a licensed facility (note that partial facility decommissioning would also be accomplished using the appropriate decommissioning group discussed in previous chapters).

15.5.2 SPECIFIC LICENSES

Typically, a specific licensee's facilities are identified in its license, and thus an amendment to the license is required prior to releasing the building or area for unrestricted release. Licensees with a single address or location of use incorporating multiple sites or buildings, such as a research facility with licensed material usage in several buildings, and who have determined that the provisions of 30.36(d) apply, may be required to develop a DP for each area of use (see Chapters 8–14 for applicable decommissioning groups). If a DP were required, or if several areas were to be decommissioned, a single plan that incorporates the decommissioning of each of these areas may be acceptable.

If only a portion (i.e., a single building or an outdoor area) of a licensed facility is to be decommissioned, the DP, if required, should address the portion of the property that will be removed from the license. In addition, the DP should incorporate measures to ensure that the area being decommissioned is separated from the area that will remain as a controlled area. For example, this may be accomplished by erecting a fence, or establishing administrative controls between the two portions of the site. The DP should address not only the decommissioning of the portion of the site but also the measures that will ensure that the decommissioned area does not become recontaminated by future licensed activities. At the completion of the decommissioning operations, the license will be amended to indicate that radioactive material use is no longer authorized in that portion of the facility that was decommissioned.

Upon final decommissioning of the site, the licensee will consider residual radioactivity and any dose contribution from all previous site releases when computing the final dose. The total dose contribution must meet the site's final release criteria. See Volume 2 of this NUREG for more information.

15.5.3 BROAD SCOPE LICENSES

Broad scope licensees pursuant to 10 CFR Part 33 are authorized to internally establish, terminate, and resume uses of licensed materials at separate locations (e.g., individual laboratories within a building). Typically, based on license conditions, these licensees are not required to notify NRC as described in 10 CFR 30.36(d), because a decision has not been made to permanently cease principal activities at the entire site or in any separate building. Broad scope licensees also have license requirements incorporated into their operational program for the release of existing and approval of new material use areas. Furthermore, broad scope licensees generally would not have to submit a DP and would not request an amendment to their license to describe changes in areas of use. Broad scope licensees who issue internal approvals would only be required to maintain records of the decommissioning for review by NRC inspectors, per 10 CFR 30.35(g).

However, two specific provisions of 10 CFR Part 30, Section 30.36, "Expiration and termination of licenses and decommissioning of sites and separate buildings or outdoor areas," 10 CFR 30.36(d)(2) and (4) apply to broad scope licensees:

- 10 CFR 30.36(d)(2) states that the licensee has decided to permanently cease principal activities, as defined in this part, at the entire site or in any separate building or outdoor area that contains residual radioactivity such that the building or outdoor area is unsuitable for release in accordance with NRC requirements; or

- 10 CFR 30.36(d)(4) states that no principal activities have been conducted for a period of 24 months in any separate building or outdoor area that contains residual radioactivity such that the building or outdoor area is unsuitable for release in accordance with NRC requirements.

A key qualifier, cited in the above subsections, is the emphasis placed on the presence of radiological contamination in excess of NRC unrestricted release limits in licensed facilities (separate buildings or areas). Therefore, broad scope licensees who remediate their facilities to meet operational radiological release limits (specified in their license) may find that information on areas where past licensed activities were conducted do not need to be provided to NRC as required in 10 CFR 30.36, and would need only to maintain records (10 CFR 30.35(g)) for NRC inspection. Since licensees must account for dose consequences for all past areas of use upon license termination, licensees who elect not to notify NRC may wish to contact NRC prior to relinquishing control of a building or area, if prior to license termination. Licensees are also encouraged to review decommissioning surveys in Section 15.4 to ensure that the operational release surveys contain sufficient information to satisfy FSS requirements at license termination.

However, just as required for a specific licensee, if a broad scope licensee were to identify a building or area in excess of NRC's unrestricted release criteria, or if the remediation would require use of procedures not approved in their license, or if the remediation would have adverse dose consequences upon workers, the public, or the environment, they would also be required to notify NRC, as well as to make a determination as to whether or not a DP is required.

15.6   SITE DECOMMISSIONING MANAGEMENT PLAN SITES

In March 1990, NRC established the Site Decommissioning Management Plan (SDMP) program to help ensure the timely cleanup of sites warranting special attention by the Commission. The SDMP program was implemented to identify and resolve the issues associated with the remediation of numerous licensed, formerly licensed, and unlicensed sites contaminated with residual radioactive material in excess of NRC criteria.

In 1992, the staff developed the Site Decommissioning Management Plan (SDMP) Action Plan to: 1) identify criteria that would be used to guide the cleanup of sites; 2) state the NRC's position on finality; 3) describe the NRC's expectation that cleanup would be completed within 3-4 years; 4) identify guidance on site characterization; and 5) describe the process for timely cleanup on a site-specific basis.

Since development of the SDMP Action Plan, the staff has addressed the issues identified in the Action Plan, as follows: (1) The criteria for site cleanup and NRC's position on finality were codified in 10 CFR Part 20, Subpart E (LTR); (2) NRC's expectations regarding the completion of site decommissioning have been codified in 10 CFR 30.36, 40.42, 70.38, and 72.54; and (3) Issues associated with site characterization have been addressed in the Multi-Agency Radiation Survey and Site Investigation Manual (MARSSIM) (NUREG-1575, Rev. 1, August 2000), and in Volume 2: Characterization, Survey, and Determination of Radiological Criteria, of the Consolidated Decommissioning Guidance (NUREG-1757, Vol. 2, September 2003).

The LTR authorized two different sets of cleanup criteria—the concentration-based SDMP Action Plan criteria and the dose-based LTR criteria. Under the provisions of 10 CFR 20.1401(b), any licensee that submitted its decommissioning plan (DP) before August 20, 1998, and received NRC approval of that DP before August 20, 1999, could use the SDMP Action Plan criteria for site remediation. In the SRM on SECY-99-195, the Commission granted an extension of the DP approval deadline, for 12 sites, to August 20, 2000. In September 2000, the staff notified the Commission that all 12 DPs were approved by the deadline. All of the sites that received approval of their DPs before August 20, 2000, are referred to as "grandfathered" sites.

On June 17, 2004, the staff announced the elimination of the SDMP designation in the *Federal Register* (69 *Federal Register* 33946). NRC now manages materials decommissioning sites as "complex sites," under a comprehensive decommissioning program. The SDMP designation is now used only to describe the cleanup criteria prior to the LTR.

In the past, in order to remove a site from the SDMP, the staff would develop an analysis of the remediation of the site, inform the Commission of its intent to remove the site from the SDMP, and await the Commission's approval. In the SRM on SECY-04-0024, the Commission approved changing the approach for the staff to notify the Commission of its intent to remove sites from the SDMP. The staff no longer seeks Commission approval to release sites that have been grandfathered and are being remediated to the concentration-based SDMP Action Plan criteria, as long as the dose from residual radioactivity at the site does not exceed the dose-based unrestricted release provisions in the LTR. However, the staff must seek specific Commission permission before releasing any grandfathered site that does not meet the dose-based unrestricted release provisions of the LTR.

A grandfathered SDMP site is not required to demonstrate that the concentration-based SDMP Action Plan cleanup criteria meet the unrestricted release provisions of the LTR (0.25 mSv/y (25 mrem/y) and ALARA). In these cases, the NRC staff is ultimately responsible for estimating the dose from residual radioactivity at the site.

15.7   NATIONAL ENVIRONMENTAL POLICY ACT COMPLIANCE

The National Environmental Policy Act (NEPA) of 1969 requires Federal Agencies, as part of their decision-making process, to consider the environmental impacts of actions under their jurisdiction. Both the Council on Environmental Quality (CEQ) and NRC have promulgated regulations to implement NEPA requirements. CEQ regulations are contained in 40 CFR Parts 1500 to 1508, and NRC requirements are provided in 10 CFR Part 51. The NEPA review (also referred to as the environmental review) process for decommissioning is initiated by a licensee's request for a license amendment to decommission. A flow chart illustrating the NEPA process is shown in Figure 15.2.

Most decommissioning actions are in Group 1 and have little, if any, significant impact on the environment; compliance with NEPA involves a determination that the action qualifies as a categorical exclusion (CATX). For decommissioning actions in Groups 2–5, an environmental assessment (EA) is needed. For Groups 6 and 7, potentially significant impacts may result from the proposed decommissioning actions, and a detailed environmental review and preparation of an EIS may be required. See Table 15.3 for a listing of NEPA actions appropriate for each decommissioning group.

NUREG–1748 (Environmental Review Guidance for Licensing Actions Associated with NMSS Programs) provides general procedures for the environmental review of licensing actions for materials facilities regulated by NMSS. The NMSS environmental guidance includes:

- whether a licensee's request is a CATX or whether the staff needs to prepare an EA or EIS;

- early planning for an EA or EIS;

- methods of using previous environmental analyses related to the proposed action;

- the EA process, including preparation and content of the EA, agencies to be consulted, and preparation of the Finding of No Significant Impact (FONSI);

- the process of preparing an EIS, from developing a project plan, through scoping, consultations, and public meetings, to preparing the Record of Decision;

- the content of the EIS; and

- environmental information that should be considered by licensees in their environmental report (ER).

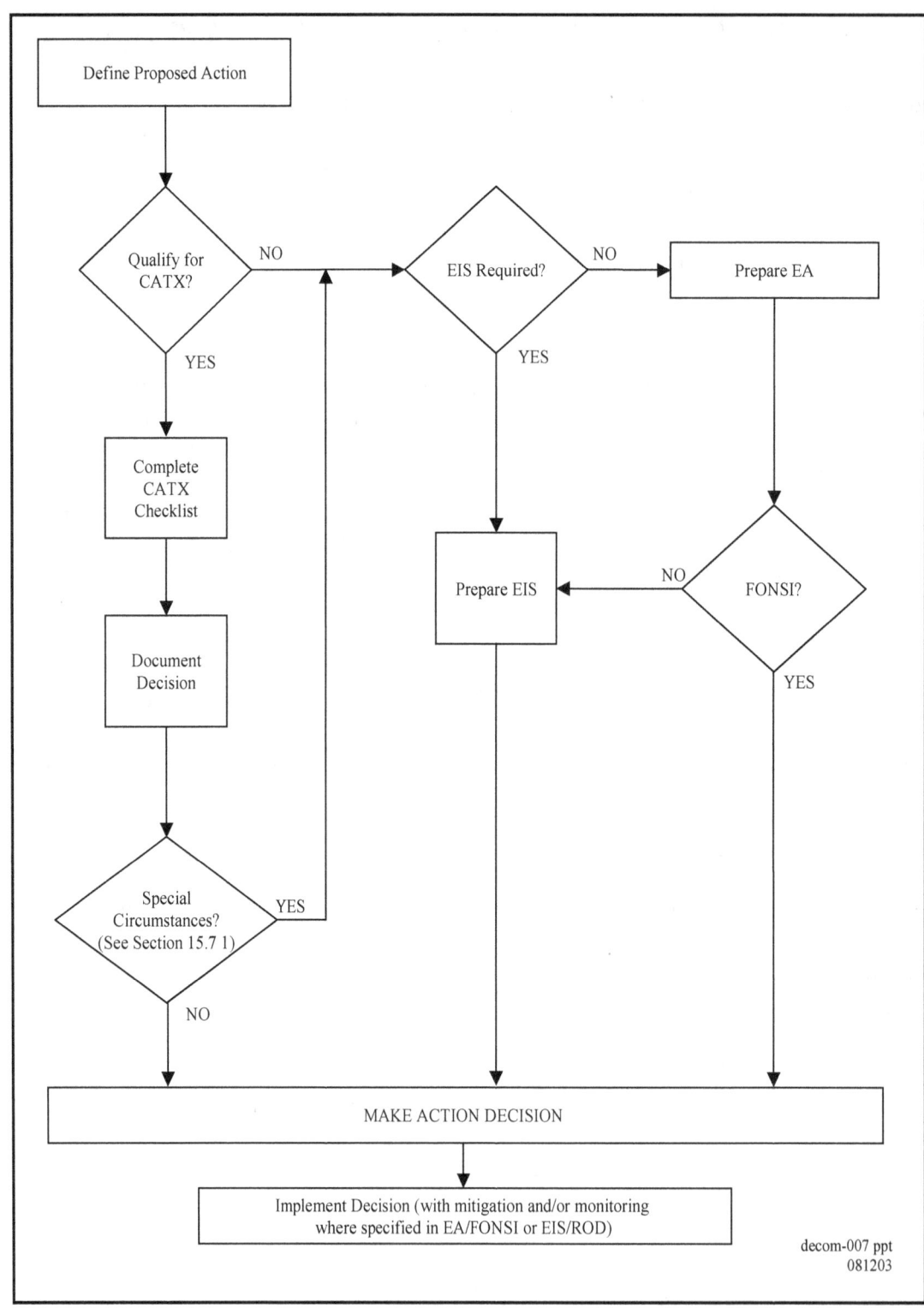

Figure 15.2   NEPA Screening Process (from NUREG–1748).

Table 15.3   Decommissioning Groups and Associated NEPA Actions

GROUP	DESCRIPTION OF GROUP	NEPA ACTION
1	Licensed material used in a manner that would preclude releases to the environment. No DP required. (Limited to sealed sources and small quantities of short half-life materials.)	An environmental assessment (EA) for termination of the license is not required, since this action is categorically excluded under 10 CFR 51.22(c)(20).
2	Would not typically be expected to result in unmonitored releases into the environment. Dose screening methodology or final status survey report (FSSR) required. No DP required.	An EA is required; consider relying on the license termination rule GEIS, as described in Section 15.7.3 of this guidance.
3	Dose screening methodology. DP required.	Same as Group 2.
4	Typically results or has resulted in releases into the environment. Volumetric contamination without existing ground water contamination, and surface and soil contamination that does not meet screening criteria. Licensee plans unrestricted use.	Same as Group 2.
5	Licensed material used in a manner that resulted in releases into the environment, including ground water contamination. Licensee plans unrestricted use.	An EA will be required. If ground water is contaminated and a FONSI cannot be determined, an EIS will be necessary.
6	Licensed material used in a manner that resulted in releases into the environment. Licensee plans restricted use.	Because the licensee plans to limit future land uses at the site, the staff should prepare an EIS. NUREG–1748 discusses the process of preparing an EIS, environmental information that should be considered by licensees in their environmental report, and the content of the EIS.
7	Licensed material used in a manner that resulted in releases into the environment. Licensee plans restricted use and requests use of the alternate criteria in 10 CFR 20.1404.	Same as Group 6.

15.7.1 CATEGORICAL EXCLUSION

When a request for decommissioning is received from a licensee, NRC first determines whether a CATX is applicable for the proposed action. CATXs are categories of actions that NRC, in consultation with CEQ, has determined do not individually or cumulatively have a significant effect on the environment. Criteria for identifying a CATX and a list of actions eligible for CATX are provided in 10 CFR 51.22. Group 1 license termination actions qualify for a categorical exclusion under 10 CFR 51.22(c)(20). An EA or EIS is only prepared when there is a potential for environmental impacts from the decommissioning of the facility.

For a CATX, the finding should be documented in the staff's safety review or in the response to the licensee. For example, the staff could state in the letter to the licensee that "an environmental assessment for this action is not required, since this action is categorically excluded under 10 CFR 51.22(c)(20), because licensed operations have been limited to the use of small quantities of short-lived radioactive materials." The proposed action is subject to no further NEPA review, but it is still evaluated for compliance with NRC radiation protection regulations and other applicable regulations.

Further guidance on CATXs, including a CATX checklist, is contained in the NMSS environmental guidance (NUREG–1748). The reviewer should complete the checklist in NUREG–1748 to ensure that no special circumstances exist that would require preparation of an EA. Special circumstances in which a CATX may not apply for decommissioning include (a) decommissioning activities that could significantly affect the natural or cultural environment, (b) activities that could generate a great deal of public interest, or (c) a high level of uncertainty about the decommissioning's environmental effects.

15.7.2 ENVIRONMENTAL ASSESSMENT

If no CATX applies, NRC typically prepares an EA (10 CFR 51.21 and 51.30). An EA is a concise, publicly available document that provides sufficient evidence and analysis for determining whether to prepare an EIS or a FONSI. EAs are prepared by project managers or license reviewers. If it is determined that no significant impacts exist, the FONSI (10 CFR 51.32 to 51.35) is prepared for publication in the *Federal Register* (10 CFR 51.119).

If the EA reveals the proposed action may significantly affect the environment and cannot be mitigated, the development of an EA is discontinued, and the process to develop an EIS is initiated. If the action under review is certain to result in significant impacts, the EA can be skipped, and the environmental review to support the action should move directly to an EIS. Figure 15.3 lists when an EA must be prepared.

> **An EA must be prepared for proposed actions that are not:**
>
> - exempt from NEPA;
>
> - categorically excluded (10 CFR 51.22);
>
> - covered in an existing EIS or other environmental analysis; or
>
> - required to have an EIS prepared (10 CFR 51.21).

Figure 15.3 When to Prepare an Environmental Assessment.

As provided in 10 CFR 51.45 and 10 CFR 51.60, certain license amendment requests are required to be accompanied by an environmental report (ER). In cases where an ER is not required, NRC staff may require that environmental information be submitted to aid the NRC staff in complying with NEPA (10 CFR 51.41). The general requirements for an ER are described in 10 CFR 51.45. When the environmental information is submitted, NRC staff should conduct an acceptance review to determine whether (a) the requested action will require an EA or EIS, and (b) the information is complete and will support the required environmental analyses.

EAs and EISs (i.e., NEPA documents) focus on the potential environmental impacts of the proposed action. NRC also prepares an SER to evaluate the safety of the proposed action and compliance with NRC regulations. (Appendix G contains an SER outline and template.) The safety and environmental reviews are conducted in parallel. Although there is some overlap between the content of an SER and the NEPA document, the intent of the documents is different. The NEPA document usually includes a summary of the SER findings to aid in the decision process. Much of the information describing the affected environment is also applicable to the SER (e.g., traffic patterns, demographics, geology, and meteorology), and NRC staff should ensure consistency between the NEPA document and the SER, preferably by references to each other.

15.7.3 ABBREVIATED ENVIRONMENTAL ASSESSMENT—RELYING ON THE LICENSE TERMINATION RULE GENERIC ENVIRONMENTAL IMPACT STATEMENT

NRC staff may be able to rely on the 1997 "Generic Environmental Impact Statement in Support of Rulemaking on Radiological Criteria for License Termination of NRC–Licensed Nuclear Facilities" (GEIS, NUREG–1496), to satisfy NEPA obligations for decommissioning sites where the licensee proposes to release the site for unrestricted use. To determine if the GEIS can be applied to a specific decommissioning site, perform the following steps:

Determine if the screening values are applicable to the decommissioning site, as discussed in Section 6.6. If the screening values can be used, the GEIS applies to the site.

If the screening values cannot be used, compare the site conditions to the models used in the GEIS. Appendix E should be used to determine if the generic analysis in the GEIS encompasses the range of environmental impacts at the site. Appendix E contains (a) checklists for structures and soil that indicate whether the GEIS is applicable and (b) tables that show the parameters used for the reference facilities studied in the GEIS.

If the GEIS does apply to the decommissioning site, NEPA compliance can be demonstrated in an abbreviated EA; Appendix E contains a sample EA. The PM can develop an abbreviated EA by performing the following six actions:

- Characterize briefly the contamination and remediation activities.

- Reference the appropriate licensee documents, and direct the reader to the licensee's DP for a more thorough description of the contamination and remediation activities.

- Describe the affected environment (including location, climate, geology, hydrology, cultural resources, and ecology) to demonstrate that NRC has looked for any site-specific impacts that are not covered by the GEIS. Special environmental or cultural issues may be associated with a decommissioning action, which may require a particular analysis.

- Add the following statement to the EA:

  "NRC staff has reviewed the decommissioning plan for the XYZ facility and examined the impacts of decommissioning. Based on its review, the staff has determined that the environmental impacts associated with the decommissioning of the XYZ facility are bounded by the impacts evaluated by [*either* "the 'Generic Environmental Impact Statement in Support of Rulemaking on Radiological Criteria for License Termination of NRC–Licensed Nuclear Facilities' (NUREG–1496)" *or* "the NRC Final EIS related to construction and operation of XYZ facility, dated __, 20_)"]. The staff also finds that the proposed decommissioning of XYZ is in compliance with 10 CFR 20.1402, "Radiological Criteria for Unrestricted Use.""

- Contact the U.S. Fish and Wildlife Service for a list of threatened and endangered species and determine the impacts, if any, of the decommissioning activities on these species. Since impacts on plant and animal populations could occur, the reviewer or the licensee will need to contact the U.S. Fish and Wildlife Service for a list of threatened and endangered species and determine the impacts, if any, of the decommissioning activities on these species. The State would also need to be consulted about possible impacts on State-listed species.

- Contact the State Historic Preservation Officer to determine if there are any historic properties that could be impacted by the decommissioning activities.

If a FONSI has been made and there is no potential for offsite impacts, environmental justice issues need not be considered (environmental justice is disproportionately high and adverse human health or environmental effects on minority and low-income populations).

15.7.4   ENVIRONMENTAL IMPACT STATEMENT

If there are potentially significant impacts, an EIS must be prepared. An EIS provides decision makers and the public with a detailed and objective evaluation of significant environmental impacts, both beneficial and adverse, likely to result from a proposed action and reasonable alternatives. In contrast to the brief analysis in an EA, the EIS includes a more detailed interdisciplinary review. The EIS provides sufficient evidence and analysis of impacts to support the final NRC action in the Record of Decision (for NRC, the issuance of the license amendment). The NMSS environmental guidance (NUREG–1748) discusses the EIS process and preparation of the ER and EIS documents. Except for rulemaking EISs, all NMSS EISs are prepared by the Environmental and Performance Assessment Directorate (EPAD). Decommissioning of facilities that plan to use the restricted release criteria (10 CFR 20.1403–1404) for license termination typically require an EIS. Figure 15.4 lists when an EIS must be prepared.

An EIS must be prepared for proposed actions that:

- are major Federal actions significantly affecting the quality of the human environment (10 CFR 51.20(a)(1));

- involve a matter which the Commission, at its discretion, has determined should be covered by an EIS (10 CFR 51.20(a)(2));

- are of the type listed in 10 CFR 51.20(b); or

- are determined to require an EIS by the NRC manager responsible for authorizing the action, based either on the results of an EA or on other information indicating potentially significant impacts.

Figure 15.4 When to Prepare an Environmental Impact Statement.

15.7.5   ENVIRONMENTAL ASSESSMENT AND ENVIRONMENTAL IMPACT STATEMENT CONSIDERATIONS SPECIFIC TO DECOMMISSIONING

For decommissioning actions, the proposed action is to remove a facility safely from service and reduce residual radioactivity to a level that permits release of the property for unrestricted use or under restricted conditions, and termination of the license. NRC's purpose is to fulfill its responsibilities under the AEA, which is to make a decision on a proposed license amendment for decommissioning that ensures protection of the public health and safety. The objective of the proposed action is to ensure that the decommissioning of the facility meets the license termination criteria in 10 CFR Part 20, Subpart E.

The EA or EIS is required to consider all reasonable alternatives, including the licensee's decommissioning proposal and the no-action alternative. Because decommissioning is required by regulation and is necessary to protect the public, the no-action alternative may not be a reasonable alternative. However, NEPA regulations require analysis of the no-action alternative because it provides a benchmark, enabling decision makers to compare the magnitude of environmental effects of the action alternatives. For decommissioning sites, the no-action alternative does not require a detailed analysis.

Reasonable alternatives could include other means to decommission the facility or decommissioning only part of it. Local, State, Tribal, or Federal laws (for example, a local law that prohibits onsite disposal of radioactive waste) do not necessarily render an alternative unreasonable, although such conflicts must be considered and discussed in the EA or EIS.

A complete list of impacts to be considered is contained in the environmental guidance (NUREG–1748). The following is a list of some typical decommissioning impacts:

- construction impacts such as fugitive dust emissions, vehicle and equipment exhaust emissions, and noise;

- hazardous and radioactive emissions;

- ground water contaminant plumes;

- doses to the public from transporting radioactive materials to disposal sites; and

- land use and aesthetic impacts from construction of a disposal cell.

15.8 NUCLEAR MATERIALS MANAGEMENT AND SAFEGUARDS SYSTEM

The Nuclear Materials Management and Safeguards System (NMMSS) serves as the U.S. Government's information system containing current and historic data on the possession and shipment of certain source and special nuclear material. NMMSS satisfies the requirements of the AEA of 1954, as amended, for

> "a program for Government control of the possession, use and production of atomic energy and special nuclear material, whether owned by the Government or others, so directed as to make the maximum contribution to the common defense and security and the national welfare, and to provide continued assurance of the Government's ability to enter into and enforce agreements with nations or groups of nations for the control of special nuclear materials."

It is also used to satisfy treaty obligations to the International Atomic Energy Agency and a variety of agreements for nuclear cooperation for a state system of accountancy of source and special nuclear materials. Transaction, Inventory and Material Balance data from over 1000 facilities, which are either operated by DOE or regulated by NRC, are reported to NMMSS.

When decommissioning a license in preparation for license termination, NRC staff should request the NRC NMMSS project manager (in the Division of Security Policy in the Office of Nuclear Security and Incident Response), to confirm that all NMMSS material has been properly accounted between the license and the NMMSS database. This process can take as little as two days to complete, and it should be conducted for all licenses that were issued to possess materials in the following minimum quantities of materials as described in Table 15.4.

Table 15.4   NMMSS Reportable Quantities

Isotope or Element	Reportable Quantity
Plutonium-238	0.1 gram
Plutonium	1 gram
Enriched Uranium	1 gram Uranium-235
Uranium-233	1 gram Uranium-233
Foreign-Origin Thorium	1 kilogram
Foreign-Origin Natural Uranium	1 kilogram
Foreign-Origin Depleted Uranium	1 kilogram

A statement to this effect should be included in the SER, either in the "Radiological Status of the Facility" or the "Radioactive Waste Management" sections.

15.9   DECOMMISSIONING CONTRACTORS

It should be noted that Group 1 and 2 licensees may consider using Decommissioning Service Contractors who are licensed to perform decommissioning activities, without amending their license, if the Service Licensee's license allows such activities. For Groups 3 and 4, Decommissioning Service Contractors may be able to perform work under their license (consult with NRC for a determination). Decommissioning Service Contractors would not typically be used under the contractor's license for Groups 5–7. These higher Group decommissioning activities typically are done under the authority of the licensee's license, because the DP requires both the public's involvement and the need for safety and environmental assessments. The site owner remains responsible for the eventual release of a site regardless of whomever the owner hires to perform specific activities.

Appendix K contains the final policy and guidance directive on licensing site remediation contractors to operate under their own license at temporary job sites. The guidance includes example license conditions for service licenses.

15.10 DECOMMISSIONING COMMUNICATION PLANNING

The regulation 10 CFR 20.1405 requires that NRC contact members of local governments and the public when a DP is received. The purpose of the contacts is to notify these entities of a licensee's plan to terminate its license under the license termination rule, and to solicit comments on the licensee's plans. The staff should send letters to local Native American associations, the State Government—usually the environmental branch—the county executive or manager, and nearby city mayors. The staff should also publish, in the *Federal Register* and in one or more local newspapers of wide circulation, a notice soliciting public comment on the licensee's plans. If the licensee is proposing alternate criteria (see 10 CFR 20.1404), staff should also notify and seek comments from the EPA.

For complex sites, the NRC reviewer should develop a communication plan following the "Policy and Procedures Guide for Developing NMSS Communication Plans" for developing individual communication plans for decommissioning activities. Communication plans cover topics such as applying public outreach tools and techniques, identifying stakeholders, and estimating costs and schedules for public outreach meetings. For simple sites, the NRC reviewer may need to develop a communication plan if there are active stakeholders.

The ADAMS Document Processing Instruction Template NRC–001, "Meeting Related Documents for NRC Staff-Level Offices," instructs the staff in the preparation of all meeting related documents. These documents include meeting notices, agenda, handouts, summaries, and so forth. NRC's Web site http://www.nrc.gov/NRC/PUBLIC/meet.html is the official forum to announce meetings open to the public. NUREG–1748 contains detailed guidance concerning public meetings associated with an EIS.

15.11 CONTROLLING THE DISPOSITION OF SOLID MATERIALS

15.11.1 CURRENT NRC APPROACH TO RELEASES OF SOLID MATERIAL

Currently, NRC staff generally addresses the release of solid material on a case-by-case basis using license conditions and existing regulatory guidance. In each case, material may be released from a licensed operation with the understanding and specific acknowledgment that the material may contain very low amounts of radioactivity, but that the concentration of radioactive material is so small that its control through licensing is no longer necessary.

The case-by-case approach includes guidance that is applicable to equipment and material with radioactivity located on the surface or within the material or equipment itself. However, there are differences in the application of this guidance between reactor licensees and materials licensees, which is explained below.

15.11.1.1 Release of Solid Materials with Surface Residual Radioactivity

All Licensees

Criteria which licensees must use in determining whether the material may be released are approved for use by the NRC staff during the initial licensing or license renewal of a facility, as part of the facility's license conditions or radiation safety program. The licensees' actions must be consistent with the requirements of 10 CFR Part 20 (e.g., Subpart F of Part 20 (10 CFR 20.1501)). Thus, the licensee performs a survey of the material prior to its release.

Reactor Licensees

Reactor licensees typically follow a policy that was established by Office of Inspection and Enforcement Circular 81-07 and Information Notice 85-92. Under this approach, reactor licensees must survey equipment and material before its release. If the surveys indicate the presence of AEA material above natural background levels, then no release may occur. If the appropriate surveys have not detected licensable material above natural background levels, the solid material in question does not have to be treated as waste under the requirements of Part 20. The fact that no radioactive material above background is detected does not mean that none is present; there are limitations on detection capability. In practice, the actual detection capability of survey instruments are typically consistent with those contained in Regulatory Guide 1.86.

Materials Licensees

For materials licensees, NRC staff usually authorizes the release of solid material through specific license conditions. One set of criteria that is used to evaluate solid materials before they are released is contained in Regulatory Guide 1.86, entitled "Termination of Operating Licenses for Nuclear Reactors." A similar guidance document is Fuel Cycle Policy and Guidance Directive FC 83-23, entitled "Guidelines for Decontamination of Facilities and Equipment Prior to Release for Unrestricted Use or Termination of Byproduct, Source or Special Nuclear Materials Licenses." Both documents contain a table of surface contamination criteria which may be applied by licensees for use in demonstrating that solid material with surface contamination can be safely released with no further regulatory control.

Although Regulatory Guide 1.86 was originally developed for nuclear power plant licensees, the surface contamination criteria have been used in other contexts for all types of licensees for many years. By setting maximum allowable limits for surface contamination, Regulatory Guide 1.86 implicitly reflects the fact that materials with surface contamination below those limits may be released without adverse effects on the public health and safety.

15.11.1.2 Release of Solid Materials with Volumetric Residual Radioactivity

In the case of volumetrically contaminated materials, NRC staff has not provided guidance like that found in Reg Guide 1.86 for surface contamination. Instead, NRC staff has treated these situations on an individual basis, typically seeking to assure, by an evaluation of doses associated with the proposed release of the material, that maximum doses are a small percentage of the Part 20 dose limit for members of the public. Thus, the NRC staff practice over the years has been to allow the release of material with slight levels of volumetric contamination based on a case-by-case evaluation. These evaluations follow guidance discussed in the June 1999 Issues Paper (NRC 1999b) and in three All-Agreement States letters (STP-00-070, STP-01-081, STP-03-003), dated August 22, 2000, November 28, 2001, and January 15, 2003, respectively. Licensees have used the process set out in 10 CFR 20.2002 to seek approval for alternate disposal methods of solid material. The release of material using the 10 CFR 20.2002 process is consistent with other disposition provisions in Part 20 that allow for the release of material (e.g., 10 CFR 20.2003 and 10 CFR 20.2005). The current guidance that would be used to evaluate doses associated with 10 CFR 20.2002 requests is NUREG-1757, Volume 2.

Reactor Licensees

For reactor licensees, the release of volumetrically contaminated materials is being implemented under the provisions of Information Notice No. 88-22: Disposal of Sludge from Onsite Sewage Treatment Facilities at Nuclear Power Stations. Certain materials may be surveyed using a representative sample and gamma spectrometry analytical methods. The provision requires that materials can be released if no licensed radioactive material above natural background levels is detected, provided the radiation survey used a detection level that is consistent with the lower limit of detection values used to evaluate environmental samples. NRC guidance states that the lower limit of detection (LLD) to be used for radiation surveys is the "operational state of the art" LLD values given in the Standard Radiological Effluent Technical Specifications for environmental samples taken as part of the licensee's radiological environmental monitoring program.

The environmental LLDs are contained in Regulatory Guide 4.8, "Environmental Technical Specifications for Nuclear Power Plants," and in a Branch Technical Position (NRC 1979). They are also contained in NUREG-1301, "Offsite Dose Calculation Manual Guidance: Standard Radiological Effluent Controls for Pressurized Water Reactors," and NUREG-1302, "Offsite Dose Calculation Manual Guidance: Standard Radiological Effluent Controls for Boiling Water Reactors." There are several different acceptable survey applications of the environmental LLDs and applications have included a variety of environmental media including soils, sediments, liquids and slurries.

Materials Licensees

For materials licensees, the release of volumetrically contaminated materials is being implemented under the provisions of the December 27, 2002, NRC Memorandum, "Update on Case-Specific Licensing Decisions on Controlled Release of Concrete from Licensed Facilities" (referenced in STP-03-003). This memorandum indicates that controlled releases of volumetrically contaminated concrete may be approved, pursuant to 10 CFR 20.2002, under an annual dose criterion of a "few mrem."

15.11.2 CASE-SPECIFIC LICENSING DECISIONS ON DISPOSITION OF SOLID MATERIALS FROM LICENSED FACILITIES

NRC staff should use the following approach for making decisions on specific licensing actions, as well as generic requests, concerning the disposition of solid materials.

Existing NRC regulations do not contain generally applicable standards for the disposition of solid materials with relatively small amounts of radioactivity in, or on, materials and equipment. Therefore, the offsite disposition of solid materials prior to license termination will continue to be evaluated on a case-by-case basis using existing guidance (e.g., application of Regulatory Guide 1.86 and its equivalent, Fuel Cycle Policy and Guidance Directive FC 83–23, for materials licensees and Office of Inspection and Enforcement Circular 81–07 and Information Notices 85–92 and 88–22 for reactor facilities).

To ensure a consistent approach for the disposition of solid materials, NRC reviews of licensee requests for the disposition of these materials using criteria other than those in existing guidance are to be coordinated with NMSS or NRR Divisions. NRC contact information is provided in Table 15.5.

Table 15.5   NRC Contact Information for Requests Concerning the Disposition of Solid Materials

NRC Office	NRC Division	Contact
Office of Nuclear Materials Safety and Safeguards	Division of Waste Management and Environmental Protection	Tel: 301–415–7437 Fax: 301–415–5397 Mail Stop: T–7J8
Office of Nuclear Materials Safety and Safeguards	Division of Industrial, Medical and Nuclear Safety	Tel: 301–415–7197 Fax: 301–415–5369 Mail Stop: T–8F5
Office of Nuclear Materials Safety and Safeguards	Division of Fuel Cycle Safety and Safeguards	Tel: 301–415–7213 Fax: 301–415–5730 Mail Stop: T–8A33
Office of Nuclear Reactor Regulation	Division of Inspection Program Management	Tel: 301–415–1004 Fax: 301–415–2220 Mail Stop: O–6E3

15.11.3 REVIEW OF RETROSPECTIVE AND PROSPECTIVE CASES INVOLVING SOIL DISPOSITION

NRC staff should use the following guidance in review of retrospective and prospective cases involving offsite soil disposition prior to license termination. Requests for approvals for the disposition of soils should be coordinated with the NRC Divisions provided in Table 15.5, on a case-by-case basis.

15.11.3.1 Retrospective Cases

For retrospective cases, if offsite soil releases have been identified, reviewed, and accepted in an approved decommissioning plan (DP) based on Site Decommissioning Management Plan (SDMP) Action Plan criteria, a 10 CFR 20.2002 disposal, or other specific license condition, previously approved offsite soil releases should be considered as final, but further examination is recommended if offsite soil releases could produce a dose of more than 1 mSv/y (100 mrem/y) to a member of the public under realistic conditions. The examination should be based on a case-specific dose assessment rather than a conservative screening assessment.

15.11.3.2 Prospective Cases

For prospective cases or cases that are not grandfathered—where a proposed disposition of offsite soil is not covered under an existing DP, a 10 CFR 20.2002 disposal, or other specific license condition—there may be approval under a criterion of a "few mrem" (pursuant to a 10 CFR 20.2002 procedure, DP, or other specific license amendment) rather than use of license termination criteria either in Subpart E of 10 CFR Part 20 or in the SDMP Action Plan.

15.11.4 CASE-SPECIFIC LICENSING DECISIONS ON DISPOSITION OF CONCRETE FROM LICENSED FACILITIES

NRC staff should use the following guidance in review of retrospective and prospective cases involving concrete disposition. Requests for approvals for the disposition of concrete should be coordinated with NRC Divisions provided in Table 15.5, on a case-by-case basis.

15.11.4.1 Retrospective Cases

For retrospective cases, if offsite concrete releases have been identified, reviewed, and accepted in an approved DP based on SDMP Action Plan criteria, a 10 CFR 20.2002 disposal, or other specific license condition, previously approved offsite concrete releases should be considered as final. However, if upon further review of information, it is estimated that offsite concrete releases could produce a dose of more than 1 mSv/y (100 mrem/y) to individual members of the public under realistic conditions, further examination is recommended. The examination should be based on a case-specific dose assessment rather than a conservative screening assessment.

15.11.4.2 Prospective Cases

For prospective cases or cases that are not grandfathered—where a proposed disposition of offsite concrete is not covered under an existing DP, a 10 CFR 20.2002 disposal, or other specific license condition—disposition of concrete with volumetric sources of contamination may be approved under a criterion of a "few mrem" (pursuant to a 10 CFR 20.2002 procedure, DP, or other specific license amendment) rather than use of license termination criteria either in Subpart E of 10 CFR Part 20 or in the SDMP Action Plan. The following guidance is provided for these types of cases:

- Licensees should assess surficial contamination of concrete based on process knowledge and should take appropriate core samples to confirm that the concrete does not contain contamination beyond a depth that can be measured by the instrumentation used for the survey. Survey instruments should be used that are appropriate for evaluating the radioactive contamination of interest and all accessible surfaces should be evaluated. The number of core samples and the method for determining the depth at which a survey instrument can measure below the surface of the concrete should be determined on a case-specific basis.

- At materials sites, based on a licensee's determination that the concrete contains either surficial or volumetric contamination:

    — Disposition of concrete with surficial contamination should be evaluated using "Guidelines for Decontamination of Facilities and Equipment Prior to Release for

Unrestricted Use or Termination of Licenses for Byproduct, Source, or Special Nuclear Material," dated April 1993, which is based on Fuel Cycle Policy and Guidance Directive 83–23. There is no upper limit on the amount of concrete with surficial contamination that can be released from a materials site if it meets criteria contained in the April 1993 guidance document.

— Disposition of concrete with volumetric contamination should be pursuant to 10 CFR 20.2002 procedures.

• Surveys for the disposition of concrete with surficial contamination should be conducted before the concrete floor or wall is broken up. If the concrete wall or floor has been broken up, then it is considered a volumetric source of contamination and 10 CFR 20.2002 procedures should be followed.

15.11.4.3 Concrete Remaining Onsite

If a licensee proposes to allow concrete with surficial or volumetric contamination to remain onsite after license termination, the concrete should be evaluated as part of the licensee's overall decommissioning approach for license termination pursuant to 10 CFR 20, Subpart E.

15.12 ONSITE DISPOSAL OF RADIOACTIVE MATERIALS UNDER 10 CFR 20.2002

15.12.1 OPTIONS FOR ONSITE DISPOSALS AT NRC LICENSED FACILITIES

NRC regulations allow onsite disposals or burial of radioactive materials under 10 CFR 20.2002. 10 CFR 20.2002 identifies the information that a licensee must include in its request for disposal and requires that the disposal result in doses that are maintained ALARA and are within the dose limits of 10 CFR Part 20. Part 20 includes the public dose limit of 1 mSv/y (100 mrem/y) and the LTR (10 CFR Part 20, Subpart E) criteria for license termination (0.25 mSv/y (25 mrem/y) and ALARA for unrestricted use). NRC staff's current practice is to approve requests for onsite disposal that result in doses not exceeding a "few millirem" per year.

The LTR requires that the dose contribution from all onsite disposals be accounted for at the time of license termination. This suggests, at a minimum, that the LTR radiological criteria of 0.25 mSv/y (25 mrem/y) and ALARA for unrestricted use would apply to an onsite disposal. Onsite disposals resulting in higher doses (up to 1 mSv/y (100 mrem/y)) would need to be remediated for a site to meet the radiological criteria for unrestricted use in the LTR. In addition, because the Timeliness Rule in 10 CFR 30.36, 40.42, 70.38, and 72.54 also applies to onsite disposals, licensees may need to remediate such onsite disposals (i.e., those approved at higher doses) prior to license termination.

The NRC staff originally examined the potential conflicts between onsite disposal under 10 CFR 20.2002 and the LTR criteria and the Timeliness Rule, in the staff's LTR Analysis (SECY-03-0069), which provided options for onsite disposal. NRC staff reevaluated these options for onsite disposal, after consideration of stakeholder comments, and this evaluation is presented in SECY-06-0143. NRC staff's evaluation and the Commission's associated direction (in SRM-SECY-06-0143) is reflected in the following guidance, which is based on the goal of preventing future legacy sites (sites with complex decommissioning problems where funding is not typically available to adequately decommission the site for unrestricted use) and allows reasonable flexibility for onsite disposals within the current regulations. NRC staff should use the following guidance to determine the acceptability of licensee proposals for onsite disposal under 10 CFR 20.2002.

NRC will continue the current practice of approving onsite disposal of radioactive materials based on a dose criterion of a "few millirem" per year. At the time of license termination, there may be multiple sources of residual radioactivity, including onsite disposals. By generally constraining doses from onsite disposals to a few millirem per year, it is likely that the entire site (including onsite disposals) will meet the LTR criteria, without remediation of the onsite disposal.

The NRC will also consider requests for onsite disposals, using dose criteria other than a few millirem per year. NRC staff's approval of these requests will be based on the goal of preventing future legacy sites.

15.12.2 ACCEPTABLE ONSITE DISPOSAL DOSE CRITERIA

Onsite disposals are allowed under 10 CFR 20.2002, which identifies, in general terms, the information that licensees must include in their requests for disposal under 10 CFR 20.2002, to allow NRC staff to determine if the proposed disposal is acceptable. This information includes: a description of the material to be disposed of; physical and chemical properties of materials; the manner and conditions of waste disposal; site characteristics used for dose modeling; and dose impacts and ALARA considerations. Licensees should submit adequate information to allow NRC reviewers to determine if adequate dose and other analyses were done.

Volume 2 of this NUREG report addresses technical aspects of onsite disposal under 10 CFR 20.2002 that need to be considered by the licensee and NRC reviewers. Requests for onsite disposals of wastes containing mobile radionuclides (e.g., H-3 and C-14), which may reach subsurface soils and potentially reach groundwater, should provide detailed information on the design of any engineered structures or barriers, if they are used. The site geology and hydrology, which is important to containing the disposal area and to the potential for radionuclide transport in the subsurface environment, should be described.

It is anticipated that an onsite disposal will occur during the conduct of licensed operations and will precede decommissioning and license termination. Licensees should consider the site life-cycle in developing an approach for onsite disposal of radioactive materials, because the entire site (including any onsite disposals) must meet radiological criteria for license termination in the LTR. Onsite disposals or burials may have to be remediated at the time of license termination or earlier, per the Timeliness Rule.

15.12.2.1 Current Practice of a Few Millirem Per Year

NRC will continue the current practice of approving onsite disposals based on a dose criterion of a few millirem per year. The few millirem per year criterion encompasses 0–0.05 mSv (0–5 mrem) per year TEDE. At the time of license termination, there may be multiple sources of residual radioactivity, including onsite disposals. By generally constraining doses from onsite disposals to a few millirem per year, it is likely that the entire site (including the contribution from onsite disposals) will meet the LTR criteria, without remediation of the onsite disposal.

Requests for onsite disposal should consider the doses from all previous onsite disposals. Thus, the few millirem per year dose criterion applies to the cumulative dose from all onsite disposals.

15.12.2.2 Other Dose Criteria

NRC will consider requests for onsite disposals of radioactive materials under 10 CFR 20.2002 that exceed a few millirem per year. The prevention of future legacy sites will be a primary consideration in the approval of these requests. Accordingly, the NRC staff will gauge the likelihood of the creation of a future legacy site by these considerations: (1) time of potential doses based on the half-lives of the material and time of license termination; (2) the mobility of the radioactive materials to be disposed; (3) the additional financial assurance that the licensee may provide to ensure necessary cleanup can be completed for license termination; and (4) other aspects that ensure that the facility will not become a future legacy site.

Onsite disposals resulting in greater than a few millirem per year may conflict with requirements of the Timeliness Rule, and licensees should consider the implications of the Timeliness Rule in their proposals for onsite disposal (see Section 15.12.3.7).

15.12.3 CONSIDERATIONS FOR REVIEW OF ONSITE DISPOSAL REQUESTS

15.12.3.1 Radiological Dose Assessment

The licensee's onsite disposal request should include an evaluation of doses to workers and to the public for the site conditions at the time of disposal. It should also include an assessment of the potential doses to critical groups of exposed persons after license termination.

NUREG-1757, Vols. 1 and 2, provide guidance on the applicable radiological dose modeling that is needed to evaluate an onsite disposal request. The radiological dose assessment for onsite disposals should include site-specific, realistic scenarios that are applicable to decommissioning and license termination.

15.12.3.2 Recordkeeping

Pursuant to 10 CFR 20.2108 (see also 10 CFR Parts 30, 40, 70, and 72), licensees are required to maintain records of disposals made under 10 CFR 20.2002. NUREG-1757, Vol. 3, provides guidance on recordkeeping requirements for licensees while they are conducting licensed operations. NRC expects that adequate records of onsite disposals or burials will facilitate decommissioning and/or remediation at license termination.

15.12.3.3 Radiological Surveying and Monitoring

Licensees should perform applicable surveys and monitoring to assure that radioactive materials in onsite disposals are contained and do not migrate from the disposal site. The radiation dose rates from an onsite disposal may not be distinguishable from natural radiation background levels or from other site operations. Therefore, instead of surveys, periodic surveillance of an onsite disposal area may be more appropriate to ensure that it is not relocated or disturbed.

15.12.3.4 Financial Assurance

NUREG-1757, Vol. 3, provides guidance on financial assurance requirements. A licensee may provide additional financial assurance to cover the cost of decommissioning or remediation of any onsite disposals that exceed 0.25 mSv/y (25 mrem/y). NRC staff should review the licensee's decommissioning cost estimate and the financial assurance available to the licensee to remediate or decommission an onsite disposal.

15.12.3.5 Licensed Materials Remaining Onsite

Typically, onsite disposals or burials occur during licensed operations, and will precede any necessary remediation or decommissioning activities and license termination. Radioactive materials disposed onsite, in accordance with 10 CFR 20.2002, may be allowed to remain in-place at license termination if the LTR radiological criteria are met for the entire site, including contributions from residual radioactivity in the onsite disposal.

License requests for onsite disposal under 10 CFR 20.2002 generally should not include requests to leave contaminated structures in-place in contemplation of license termination. Those actions should be addressed in the decommissioning plan and in the license termination process.

15.12.3.6 Compliance with Environmental and Health Protection Regulations

An NRC licensee must comply with all applicable local, State, and Federal regulations governing any other toxic or hazardous properties of materials that may be disposed under 10 CFR 20.2002.

15.12.3.7 Timeliness Rule

Onsite disposals are subject to the Timeliness Rule, and licensees should consider the implications of the Timeliness Rule and the potential need to remediate the disposal in their requests for onsite disposal.

For requests of onsite disposals resulting in greater than 25 millirem per year, materials licensees would be subject to the two-year time frame discussed in the Timeliness Rule, unless the NRC has approved an alternate schedule for completing the decommissioning of the onsite disposal. Materials licensees that intend to dispose of material onsite in accordance with 10 CFR 20.2002 should consider whether an alternate decommissioning schedule is needed, as appropriate, and should use the guidance on alternate schedule, in NUREG-1757, Vol. 3, Section 2.2.

15.12.3.8 Transparency in the 10 CFR 20.2002 Process

NRC reviewers should be familiar with SECY-06-0056, "Improving Transparency in the 10 CFR 20.2002 Process," and the associated SRM, which provides direction on the information to be provided to the public on 10 CFR 20.2002 disposals. NRC staff reviewers should contact the Environmental & Performance Assessment Directorate, Division of Waste Management and Environmental Protection, Office of Nuclear Material Safety and Safeguards, for current information and applicable policies related to transparency in the 10 CFR 20.2002 process.

15.13 USE OF INTENTIONAL MIXING OF CONTAMINATED SOIL

15.13.1 INTRODUCTION

As part of the LTR analysis, NRC staff examined the use of intentional mixing of contaminated soil to meet the LTR release criteria as an option to provide flexibility in achieving the goals of the LTR (10 CFR Part 20, Subpart E). The results of the staff's analysis of this issue are in SECY-04-0035 (NRC 2004a). The staff analyzed the possible ways that a licensee could intentionally mix soil to lower its concentration and identified which of these scenarios should be considered further in the analysis. Using these scenarios, the staff evaluated options for allowing intentional mixing[20]. The analysis considered a wide range of relevant information and experience from the NRC and other domestic and international sources.

In SRM-SECY-04-0035 (NRC 2004b), the Commission approved the use of intentional mixing of contaminated soil to meet the LTR release criteria, in limited circumstances, on a case-by-case basis, while continuing the current practice of allowing intentional mixing for meeting waste acceptance criteria (WAC) of offsite disposal facilities and for limited onsite waste disposals at operating facilities (approved under 10 CFR 20.2002).

Intentional mixing has been approved by the NRC staff where homogenous waste streams (for example, soil from two areas of a facility contaminated by similar waste from two different processes) have been mixed to meet the WAC of a disposal facility, as long as the classification of the waste, as determined by the requirements of 10 CFR 61.55, is not altered. NRC staff will continue to consider proposals from decommissioning sites for intentionally mixing contaminated soil (and other homogeneous waste streams) to meet WAC of offsite disposal facilities to aid in the completion of remedial actions at sites undergoing decommissioning.

Intentional mixing also has been approved by the NRC staff for limited onsite disposals approved under 10 CFR 20.2002. A decommissioning licensee will normally not seek approval under 10 CFR 20.2002 for an onsite burial (although 10 CFR 20.2002 may be used for disposal at an offsite location). Licensees should be aware that if an onsite disposal under 10 CFR 20.2002 is approved during operations, the onsite disposal will need to be readdressed at the time of license termination, in the evaluation of whether the dose criteria of the LTR are met (see guidance in Section 15.12 of this volume).

This guidance implements the Commission's policy decisions on the use of intentional mixing of contaminated soil and other homogeneous waste streams from decommissioning sites to meet WAC of offsite disposal facilities and for intentional mixing of soil that remains at the decommissioning site to meet the LTR release criteria.

15.13.2 REVIEW PROCEDURES

The NRC staff will consider proposals to use intentional mixing of contaminated soil (or other homogeneous waste streams) to meet the WAC of an offsite disposal facility to facilitate completion of decommissioning. Licensees should be aware that local and/or State requirements may also apply to waste that is transported to a disposal facility away from the decommissioning site, and that these requirements will have to be met. Approval of a process for a waste stream by the NRC does not imply approval for disposal by the local or State regulators with jurisdiction over the disposal facility. Decisions on approving the use of intentional mixing to

---

[20] NRC staff recognizes that some incidental mixing of contaminated soil and non-contaminated soil may occur as a result of excavation and other earth-moving activities. This mixing that occurs from the use of excavating and earth-moving equipment in normal activities associated with site decommissioning is not considered "intentional" mixing for the purposes of this guidance.

meet the WAC of an offsite disposal facility will be performance-based, using the appropriate criteria of 10 CFR Part 20 or other NRC regulations, if they apply.

The NRC staff will consider the use of intentional mixing of soil to meet the LTR release criteria (where the mixed soil will be left on the site) only in cases in which an "overall approach" to site cleanup is proposed that includes soil mixing and ALARA principles. Proposals to use intentional mixing should be part of an overall plan for decontamination and decommissioning (presented in a DP or LTP) of a licensee's property, that seeks to achieve unrestricted use of the site [21] and renders doses ALARA, which may include: (1) removal and disposal of contaminated components and equipment; (2) decontamination (and demolition, if appropriate) of buildings; (3) removal and disposal of waste streams remaining onsite from past operations; and (4) excavation and removal of large areas of soil contamination as waste. Intentional mixing should not be proposed as the sole means to achieve the license termination dose criteria, unless it is the only practical means to meet the LTR criteria.

The NRC staff will consider only cases in which this overall approach to site cleanup demonstrates that the removal of soil would not be reasonably achievable. The NRC will consider the same criteria used to determine the eligibility of a site for restricted use (see 10 CFR 20.1403(a)) for determining when removal of soil is not reasonably achievable (i.e., a demonstration that further removal of contaminated soil would result in net public or environmental harm or leaving the soil in place is ALARA). Licensees also should include other considerations (e.g., distance to disposal facility, efficient utilization of available disposal capacity at the offsite facility, unavailability of required treatment options, lack of disposal options other than leaving the contaminated soil onsite, and the need to use funds for remediation of non-radioactive hazards at the same site) in proposals for intentional mixing, if they are applicable and appropriate to a determination of whether the removal of soil for offsite disposal is reasonably achievable.

Decisions on approving the use of intentional mixing of contaminated soil to meet the LTR will be performance-based using the dose criteria of the LTR. Therefore, licensees have flexibility in how intentional mixing may be used together with other remediation activities to achieve the dose criteria. In addition, staff will base the approval decisions using a risk-informed approach. In their proposal to use intentional mixing of soil, licensees should include all relevant information concerning the risks of using the approach versus other remediation alternatives.

---

[21] The NRC's staff preferred option for decommissioning is to achieve license termination for unrestricted use of sites where possible. NRC may consider remedies that include intentional mixing of contaminated soil to achieve unrestricted use of a site, when other remedies alone would result in restricted use. (For example, NRC staff could consider intentional mixing that uses additional uncontaminated soil from outside the footprint if it will achieve unrestricted use). Intentional mixing also may be used to achieve the restricted use or alternate criteria of the LTR.

15.13.3 ACCEPTANCE CRITERIA

Information to be Submitted

The information supplied by the licensee should be sufficient to allow the staff to determine that the information adequately describes how the intentional mixing operation will be carried out and that the conditions for approving the use of intentional mixing have been met. In the case where intentional mixing will be used to meet the LTR criteria, the information supplied by the licensee should be sufficient to allow the staff to determine that the limited circumstances, for which mixing will be considered, are present.

Intentional Mixing to Meet Waste Acceptance Criteria

The staff's review should verify that the following information is included in the sections of the DP, corresponding to the sections of the Volume 1 of this NUREG report (indicated in parentheses), for decommissioning sites proposing to use intentional mixing to meet the WAC of an offsite disposal facility:

- Information on the intentional mixing activities to be conducted by the licensee or contractors, including the machinery to be used and the methods to be employed with the equipment to achieve a homogeneous mix of soil. Information should be included on important features and parameters of machinery operation that control the homogeneity of the resultant mix, such as mixing time, discharge time, number of mixing blades or paddles, and the maximum particle size. (Section 17.1.3)

- Information on any slag or other larger non-soil like waste materials that will be included in the soil that is intentionally mixed, and how it will be rendered compatible with the mixing machinery (e.g., maximum particle size), if necessary. Information should also be included on non-soil like waste materials that are included in the mixed soil, but which are not compatible with the mixing machinery, and how it is compatible with the WAC of the disposal facility. (Section 17.1.3)

- Information on the method to be used to ensure that the mixing operation has resulted in a sufficiently homogeneous mixture to achieve the requirements of the disposal facility. This should include any instrumentation that may be used in support of the machinery used for mixing, as well as any proposed surveying and/or sampling and analysis that is employed. (Sections 17.1.3 and 17.3.1.7)

- Information on how soil following intentional mixing is controlled (e.g., temporary storage), in accordance with the licensee's program for management of volumetrically contaminated materials to ensure it maintains its required properties, if appropriate. (Section 17.5.1)

- Information on how the soil following the intentional mixing operation will meet the WAC of the disposal facility. (Section 17.5.1)

Intentional Mixing to Meet the License Termination Rule

The staff's review should verify that the following information is included in the sections of the DP, corresponding to the sections of the Volume 1 of this NUREG report (indicated in parentheses), for sites proposing to use intentional mixing to meet the release criteria of the LTR:

- A summary discussion of the overall decommissioning of the site that includes the use of intentional mixing in a comprehensive cleanup approach, including how the licensee will complete interrelated decommissioning activities and the timeframes for completing the activities. This discussion should describe how the intentional mixing proposed helps achieve the goal of unrestricted use, how it is risk-informed, and the reasons that removal of all contaminated soil is not reasonably achievable. (Section 17.1)

- Information on the locations of surface and subsurface contamination that define the areas of contamination for which intentional mixing will be utilized. (Section 16.4.3 and 16.4.4)

- Information on the configuration of the "footprint" of the areas of contamination prior to the mixing operation and the final area comprised of the intentionally mixed soil. (Section 17.1.3)

- Information on any locations of uncontaminated surface or subsurface soil that will be incorporated into the footprint. (Sections 16.4.3 and 16.4.4)

- Information on the intentional mixing activities to be conducted by the licensee or contractors, including the machinery to be used and the methods to be employed with the equipment to achieve a homogeneous mix of soil. Information should be included on important features and parameters of machinery operation that control the homogeneity of the resultant mix, such as mixing time, discharge time, number of mixing blades or paddles, and the maximum particle size. (Section 17.1.3)

- Information on any slag or other larger non-soil like waste materials that will be included in the soil that is intentionally mixed, and how it will be rendered compatible with the mixing machinery (e.g., maximum particle size), if necessary. Information should also be included on non-soil like waste materials that are included in the mixed soil but which are not compatible with the mixing machinery and how it contributes to the overall plan for decommissioning. (Section 17.1.3)

- Information on the method to be used to ensure that the mixing operation has resulted in a sufficiently homogeneous mixture to achieve the goals of the decommissioning project. This should include any instrumentation that may be used in support of the machinery used for mixing, as well as any proposed surveying and/or sampling and analysis that is employed. (Sections 17.1.3 and 17.3.1.7)

- Information on the final configuration and design attributes of the area containing the intentionally mixed soil, including a soil cap if it is employed. (Section 17.1.3)

- Results of and information that contributes to the ALARA analysis relating to the use of intentional mixing, considering the criteria used to determine the eligibility of a site for restricted use (see 10 CFR 20.1403(a)). (Section 17.4.1)

- Information on how soil following intentional mixing is controlled (e.g., temporary storage) in accordance with the licensee's program for management of volumetrically contaminated materials to ensure it maintains its required properties, if appropriate. (Section 17.5.1)

- If intentional mixing is used to meet the restricted use criteria, information on advice from affected parties concerning the use of intentional mixing as part of the remediation of a site. (Sections 17.7.5 and M.6)

15.13.4 EVALUATION FINDINGS

Approval Conditions

The NRC staff will consider approval of proposals to use intentional mixing from decommissioning sites to meet the WAC of offsite disposal facilities. For these cases, the mixture should be comprised of soil or other homogeneous waste streams and should not result in lowering the classification of the wastes (in accordance with 10 CFR 61.55). Proposals to use mixing to meet WAC of an offsite disposal facility should not use clean soil or non-contaminated materials similar to the waste stream to lower the concentrations of a mixture.

NRC staff will consider approval of intentional mixing to meet the release criteria of the LTR for soils left onsite, in which:

1. The intentional mixing is part of the proposed overall approach to site cleanup. The overall approach also includes application of the ALARA principle.

2. The area containing the mixed contaminated soil after license termination will be equal to or smaller than the footprint of the zones of contamination before decommissioning begins.

3. Clean soil, from outside the footprint of the area containing the contaminated soil, should generally not be mixed with contaminated soil to lower concentrations. Staff will consider use of clean soil only in cases where the licensee has demonstrated that: (a) the only viable approach to achieving the dose criteria of the LTR is to use clean soil from outside the contaminated area footprint; or (b) the only viable approach to achieving the unrestricted use criteria (when other remedies would only achieve the restricted use criteria) is to use clean soil from outside the contaminated area footprint.

Proposals to use intentional mixing of soil to meet the LTR criteria will be approved only if the area of land containing the intentionally mixed soil following remediation is no larger than the total of the areas of contaminated soil before remedial actions began. It is reasonable to include some portions of uncontaminated land within the footprint of contaminated areas, where an area encompassing several "zones" of contamination is designated as the footprint to be mixed. To include them, however, the uncontaminated areas should be small in comparison to the areas that are contaminated.

The NRC staff analysis of the use of intentional mixing contemplated circumstances where a contaminated soil was mixed with a contaminated soil of lower concentrations to achieve a mixture that allowed the dose criteria of the LTR to be met. The use of clean soil to achieve the goals of intentional mixing should be limited to the circumstances just described. Any uncontaminated soil that is utilized in the mixing operation should normally be included within the footprint of the contaminated zones that are to be mixed. Staff will consider the inclusion of uncontaminated soil that comes from outside of the footprint of the contaminated zones only in cases where its use is the only viable approach for meeting the dose criteria of the LTR. If a licensee proposes intentional mixing using offsite clean soil to meet the LTR criteria, the NRC staff will consult with the Commission on the acceptability of the proposal.

The staff will also consider the inclusion of uncontaminated soil that comes from below the contaminated zones within the footprint as long as it is consistent with the overall approach described for achieving license termination and considers the impacts associated with an increased depth of disposal (e.g., affect on groundwater).

The staff will consider the inclusion of a limited volume of non-soil materials (e.g., slag or concrete rubble) within the mixed soil as part of remediation, as long as analysis is presented demonstrating that the release criteria of the LTR are met and that inclusion in the mixed soil is consistent with the overall approach to site cleanup in the DP or LTP. In order to be consistent with the overall approach, the non-soil materials to be included in the mixed soil should be incidental to the excavation and removal of buildings, equipment, and major waste streams to be managed at the decommissioning site.[22] Intentionally mixing a significant non-soil like waste stream resulting from the activities that were conducted at the site during operations (e.g., slag) that is easily removed from the site (e.g., in a pile on the soil surface) should not be included in a proposal for intentional mixing to meet the LTR release criteria.

Evaluation Criteria

The staff should verify that the information summarized under "Information to be Submitted," above, is included in the licensee's descriptions of the surface and subsurface soil contamination, the soil decommissioning activities, instrumentation, control of contaminated material, ALARA evaluation, and stakeholder involvement (if necessary). The staff should verify that intentional mixing of contaminated soil is part of an overall approach to site remediation in which it is demonstrated that removal of the soil to be mixed is not reasonably achievable. The staff should verify that the descriptions of the mixing operation, the use of machinery, and the methodology

---

[22] Staff would consider non-soil materials to be incidental if, for example, a few pieces of small equipment, building rubble, or non-soil waste (e.g., slag) were discovered that required disposal following completion of waste shipping campaigns, or where a waste were most effectively managed (e.g., to avoid a technical difficulty that would increase worker dose) if it were included in the mixed soil.

for ensuring that the mixture is homogeneous are sufficiently detailed to allow the staff to understand the manner in which the licensee will ensure that the expected properties of the mixed soil have been achieved. The staff should ensure that the area containing mixed soil is no greater than the footprint of contaminated areas defined at the start of remediation. The staff should also ensure that the use of uncontaminated soil in mixing is limited only to cases where it is the only viable approach to meeting the LTR criteria. If a licensee proposes intentional mixing using offsite clean soil to meet the LTR criteria, the NRC staff will consult with the Commission on the acceptability of the proposal. The staff should ensure that any operation to mix contaminated soil to meet WAC of an offsite disposal facility does not result in lowering the classification of the waste in accordance with 10 CFR 61.55.

Sample Evaluation Findings

None required. The staff should combine the assessment of a DP proposing the use of intentional mixing with the findings on the Sections corresponding to the sections in parentheses above.

References

- NRC 2004a. SECY-04-0035, "Results of the License Termination Rule Analysis of the Use of Intentional Mixing of Contaminated Soil," March 1, 2004.

- NRC 2004b. SRM-SECY-04-0035, "Staff Requirements – SECY-04-0035 – Results of the License Termination Rule Analysis of the Use of Intentional Mixing of Contaminated Soil," May 11, 2004.

PART II: DECOMMISSIONING PLANS

16   DECOMMISSIONING PLANS: SITE DESCRIPTION

NRC regulations require that a licensee must submit a DP to support the decommissioning of its facility when it is required by license condition, or if NRC has not approved the procedures and activities necessary to carry out the decommissioning and these procedures could increase the potential health and safety impacts to the workers or the public. Chapters 16 through 18 provide a description of the contents of specific DP modules, as well as evaluation and acceptance criteria for use in reviewing DPs and other information submitted by licensees to demonstrate that the facility is suitable for release in accordance with NRC requirements.

This chapter addresses the general description of the site and its current radiological condition; the next chapter is the decommissioning process (e.g., activities, management, and QA); and the third is devoted to changes after submission of a DP. Discussions of dose modeling, ALARA, surveys, and financial assurance are found in Volumes 2 and 3 of this NUREG. Complete NEPA guidance can be found in NUREG–1748.

The topical contents of Chapters 16–18 are as follows:

Chapter 16: Decommissioning Plans: Site Description

16.1   Decommissioning Plan: Executive Summary

16.2   Decommissioning Plan: Facility Operating History

16.3   Decommissioning Plan: Facility Description

16.4   Decommissioning Plan: Radiological Status of the Facility

Chapter 17: Decommissioning Plans: Program Organization

17.1   Planned Decommissioning Activities

17.2   Decommissioning Plan: Project Management and Organization

17.3   Decommissioning Plan: Radiation Safety and Health Program During Decommissioning

17.4   Decommissioning Plan: Environmental Monitoring and Control Program

17.5   Decommissioning Plan: Radioactive Waste Management Program

17.6   Decommissioning Plan: Quality Assurance Program

17.7   Restricted Use

17.8   Alternate Criteria

Chapter 18: Decommissioning Plans: Modifications to Decommissioning Programs and Procedures

Licensee Procedures:

NRC staff DP reviews do not typically require evaluation of a licensee's detailed procedures, which are normally evaluated during the inspection process (see Section 15.3). However, NRC staff will request and review operating and decommissioning procedures when necessary to ensure safety and regulatory compliance.

DISCLAIMER

This guidance is being issued to describe and make available to the public methods acceptable to NRC staff in implementing specific parts of the Commission's regulations, to delineate techniques and criteria used by the staff in evaluating DPs, and to provide guidance to licensees. This guidance is not a substitute for regulations, and compliance with it is not required. Methods and solutions different from those set out in this guidance will be acceptable, if they provide a basis for concluding that the DP is in compliance with the Commission's regulations.

16.1   DECOMMISSIONING PLAN: EXECUTIVE SUMMARY

NRC staff should review the general information supplied by the licensee to determine if the decommissioning objective and general decommissioning schedule comply with NRC requirements. Expected contents of the DP are listed in Section 16.1.2.

The purpose of review by NRC staff of the "Executive Summary" is to determine, in a general manner, whether the licensee's submitted DP provides an adequate demonstration that the licensee understands, and has complied with, the requirements of 10 CFR 20.1400–1404, 30.36, 40.42, 70.38, and 72.54 for decommissioning and license termination. The staff should not perform a technical review of any information in the "Executive Summary."

16.1.1   REVIEW PROCEDURES

SAFETY EVALUATION

The material to be reviewed is informational in nature, and no specific detailed technical analysis is required. NRC staff should verify that the specific information (e.g., licensee's name and address) is correct. NRC staff should make a qualitative assessment as to (a) the licensee's compliance with the requirements of 10 CFR 20.1402, regarding the estimated dose to the public from residual radioactive material at the completion of decommissioning and the method that the estimated dose from residual radioactivity was determined; (b) the requirements of 10 CFR 20.1403 or 20.1404, if the decommissioning alternative proposed by the licensee is license termination under restricted conditions or using alternate criteria, and; (c) if the decommissioning schedule summary is reasonable.

16.1.2 ACCEPTANCE CRITERIA

REGULATORY REQUIREMENTS

10 CFR 20.1400–1404, 30.36, 40.42, 70.38, 72.54

INFORMATION TO BE SUBMITTED

The information supplied by the licensee should provide contributory evidence as to the licensee's understanding of the technical and institutional requirements for the decommissioning of licensed nuclear facilities. NRC staff review should verify that the following information is included in the "Executive Summary":

- the name and address of the licensee or owner of the site;

- the location and address of the site;

- a brief description of the site and immediate environs;

- a summary of the licensed activities that occurred at the site, including the number and type of license(s); when the facility began and ceased using licensed material, and the types and activities of licensed material authorized and used under the license(s);

- the nature and extent of contamination at the site;

- the decommissioning objective proposed by the licensee (i.e., restricted or unrestricted use);

- the DCGLs for the site, the corresponding doses from these DCGLs, and the method by which the DCGLs were determined;

- a summary of the ALARA evaluations performed to support the decommissioning;

- if the licensee requests license termination under restricted conditions, the restrictions the licensee intends to use to limit doses as required in 10 CFR 20.1403 or 20.1404, and a summary of institutional controls and financial assurance arrangements for the site;

- if the licensee requests license termination under restricted conditions, or using alternate criteria, a summary of the public participation activities undertaken by the licensee to comply with 10 CFR 20.1403(d) or 20.1404(a)(4);

- the proposed initiation and completion dates of decommissioning;

- any post-remediation activities (such as groundwater monitoring) that the licensee proposes to undertake prior to requesting license termination; and

- a statement that the licensee is requesting that its license be amended to incorporate the DP.

16.1.3 EVALUATION FINDINGS

EVALUATION CRITERIA

NRC staff should verify that the information summarized in "Information to be Submitted," above, is included in the Executive Summary. The staff's review should verify that the decommissioning alternative and activities proposed by the licensee are or will be in compliance with the requirements of 10 CFR 20.1402 or 20.1403 as appropriate and that the decommissioning timeframe appears to be reasonable.

16.2 DECOMMISSIONING PLAN: FACILITY OPERATING HISTORY

Licensees who must provide a DP to NRC should submit information to determine if the description of the operating history of the facility is adequate to allow NRC to fully understand the types of radioactive material (and for Part 70 licenses, the hazardous chemicals produced from radioactive material) used at the site, the nature of the authorized use of radioactive materials at the site, and the activities at the site that could have contributed to residual radioactive material being present at the site. This information should include the license number(s) and status of the license(s) held by the licensee descriptions of:

- the activities authorized under the current license;

- past authorized activities using licensed radioactive material at the site;

- all previous decommissioning or remedial activities at the site;

- descriptions the locations of all spills and releases of radioactive material at the site; and,

- all previous burials of radioactive material, including those where the material was subsequently exhumed.

NRC staff should verify that the specific information (license numbers, status and current authorized activities) is correct. In some instances the information described in the following sections may not be available, especially for older facilities. Lack of complete information on the past facility operations would not generally be sufficient justification for rejecting the DP. Rather, the staff should make a qualitative assessment as to whether the licensee's descriptions of authorized activities, past operating activities, spills, and previous burials are adequate to serve as the basis for evaluating the accuracy of the descriptions of the radiological status of the facility and if the decommissioning activities proposed by the licensee to remediate the facility can be conducted safely.

16.2.1 LICENSE NUMBER/STATUS/AUTHORIZED ACTIVITIES

The need for the licensee to determine and evaluate past license activities is to verify that the number and types of licenses and the status of each license are accurate and to insure that the licensee is confident with their past use of radioactive material at the site. This will allow NRC staff to evaluate the licensee's determination of the radiological status of the facility and the licensee's planned decommissioning activities, to ensure that the decommissioning can be conducted in accordance with NRC requirements.

INFORMATION TO BE SUBMITTED

The information supplied by the licensee should be sufficient to enable NRC staff to fully understand what licensed activities are currently being performed by the licensee. NRC staff's review should verify that the following information is included in the "Authorized Activities" section of the DP:

The radionuclides and maximum activities and quantities of radionuclides authorized and used under the current license;

- the chemical forms of the radionuclides authorized and used under the current license;

- a detailed description of how the radionuclides are currently being used at the site;

- the location(s) of use and storage of the various radionuclides authorized under current licenses;

- a scale drawing or map of the building or site and environs showing the current locations of radionuclide use at the site; and

- a list of amendments to the license since the last license renewal.

NRC EVALUATION FINDINGS

NRC staff review should verify that the number and type of licenses and the status of each license are accurate by comparing the information presented in the DP with current NRC license and past inspection information. The staff should verify that the information summarized under "Information to be Submitted," above, is included in the licensee's description of the authorized activities under the license. The staff should verify that this information is correct by comparing it with current NRC license information.

16.2.2 LICENSE HISTORY

As indicated above, the purpose of the development of a detailed history is to ensure that the licensee has thoroughly evaluated and documented previous uses of radioactive material at the site, so that NRC staff can evaluate whether the licensee's determination of the radiological status of the facility is adequate and that the licensee's planned decommissioning activities are appropriate to ensure that the decommissioning can be conducted in accordance with NRC requirements.

INFORMATION TO BE SUBMITTED

The information supplied by the licensee should be sufficient to enable the staff to fully understand what licensed activities were performed by the licensee in the past. The staff's review should verify that the following information is included in the license history section of the facility DP:

- the radionuclides and maximum activities of radionuclides authorized and used under all previous licenses;

- the chemical forms of the radionuclides authorized and used under all previous licenses;

- a detailed description of how the radionuclides were used at the site;

- the location(s) of use and storage of the various radionuclides authorized under all previous licenses as described in 10 CFR 30.35(g), 40.36(f), 70.25(g), 72.30(d); and

- a scale drawing or map of the site, facilities and environs showing previous locations of radionuclide use at the site as described in 10 CFR 30.35(g), 40.36(f), 70.25(g), 72.30(d).

16.2.3 PREVIOUS DECOMMISSIONING ACTIVITIES

The purpose of the review of the license's previous decommissioning activities is to provide NRC staff with information that will aid the staff in evaluating the licensee's determination of the radiological status of the facility and whether previous decommissioning activities are sufficient to comply with current NRC criteria for license termination.

INFORMATION TO BE SUBMITTED

The information supplied by the licensee should be sufficient to enable the staff to fully understand what decommissioning activities were performed by the licensee in the past. The staff's review should verify that the following information is included in the previous decommissioning activities section of the DP:

- a list or summary of areas at the site that were remediated in the past;

- a summary of the types, forms, activities and concentrations of radionuclides that were present in previously remediated areas;

- the activities that caused the areas to become contaminated;

- the procedures used to remediate the areas and the disposition of radioactive material generated during the remediation;

- a summary of the results of the final radiological evaluation of the previously remediated area, including the locations and average radionuclide concentrations in the previously remediated areas; and

- a scale drawing or map of the site, facilities, and environs showing the locations of previous remedial activity.

16.2.4 SPILLS

The purpose of the review of the licensee's description of spills that have occurred at the site is to provide NRC staff with information that will aid in the staff's evaluation of the licensee's determination of the radiological status of the facility. In this context, a "spill" is defined as an uncontrolled release of radioactive material at the site that results in radioactive material being present in the site environs or any unusual occurrences involving the spread of contamination in and around the facility, equipment, or site. Note that controlled releases, such as liquid effluents released to surface water bodies in accordance with 10 CFR 20 Appendix B, would not be considered a "spill." However, the point of release may need to be evaluated to determine the radiological status of the point of release as well as the radiological status of surrounding environs.

INFORMATION TO BE SUBMITTED

The information supplied by the licensee should be sufficient to enable the staff to determine whether spills that have occurred at the facility in the past could impact on the current radiological status of the facility. The staff's review should verify that the following information is included in the spills section of the DP (note that this information may be presented with the information discussed in Section 16.2.3):

- a summary of areas at the site where spills (or uncontrolled releases) of radioactive material occurred in the past;

- the types, forms, activities and concentrations of radionuclides involved in the spill or uncontrolled release, and

- a scale drawing or map of the site, facilities, and environs showing the locations of spills.

16.2.5 PRIOR ONSITE BURIALS

The purpose of the review of the licensee's description of prior onsite burials is to provide the staff with information that will aid in the staff's evaluation of the licensee's determination of the radiological status of the facility.

ACCEPTANCE CRITERIA

Regulatory Requirements

- 10 CFR 20.2002, 30.35(g)(3)(iii), 40.36(f)(3)(iii), 70.25(g)(3)(iii);

- 10 CFR 30.36(g)(4)(i), 40.42(g)(4)(i), 70.38(g)(4)(i) and 72.54(g)(1)

Information to be Submitted

The information supplied by the licensee should be sufficient to enable the staff to determine whether previous burials at the facility could impact on the current radiological status of the facility. Note that all radioactive material at the site would be included in the staff's evaluation of the doses from residual radioactive material and as such would be included in any dose assessments that are performed for the facility. The staff's review should verify that the following information is included in the previous burials section of the DP:

- a summary of areas at the site where radioactive material has been buried in the past;

- the types, forms, activities and concentrations of waste and radionuclides in the former burial(s); and

- a scale drawing or map of the site, facilities, and environs showing the locations of former burials.

16.2.6   EVALUATION FINDINGS

The staff should verify that the information summarized under "Information to be Submitted," above, is included in the licensee's description of former burials at the site. The staff should verify that this information is correct by comparing it to historical NRC license information, as well as information submitted pursuant to 10 CFR 20.302, 20.304, 20.2002, 30.35(g)(3)(iii), 40.36(f)(3)(iii), 70.25(g)(3)(iii) and NUREG–1101, Volume 1 ("On-site Disposal of Radioactive Waste," March 1986). Note that the information required pursuant to 30.35(g)(3)(iii), 40.36(f)(3)(iii), and 70.25(g)(3)(iii) may not be submitted to NRC until license termination. However, the licensee should include or use a summary of this information in developing this section of the DP.

16.3   DECOMMISSIONING PLAN: FACILITY DESCRIPTION

SITE COMPLEXITY

This section of the decommissioning guidance was developed to provide guidance on the types of information that would be required for the most complex decommissioning sites. These sites could require complicated site-specific dose modeling, contain residual radioactivity at depths exceeding 15–30 cm, and/or have onsite disposal cells for radiologically contaminated waste. Less complex sites would not need to include all of the information described below in their DPs. Note that some of this information overlaps information required for the environmental review (see NUREG–1748). NRC staff and the licensee should work together to establish the amount and type of information needed to support the DP for each individual facility and the best method to provide NRC with the information.

The licensee should supply information to support staff analysis of the description of the facility and environs. This information should allow NRC staff to evaluate the licensee's estimation of (a) the doses to onsite and offsite populations during and at the completion of decommissioning; (b) the impacts of the proposed decommissioning activities for the site, and its surrounding areas, and for restricted-release sites; and (c) the impacts of the environment on the site (e.g., in the event of floods, tornadoes and earthquakes). This information should include all of the following:

- a description of the site and environs;

- a description of the current population distribution;

- a summary of current and potential future uses of land in and around the site;

- descriptions of the site meteorology, geology, seismology, climatology, surface and groundwater hydrology, geotechnical characteristics; and

- descriptions of the natural and water resources at the site.

In addition, a description of the ecology of the site, a description of minority and low-income populations, and a summary of all endangered species at the site, may be required for NRC staff to complete the NEPA analysis (see Section 15.7 of this NUREG and NUREG–1748 for additional guidance on NEPA).

REVIEW PROCEDURES

Safety Evaluation

The staff should verify that the provided site-specific information is complete and accurate. The staff should make a qualitative assessment as to whether the licensee's descriptions of the site and environs and summary of current and potential future land uses are adequate to serve as the bases for evaluating the licensee's estimated dose.

16.3.1   SITE LOCATION AND DESCRIPTION

The purpose of the review of the description of the site location and description is to verify that sufficient information is presented to allow NRC staff to understand the physical characteristics of the site and relationship of the site to surrounding areas. This will aid the staff in evaluating the licensee's dose estimates and planned decommissioning activities to ensure that the decommissioning can be conducted in accordance with NRC requirements.

ACCEPTANCE CRITERIA

Regulatory Requirements

10 CFR 30.36(g)(4)(i), 40.42(g)(4)(i), 70.38(g)(4)(i) and 72.54(g)(1)

Information to be Submitted

The information supplied by the licensee should be sufficient to allow the staff to fully understand the physical characteristics of the site. The staff's review should verify that the following information is included in the description of the site description and location section of the DP:

- the size of the site in acres or square meters;

- the State and county in which the site is located;

- the names and distances to nearby communities, towns and cities;

- a description of the contours and natural features of the site;

- the elevation of the site;

- a description of the man-made features of the site, such as buildings, roads, and settling ponds;

- a description of property surrounding the site, including the location of all offsite wells used by nearby communities or individuals;

- the location of the site relative to prominent features such as rivers and lakes. To facilitate presentation of this information, U.S. Geological Survey (USGS) topographic maps may be provided;

- a map that shows the detailed topography of the site using a contour interval (such as 2 feet or 1 meter) and including plot plans, the locations of characterization borings and monitoring wells, and the positions and types of geologic characterization activities;

- the location of the nearest residences and all significant facilities or activities near the site; and

- a description of the facilities (e.g., buildings, parking lots, and fixed equipment) at the site.

EVALUATION FINDINGS

The staff should verify that the information summarized under "Information to be Submitted," above, is included in the licensee's description of the site and environs. The staff's review should verify, to the maximum extent practicable, that the information supplied by the licensee is accurate by comparing it with licensing and inspection information maintained in NRC files.

16.3.2   POPULATION DISTRIBUTION

The purpose of the review of the description of the population distribution is to determine if the licensee has supplied sufficient information on the makeup and distribution of the population in the vicinity of the site to allow NRC staff to evaluate the licensee's estimate of doses to offsite individuals during and at the completion of decommissioning.

ACCEPTANCE CRITERIA

Regulatory Requirements

10 CFR 30.36(g)(4)(i), 40.42(g)(4)(i), 70.38(g)(4)(i) and 72.54(g)(1)

Information to be Submitted

The information supplied by the licensee should be sufficient to allow the staff to determine the population makeup and distribution in the vicinity of the site. The staff's review should verify that a summary of current and projected populations in the vicinity of the site, by principal compass sectors, is included in the DP. This summary should be sufficiently detailed to allow the determination of doses to offsite individuals via atmospheric pathways. The DP should include the following:

- a summary of the current population in and around the site, by compass vectors; and

- a summary of the projected population in and around the site, by compass vectors.

EVALUATION FINDINGS

The staff should verify that the information summarized under "Information to be Submitted," above, is included in the licensee's description of the distribution of populations around the site. The staff should verify the licensee's population data against available independent population data (e.g., information from the Census Bureau including any special census that may have been conducted, local and State Agencies, and regional Councils of Government). The staff should evaluate projected population information by comparing it to projections made by local planning boards or offices.

16.3.3 CURRENT/FUTURE LAND USE

The purpose of the description of current and future land use is to provide the staff with information that will aid in evaluating the licensee's estimates of doses to onsite and offsite individuals during and at the completion of decommissioning.

ACCEPTANCE CRITERIA

Regulatory Requirements

10 CFR 30.36(g)(4)(i), 40.42(g)(4)(i), 70.38(g)(4)(i) and 72.54(g)(1)

Information to be Submitted

The information supplied by the licensee should be sufficient to allow the staff to understand what current land uses are and what local, regional, or State planning boards or offices anticipate the future land uses will be at the site. The staff's review should verify that the licensee has used all available data on land use, plans and trends in land use, land use controls (such as zoning), potential for growth, or other factors likely to inhibit or stimulate growth in the area by comparing it with publicly available information from local, regional or State land use planning boards or offices. The DP should include a description of the current land uses in and around the site and a summary of anticipated land uses.

EVALUATION FINDINGS

The staff should verify that the information summarized under "Information to be Submitted," above, is included in the licensee's discussion of current and future land use. The staff should verify, to the extent practicable, that this information is correct by comparing it to publicly available information on current land use in the vicinity of the site, land use trends in and around the site, and expected future uses of the land in and around the site.

16.3.4 METEOROLOGY AND CLIMATOLOGY

The purpose of the review of the licensee's description of meteorology and climatology is to determine if the licensee has provided sufficient information to allow NRC staff to evaluate the licensee's estimations of doses to onsite and offsite individuals during and at the completion of decommissioning operations.

ACCEPTANCE CRITERIA

Regulatory Requirements

10 CFR 30.36(g)(4)(i), 40.42(g)(4)(i), 70.38(g)(4)(i) and 72.54(g)(1)

Regulatory Guidance

RG 1.23 "Onsite Meteorological Programs" (Safety Guide 23)

Information to be Submitted

The information supplied by the licensee should be sufficient to allow the staff to determine how local weather patterns will affect the estimation of doses to onsite and offsite individuals during and at the completion of decommissioning operations. The staff's review should verify that the following information is included in the climatology and meteorology section of the DP:

- a description of the general climate of the region with respect to types of air masses, synoptic features (high- and low-pressure systems and frontal systems), general air-flow patterns (wind direction and speed), temperature and humidity, precipitation, and relationships between synoptic-scale atmospheric processes and local meteorological conditions;

- seasonal and annual frequencies of severe weather phenomena, including tornadoes; water spouts, thunderstorms, lightning, hail, and high air pollution potential;

- weather-related radionuclide transmission parameters, including average and extreme wind vectors, and average and extreme duration and intensity of precipitation events;

- routine weather-related site deterioration parameters, including precipitation intensity and duration, wind vectors, and temperature and pressure gradients;

- extreme weather-related site deterioration parameters, including tornadoes, water spouts, thunderstorms, hail, and extreme air pollution (from offsite sources);

- a description of the local (site) meteorology in temperature, atmospheric water vapor, precipitation, fog, atmospheric stability and air quality; and

- the National Ambient Air Quality Standards Category of the area in which the facility is located, and, if the facility is not in a Category 1 zone, the closest and first downwind Category 1 Zone.

EVALUATION FINDINGS

The staff should review the licensee's description of the site climatology and meteorology for completeness and adequacy of basic data. The wind and atmospheric stability data should be based on onsite data. The other summaries should be based on nearby representative stations with long record retention periods. When offsite data are used, the staff should determine how well the data represent site conditions and whether more representative data are available. The staff should use National Oceanic and Atmospheric Administration (NOAA) (U.S. Department of Commerce) State meteorological summaries ("State Climatological Summary"), local climatological data ("Local Climatological Data Annual Summary with Comparative Data"), and NOAA Environmental Data Service summaries pertinent to the site to evaluate the representativeness of stations and periods of record. The staff should be familiar with all primary meteorological data collection locations. The staff should ensure that all topographic maps and topographic cross-sections presented by the licensee are legible and well-labeled so that the information needed during the review can be readily extracted. Points of interest such as facility structures, site boundary, and buffer zone should be marked on all maps and diagrams.

The staff should compare the licensee's assessment of the effect of topography with standard assessments such as those presented in "Meteorology and Atomic Energy – 1968" (Slade, 1968) and decide whether the standard regulatory atmospheric diffusion models are appropriate for this site. The staff should review for completeness and authenticity the general climatic description of the region in which the site is located. Climatic parameters such as air masses, general air flow, pressure patterns, frontal systems, and temperature and humidity conditions reported by the licensee should be checked against standard references (Thom, 1968; U.S. Department of Commerce, 1968) for appropriateness with respect to location and period of record. The staff should verify the licensee's description of the role of synoptic-scale atmospheric processes on local (site) meteorological conditions against the descriptions provided in "Climatic Atlas of the United States" and "Local Climatological Data–Annual Summary With Comparative Data" (both published by the U.S. Department of Commerce).

Because meteorological averages and extremes can only be obtained from stations in the region of the site that have long record retention periods, and the stations are not usually very close to the site, the staff should first determine the representativeness of the data to site conditions and then ascertain the adequacy of the stations and their data. The staff should verify (a) recorded meteorological averages and extremes using standard publications such as "Storm Data," published by the U.S. Department of Commerce; (b) other averages and extremes using "State Climatological Summaries" and "Storm Data," published by the U.S. Department of Commerce; (c) the potential for high air pollution; (d) extreme winds and their distribution using Regulatory Guide 1.23 (RG 1.23) and "Meteorology and Atomic Energy–1968" (Slade, 1968); and (e) gust factors using RG 1.23.

16.3.5   GEOLOGY AND SEISMOLOGY

The purpose of the review of the licensee's description of the site geology and seismology is to determine if the licensee has provided sufficient information to allow NRC staff to evaluate the licensee's estimations of doses to onsite and offsite individuals during and at the completion of decommissioning operations.

ACCEPTANCE CRITERIA

Regulatory Requirements

10 CFR 30.36(g)(4)(i), 40.42(g)(4)(i), 70.38(g)(4)(i) and 72.54(g)(1)

Information to be Submitted

The information supplied by the licensee should be sufficient to allow the staff to determine how site geological and seismological characteristics will affect the estimation of doses to onsite and offsite individuals during and at the completion of decommissioning operations and the potential effects of geological processes (e.g., earthquakes, erosion, and landslides) on restricted-release sites. The staff's review should verify that the geology and seismology section of the DP contains the following information:

Geology

- A detailed description of the geologic characteristics of the site and the region around the site.

- A discussion of the tectonic history of the region, regional geomorphology, physiography, stratigraphy, and geochronology. All tectonic structures should be identified, in particular folds and faults in the region around the site, and their geologic and structural history should be discussed. The relationship between seismicity and tectonic structures and the earthquake-generating potential of any active structures should be discussed.

- A regional tectonic map showing the site location and its proximity to tectonic structures should be provided. Appropriate references or supporting documents should be provided with regional physiographic and topographic maps, geologic and structure maps, fault maps, stratigraphic sections, boring logs, and aerial photographs.

- A description of the structural geology of the region and its relationship to the site geologic structure should be discussed. Any faults, folds, open jointing, fractures, and shear zones in the region must be identified, and their significance to the facility should be discussed.

- A description of any crustal tilting, subsidence, karst terrain, landsliding, and erosion.

- A description of the surface and subsurface geologic characteristics of the site and its vicinity. The description should include local stratigraphic units and their accepted names, ages, genetic relationships, and lithologies. To facilitate the presentation, these descriptions should be accompanied by appropriately scaled geologic maps. Descriptions of mineralogy, particle size, organic materials, degree of cementation, zones of alteration, and depositional environment of unconsolidated strata should be included.

- A description of the geomorphology of the site, including USGS topographic maps that emphasize local geomorphic features pertinent to the site. A description of the geomorphic processes affecting the present-day topography of the disposal site and vicinity should be included. Information should include descriptions of processes such as mass wasting, erosion, slumping, landsliding, and weathering where appropriate. The discussion of relevant geomorphic processes should include their rates, frequencies of occurrence, and controlling mechanisms or factors.

- A description of the location, attitude, and geometry of all known or inferred faults in the site and vicinity. Fault displacements should be identified and potential recurrence intervals addressed.

- A discussion of the nature and rates of deformation such as folding within the site and their relation to the local stress regime. Any joint sets within the site, including their densities and orientations, should be described, and their relative ages discussed. Remineralization and mineralization history of the various joint sets should also be discussed. Solution cavities and crevices in the bedrock should be described and discussed, if applicable.

- A description of any man-made geologic features, such as mines or quarries.

Seismology

- A description of the seismicity, tectonic characteristics of the site and region, correlation of earthquake activity with geologic structures and tectonic provinces, maximum earthquake potential, seismic wave transmission characteristics of the site, design earthquake, settlement and liquefaction, and geophysical methods for site characterization.

- A complete list of all historical earthquakes that have a magnitude of 3 or more or a modified Mercalli intensity of IV or more within 320 kilometers (200 miles) of the site. The listing should include all available information about the earthquakes such as epicenter coordinates, depth of focus, origin time, intensity, and magnitude, augmented by a map showing the locations of these earthquakes. The references from which the information was obtained should be indicated. In addition, any earthquake that induced geologic hazard (e.g., landsliding or liquefaction) should be identified, and the acceleration that caused the hazard should be provided.

EVALUATION FINDINGS

The staff should review for completeness the information on geologic site characterization in the DP. If the information reflects the results of a thorough literature search and an adequate reconnaissance and physical examination of the regional and site conditions by the licensee, the DP will be considered acceptable. Consultations with commercial companies and Federal, State, and local Government Agencies that may have had occasion to characterize the site will help ensure the adequacy of the characterization in the DP. The review can be completed quickly if the DP contains sufficient information to allow the staff to make an independent assessment of the licensee's assumptions, analyses, and conclusions.

16.3.6   SURFACE WATER HYDROLOGY

ACCEPTANCE CRITERIA

The purpose of the review of the licensee's description of the surface water hydrology at the site is to determine if the licensee has provided sufficient information to allow NRC staff to evaluate the licensee's estimations of doses to onsite and offsite individuals during and upon completion of decommissioning operations.

Regulatory Requirements

10 CFR 30.36(g)(4)(i), 40.42(g)(4)(i), 70.38(g)(4)(i) and 72.54(g)(1)

Information to be Submitted

The information supplied by the licensee should be sufficient to enable the staff to determine whether surface water characteristics could impact the doses to onsite or offsite individuals during or at the completion of decommissioning. For restricted-release sites, staff would also analyze the potential dose impact of atypical surface waters conditions, such as floods. The staff's review should verify that the following information is included in the surface water hydrology section of the DP:

- a description of site drainage and surrounding watershed fluvial features, including important water users;

- water resource data, including maps, hydrographs, and stream records from other agencies (e.g., USGS and USACE);

- topographic maps of the site that show natural drainages and man-made features;

- a description of the surface water bodies at the site and surrounding areas, including the location, size, shape, and other hydrologic characteristics of all streams, lakes, or coastal areas;

- a description of existing and proposed water control structures and diversions (both upstream and downstream) that may influence the site;

- flow-duration data that indicate minimum, maximum, and average historical observations for surface water bodies in the site areas;

- aerial photography and maps of the site and adjacent drainage areas identifying features such as drainage areas, surface gradients, and areas of flooding;

- an inventory of all existing and planned surface water users, whose intakes could be adversely affected by migration of radionuclides from the site (the inventory should include the owner, location, type, and amount of use; source of supply; type of intake; and surface water quality data);

- topographic and/or aerial photographs that delineate the 100-year floodplain at the site; and

- a description of any man-made changes to the surface water hydrologic system that may influence the potential for flooding at the site (such changes may include construction of reservoirs, urban development, strip mining, and lumbering) (the description of these changes should include the proximity of the affected area to the site, the surface water bodies affected, the size of the area affected, and the potential effects at the site).

EVALUATION FINDINGS

The staff should verify that the information summarized under "Information to be Submitted," above, is included in the licensee's description of the surface water features at the site. Acceptance of the information in the DP will be based in part on a qualitative evaluation of the completeness and adequacy of the information and of maps. Descriptions and evaluations of structures and facilities are adequate if they are sufficiently complete to allow independent evaluations of the effects of flooding and intense rainfall. Site topographic maps are acceptable if they are of good quality, legible, and adequate in coverage to substantiate applicable data and analyses. The descriptions of the hydrologic characteristics of surface water features and water use are acceptable if they are detailed and generally correspond to those of the USGS, NOAA, Soil Conservation Service, USACE, or appropriate State and river basin Agencies. Descriptions of existing or proposed reservoirs and dams that could influence conditions at the site should be based on reports of the USGS, U.S. Bureau of Reclamation, USACE, and others; these reports normally include tabulations of drainage areas, types of structures, appurtenances, ownership, seismic and spillway design criteria, elevation-storage relationships, and short- and long-term storage allocations.

16.3.7   GROUND WATER HYDROLOGY ACCEPTANCE CRITERIA

The purpose of the review of the license's description of the ground water hydrology section of the DP is to determine if the licensee has provided sufficient information to allow NRC staff to evaluate the licensee's estimations of doses to onsite and offsite individuals during and at the completion of decommissioning operations.

ACCEPTANCE CRITERIA

Regulatory Requirements

10 CFR 30.36(g)(4)(i), 40.42(g)(4)(i), 70.38(g)(4)(i) and 72.54(g)(1)

Information to be Submitted

The information supplied by the licensee should be sufficient to allow the staff to determine how the groundwater characteristics of the site affect the doses to onsite or offsite individuals during or at the completion of decommissioning. The staff's review should verify that the following information is included in the groundwater hydrology section of the DP:

- A description of the saturated zone including all potentially affected aquifers, the lateral extent, thickness, water-transmitting properties, recharge and discharge zones, groundwater flow directions and velocities, and other information that can be used to create an adequate conceptual model of the saturated zone.

- Descriptions for monitor wells, including location, elevation, screened intervals, depths, construction and completion details, and hydrogeologic units monitored. The description should include domestic, industrial and/or municipal wells or other monitoring devices, if applicable, and any construction and completion details for these devices, when available. Descriptions of all aquifer tests should also be provided, including test data and a discussion of the assumptions, analysis, and test procedures used.

- Physical parameters such as storage coefficients, transmissivities, hydraulic conductivities, porosities, and intrinsic permeabilities should be included.

- A description, to the extent practicable, of groundwater flow directions and velocities (horizontal and vertical) for each potentially affected aquifer. When applicable, the groundwater hydrology should be described by making use of hydrogeologic columns, cross-sections, and water table and/or potentiometric maps.

- A description of the unsaturated zone including descriptions of the lateral extent and thickness of permeable and impermeable zones, potential conduits of anomalously high flux, and direction and velocity of unsaturated flow.

- Information on all monitor stations, including location and depth.

- A description of physical parameters including the spatial and stratigraphic distribution of the total and effective porosity; water content variations with time; saturated hydraulic conductivity; characteristic relationships between water content, pressure head, and hydraulic conductivity; and hysteretic behavior during wetting and drying cycles, especially during extreme conditions.

- A description of the numerical analysis techniques used to characterize the unsaturated and saturated zones including, the model type, justification, documentation, verification, calibration and other associated information. In addition, the description should include the input data, data generation or reduction techniques, and any modifications to these data.

- The distribution coefficients of the radionuclides of interest at the site.

EVALUATION FINDINGS

The staff should verify that the information summarized under "Information to be Submitted," above, is included in the licensee's description of the groundwater hydrology at the site. The staff should review the information on the saturated zone by evaluating the testing and monitoring program and sample collection procedure. The staff should evaluate the rationale for choosing particular sampling locations and verify that they are commensurate with the complexity of the saturated zone. The staff should confirm that acceptable procedures were used by the licensee to collect, preserve, and analyze samples. Staff should determine that adequate quality control was used for the collection, preservation, and laboratory analyses of samples. The staff should evaluate the adequacy of non-licensee-constructed monitoring devices used in the characterization (including the characterization of seeps, springs, and private, municipal, or industrial wells in the vicinity of the proposed site). The staff should evaluate aquifer tests performed by the licensee to ensure that applicable test methods incorporate proper assumptions, analyses, and test procedures. The staff should assess the accuracy of the transmissivity, storativity, and hydraulic conductivity results derived from testing. The staff should determine if groundwater will discharge to the surface within the site boundary and if fluctuations in the water table will result in interactions of groundwater with the residual radioactive material. Staff should confirm the description of major hydrologic parameters, aerial extent of aquifers, recharge-discharge zones, flow rates and directions, and travel times, including seasonal fluctuations and long-term trends.

The staff should review the licensee's information on the unsaturated zone by evaluating the monitoring program and sample collection procedure. The staff should evaluate the rationale for choosing particular sampling locations and verify that they are commensurate with the complexity of the unsaturated zone. The staff should confirm that the description of the unsaturated zone incorporates the necessary field and laboratory data, including seasonal fluctuations and long-term trends. The staff should review the licensee's analysis of the likelihood of the development of perched aquifers and perform independent analyses, using accepted methods, to determine the adequacy of the description.

The evaluations described in the following paragraphs may be included in the groundwater hydrology portion or dose modeling sections of the DP.

The staff should evaluate the licensee's conceptual model that describes, to the extent practicable, all hydrogeologic processes and features, including the potential for deep percolation, recharge/discharge zones, areas of anomalous physical parameters affecting regional processes, extent of aquifers and confining layers, interactions between aquifers, and movement of groundwater in the saturated and unsaturated zone. The staff should review this model to determine its defensibility, conservatism, and adequate incorporation of data into a unified conceptual model.

The staff should evaluate the numerical analyses of groundwater data collected by the licensee for the site and vicinity. This will normally involve analytical or numerical modeling. The staff should verify that the model type chosen for analysis is properly documented, verified, and calibrated and adequately simulates the physical system of the site and vicinity. The staff should review the modeling strategy used by the licensee to assure that it is logical and defensible. The staff should review the adequacy of the model input data generation and reduction techniques. Modifications of input data required for calibration should be reviewed to ensure that the new values are realistic and defensible.

Following its review of this information, the staff should determine whether the licensee's conclusions are adequate. If the staff conducts an independent analysis, it should compare the results with those derived by the licensee to determine if the licensee's results are adequate.

16.3.8   NATURAL RESOURCES

The purpose of the review of the license's description of natural resources at the site is to aid the staff in evaluating the impacts that the decommissioning alternative chosen by the licensee may have on these resources and to evaluate whether the exploitation of these resources could impact the licensee's dose estimates for the site.

ACCEPTANCE CRITERIA

Regulatory Requirements

10 CFR 30.36(g)(4)(i), 40.42(g)(4)(i), 70.38(g)(4)(i) and 72.54(g)(1)

Information to be Submitted

The information supplied by the licensee should be sufficient to allow the staff to determine what natural resources are present at and in the vicinity of the site. The staff's review should verify that the following information is included in the natural resources section of the DP:

- a description of the natural resources occurring at or near the site, including metallic and nonmetallic minerals and ores; fuels, such as peat, lignite, and coal; hydrocarbons, including gas, oil, tar sands, and asphalt; geothermal resources; industrial mineral deposits, such as sand and gravel, clays, aggregate sources, shales, and building stone; timber; agricultural lands; and waters in the form of brines;

- a description of potable, agricultural, or industrial ground or surface waters including information on resource type, occurrence, location, extent, net worth, recoverability, and current and projected use;

- a description of economic, marginally economic, or subeconomic known or identified natural resources as defined in U.S. Geological Survey Circular 831; and

- mineral, fuel, and hydrocarbon resources near and surrounding the site which, if exploited, would affect the licensee's dose estimates.

EVALUATION FINDINGS

The staff should verify that the information summarized under "Information to be Submitted," above, is included in the licensee's description of the natural resources at the site.

The staff should determine if the licensee has identified known resources as described in U.S. Geological Survey Circular 831. The staff should verify that the DP describes economic, marginally economic, and subeconomic known resources as defined in U.S. Geological Survey Circular 831. On the basis of these data, the staff should evaluate the licensee's estimation of potential future exploitation, considering market values and current and projected demand for the resource in question. On the basis of the resources identified, the staff should examine the potential for site disruption resulting from exploration and exploitation techniques including, but not limited to, augering, drilling, shaft mining, strip mining, bulldozing and other excavation, quarrying, bore-hole injection and pumping, uprooting of vegetation, blasting, stream diversion, and dam construction. These techniques are considered for the possibility of direct site intrusion as well as indirect effects such as alteration of groundwater tables or increase in erosion.

16.4 DECOMMISSIONING PLAN: RADIOLOGICAL STATUS OF FACILITY

The licensee will provide a description of the current radiological status of the facility. This information will allow NRC staff to fully understand the types and levels of radioactive material contamination and the extent of radioactive material contamination at the facility. This information will be used by the staff during its review of the licensee's decommissioning activities, to evaluate the cost estimates for decommissioning, and decommissioning health and safety plans. This information should include summaries of the types and extent of radionuclide contamination in all media at the facility including buildings, systems and equipment, surface and subsurface soil, surface water, and groundwater.

Information presented in this section should be developed based on the methodologies and procedures described in Section 15.4 of this guidance and in Volume 2 of this NUREG. Information describing how the licensee developed the information presented in this section should be presented in the "Facility Radiation Surveys" section of the DP. Licensees who report the results of the characterization survey in the "Radiological Status of Facility" portion of the DP do not need to report it in the "Facility Radiation Surveys" portion. Similarly, licensees may combine the information required in this section of this NUREG with that described in Chapter 4 from Volume 2 of this NUREG, as long as the information discussed in both sections is included in the DP.

16.4.1 CONTAMINATED STRUCTURES

The purpose of the review of the description of the contaminated structures is to evaluate whether the licensee has fully described the types and activity of radioactive material contamination in the structures, as well as the extent of this contamination. This information should be sufficient to allow NRC staff to evaluate the potential safety issues associated with remediating the structures, whether the remediation activities and radiation control measures proposed by the licensee (described in Sections 17.1 and 17.3 of this NUREG) are appropriate for the type of radioactive material present in the structure, whether the licensee's waste management practices are appropriate, and whether the licensee's cost estimates are plausible, given the amount of contaminated material that will need to be removed or remediated.

In some instances, licensees may choose to dismantle contaminated structures and dispose of the building debris as radioactive waste in lieu of decontaminating the building. Similarly, licensees may choose to decontaminate portions of buildings to levels appropriate for unrestricted use and dismantle portions of the building to gain access to areas where contamination has migrated, such as floor/wall joints. In these instances, all of the information described below may not need to be included in the DP. NRC staff should discuss these activities with licensees to ensure that adequate information is provided in the DP to allow the staff to perform the required evaluations, without requiring the licensee to expend substantial resources characterizing the structures.

INFORMATION TO BE SUBMITTED

The information supplied by the licensee should be sufficient to allow the staff to fully understand the types and activity of radioactive material contamination in the structure, as well as the extent of this contamination. The staff's review should verify that the following information is included in the contaminated structures section of the facility DP:

- a list or description of all structures at the facility where licensed activities occurred that contain residual radioactive material in excess of site background levels;

- a summary of the structures and locations at the facility that the licensee has concluded have not been impacted by licensed operations and the rationale for the conclusion;

- a list or description of each room or work area within each of these structures;

- a summary of the background levels used during scoping or characterization surveys;

- a summary of the locations of contamination (e.g., walls, floors, wall/floor joints, structural steel surfaces, and ceilings) in each room or work area;

- a summary of the radionuclides present at each location, the maximum and average radionulide activities in disintegrations per minute per 100 square centimeters (dpm/100cm^2), the chemical form of the radionuclide, and, if multiple radionuclides are present, the radionuclide ratios;

- the mode of contamination for each surface (i.e., whether the radioactive material is present only on the surface of the material or if it has penetrated the material);

- the maximum and average radiation levels in millisievert per hour (mSv/hr) or microsievert per hour (μSv) (millirem per hour (mrem/hr) or microrem per hour (μrem/hr)), as appropriate, in each room or work area; and

- a scale drawing or map of the rooms or work areas showing the locations of radionuclide material contamination and radiation levels. All maps should include compass direction indicators.

NRC EVALUATION FINDINGS

The staff should verify that the information summarized under "Information to be Submitted," above, is included in the licensee's description of the contaminated structures. The staff's review should verify that the licensee has fully described the types and activity of radioactive material contamination in facility structures, as well as the extent of this contamination. These descriptions should be sufficient to allow NRC staff to evaluate the potential safety issues associated with remediating the structures, whether the remediation activities and radiation control measures proposed by the licensee are appropriate for the type of radioactive material present in the structures, whether the licensee's waste management practices are appropriate, and whether the licensee's cost estimates are plausible, given the amount of contaminated material that will need to be removed or remediated.

16.4.2 CONTAMINATED SYSTEMS AND EQUIPMENT

The purpose of the review of the description of the contaminated systems and equipment at the facility is to evaluate whether the licensee has fully described the types and activity of radioactive material contamination in facility systems or on equipment, as well as the extent of this contamination. This information should be sufficient to allow NRC staff to evaluate the potential safety issues associated with remediating the systems or equipment, whether the remediation activities and radiation control measures proposed by the licensee (described in Sections 17.1 and 17.3 of this NUREG) are appropriate for the type of radioactive material present in the systems or equipment, whether the licensee's waste management practices are appropriate, and whether the licensee's cost estimates are plausible, given the amount of contaminated material that will need to be removed or remediated.

Note that, in some instances, licensees may choose to remove and dispose (either as radioactive waste or as usable equipment in another radiation area) of contaminated systems and/or equipment, in lieu of decontaminating the system or equipment. In these instances, all of the information described below may not necessarily need to be included in the DP. NRC staff should discuss these activities with licensees to ensure that adequate information is provided in the DP to allow the staff to perform the evaluations described above, without requiring the licensee to expend substantial resources characterizing the equipment or system.

INFORMATION TO BE SUBMITTED

The information supplied by the licensee should be sufficient to allow the staff to fully understand the types and activity of radioactive material contamination present in systems or on equipment, as well as the extent of this contamination. The staff's review should verify that the following information is included in the contaminated systems and equipment section of the facility DP:

- a list or description and the location of all systems or equipment at the facility that contain residual radioactive material in excess of site background levels;

- a summary of the radionuclides present in each system or on the equipment at each location, the maximum and average radionulide activities in dpm/100cm^2, the chemical form of the radionuclide, and, if multiple radionuclides are present, the radionuclide ratios;

- the maximum and average radiation levels in mSv/hr or μSv/hr(mrem/hr or μrem/hr), as appropriate, at the surface of each piece of equipment;

- a summary of the background levels used during scoping or characterization surveys; and

- a scale drawing or map of the rooms or work areas showing the locations of the contaminated systems or equipment. All maps should include compass direction indicators.

16.4.3   SURFACE SOIL CONTAMINATION

The purpose of the review of the description of surface soil (i.e., soil within the top 15–30 centimeters (cm) of the soil column) contamination is to determine if the licensee has fully described the types and activity of radioactive material contamination in the surface soil, as well as the extent of this contamination. This information should be sufficient to allow NRC staff to evaluate the potential safety issues associated with remediating the surface soil, whether the remediation activities and radiation control measures proposed by the licensee (described in Sections 17.1 and 17.3 of this NUREG) are appropriate for the type of radioactive material present in the surface soil, whether the licensee's waste management practices are appropriate, and whether the licensee's cost estimates are plausible, given the amount of contaminated soil that will need to be removed or remediated.

INFORMATION TO BE SUBMITTED

The information supplied by the licensee should be sufficient to allow the staff to fully understand the types and activity of radioactive material in surface soil, as well as the extent of this contamination. The staff's review should verify that the following information is included in the description of contaminated surface soil in the facility DP:

- a list or description of all locations at the facility where surface soil contains residual radioactive material in excess of site background levels;

- a summary of the background levels used during scoping or characterization surveys;

- a summary of the radionuclides present at each location, the maximum and average radionuclide activities in becquerel per gram (Bq/gm) (picocuries per gram (pCi/gm)), the chemical form of the radionuclide, and, if multiple radionuclides are present, the radionuclide ratios;

- the maximum and average radiation levels in mSv/hr (mrem/hr) at each location; and

- a scale drawing or map of the site showing the locations of radionuclide material contamination in surface soil. All maps should include compass direction indicators.

16.4.4   SUBSURFACE SOIL CONTAMINATION

The purpose of the review of the description of subsurface soil (i.e., soil below the top 15–30 cm of soil in the soil column) contamination is to determine if the licensee has fully described the types and activity of radioactive material contamination in the subsurface soil, as well as the extent of this contamination. This information should be sufficient to allow NRC staff to evaluate the potential safety issues associated with remediating the subsurface soil, whether the remediation activities and radiation control measures proposed by the licensee (described in Sections 17.1 and 17.3 of this NUREG) are appropriate for the type of radioactive material present in the subsurface soil, whether the licensee's waste management practices are appropriate and whether the licensee's cost estimates are plausible, given the amount of contaminated soil that will need to be removed or remediated.

INFORMATION TO BE SUBMITTED

Information should be sufficient to allow NRC staff to fully understand the types and activity of radioactive material in subsurface soil, as well as the extent of this contamination. The staff's review should verify that the following information is included in the description of contaminated subsurface soil in the facility DP:

- a list or description of all locations at the facility where subsurface soil contains residual radioactive material in excess of site background levels;

- a summary of the background levels used during scoping or characterization surveys;

- a summary of the radionuclides present at each location, the maximum and average radionulide activities in Bq/gm (pCi/gm), the chemical form of the radionuclide, and, if multiple radionuclides are present, the radionuclide ratios;

- the depth of the subsurface soil contamination at each location; and

- a scale drawing or map of the site showing the locations of subsurface soil contamination. All maps should include compass direction indicators.

NRC EVALUATION FINDINGS

The staff should verify that the information summarized under "Information to be Submitted," above, is included in the licensee's description of the subsurface soil contamination at the facility. The staff's review should verify that the licensee has fully described the types and activity of radioactive material contamination in the subsurface soil at the facility, as well as the extent of this contamination. These descriptions should be sufficient to allow NRC staff to evaluate the potential safety issues associated with remediating the subsurface soil, whether the remediation activities and radiation control measures proposed by the licensee are appropriate for the type of radioactive material present in the subsurface soil, whether the licensee's waste management practices are appropriate, and whether the licensee's cost estimates are plausible, given the amount of contaminated material that will need to be removed or remediated.

16.4.5 SURFACE WATER

The purpose of the review of the description of contaminated surface water is to evaluate whether the licensee has fully described the types and activity of radioactive material present in surface water bodies at the facility, as well as the extent of this contamination. This information should be sufficient to allow NRC staff to evaluate potential safety issues associated with remediating the surface water, whether the remediation activities and radiation control measures proposed by the licensee (described in Sections 17.1 and 17.3 of this NUREG) are appropriate for the type of radioactive material present in the surface water, whether the licensee's waste management practices are appropriate, and whether the licensee's cost estimates are plausible, given the amount of contaminated water that will need to be removed or remediated.

INFORMATION TO BE SUBMITTED

The information supplied by the licensee should be sufficient to allow the staff to fully understand the types and activity of radioactive material contamination in surface water at the facility, as well as the extent of this contamination. NRC review should verify that the following information is included in the description of surface water contamination in the DP:

- a list or description and map of all surface water bodies at the facility that contain residual radioactive material in excess of site background levels;

- a summary of the background levels used during scoping or characterization surveys; and

- a summary of the radionuclides present in each surface water body and the maximum and average radionuclide activities in becquerel per liter (Bq/L) (picocuries per liter (pCi/L)).

NRC EVALUATION FINDINGS

The staff should verify that the information summarized under "Information to be Submitted," above, is included in the licensee's description of the surface water contamination at the site. The staff's review should verify that the licensee has fully described the types and activity of radioactive material contamination in the surface water at the site, as well as the extent of this contamination. These descriptions should be sufficient to allow NRC staff to evaluate potential safety issues associated with remediating the surface water, whether the remediation activities and radiation control measures proposed by the licensee are appropriate for the type of radioactive material present in the water, whether the licensee's waste management practices are appropriate and whether the licensee's cost estimates are plausible, given the amount of contaminated material that will need to be removed or remediated.

16.4.6   GROUND WATER

The purpose of the review of the description of contaminated ground water is to evaluate whether the licensee has fully described the types and activity of radioactive material present in groundwater at the facility, as well as the extent of this contamination. This information should be sufficient to allow NRC staff to evaluate potential safety issues associated with remediating the groundwater, whether the remediation activities and radiation control measures proposed by the licensee (described in Sections 17.1 and 17.3 of this NUREG) are appropriate for the type of radioactive material present in the groundwater, whether the licensee's waste management practices are appropriate, and whether the licensee's cost estimates are plausible, given the amount of contaminated water that will need to be removed or remediated.

INFORMATION TO BE SUBMITTED

The information supplied by the licensee should be sufficient to allow the staff to fully understand the types and activity of radioactive material contamination in groundwater at the facility, as well as the extent of this contamination. The staff's review should verify that the following information is included in the description of groundwater contamination in the DP:

- a summary of the aquifer(s) at the facility that contain residual radioactive material in excess of site background levels;

- a summary of the background levels used during scoping or characterization surveys; and

- a summary of the radionuclides present in each aquifer and the maximum and average radionulide activities in becquerel per liter (Bq/L) (picocuries per liter (pCi/L)).

NRC EVALUATION FINDINGS

The staff should verify that the information summarized under "Information to be Submitted," above, is included in the licensee's description of the groundwater contamination at the site. The staff's review should verify that the licensee has fully described the types and activity of radioactive material contamination in the groundwater at the site, as well as the extent of this contamination. These descriptions should be sufficient to allow NRC staff to evaluate the potential safety issues associated with remediating the groundwater, whether the remediation activities and radiation control measures proposed by the licensee are appropriate for the type of radioactive material present in the groundwater, whether the licensee's waste management practices are appropriate, and whether the licensee's cost estimates are plausible, given the amount of contaminated material that will need to be removed or remediated.

17 DECOMMISSIONING PLANS: PROGRAM ORGANIZATION

When a licensee is preparing the DP, the licensee should contact the staff about site-specific actions and conditions to confirm that both the specific information necessary and the planned actions will be effective and timely.

17.1 PLANNED DECOMMISSIONING ACTIVITIES

OVERVIEW

The staff should review the information supplied by the licensee to determine if the description of the planned decommissioning activities is adequate to allow the staff to fully understand the methods and procedures the licensee intends to use to remove residual radioactive material at the site to levels that allow for release of the site in accordance with NRC requirements. This information should include descriptions of how the licensee intends to remediate structures, systems and equipment, surface and subsurface soil, and surface and ground water at the site. In addition, the licensee should provide a schedule that demonstrates how the licensee will complete the interrelated decommissioning activities and the timeframes for completing the decommissioning. The licensee should also summarize which activities are being performed by licensee staff and which are being performed by decommissioning contractors, including which activities are being performed under the licensee's license and which are being performed under the contractor's license.

REVIEW PROCEDURES

Safety Evaluation

The material to be reviewed is informational in nature, and no specific detailed technical analysis is required. The staff should make a qualitative assessment as to whether the licensee's descriptions of planned decommissioning activities are adequate to serve as the basis for evaluating the licensee's methods and procedures for remediating the site and whether the decommissioning activities proposed by the licensee to remediate the facility can be conducted safely. In addition, the staff should ensure that the licensee's proposed schedule for completing the decommissioning complies with NRC requirements under 10 CFR 30.36(h), 10 CFR 40.42(h), 70.38(h), or 72.54(j).

Finally, the staff should ensure that the licensee and contractor are already authorized to perform the decommissioning procedures described in the DP or that the licensee has described the decommissioning procedures sufficiently to allow the staff to incorporate them into the license.

17.1.1   CONTAMINATED STRUCTURES

The purpose of the review of the planned decommissioning activities for contaminated structures is to allow the staff to fully understand what methods and procedures the licensee will undertake to remediate the contaminated structure. This will allow the staff to evaluate the licensee's methods and procedures to qualitatively assess if they can be performed safely and in compliance with NRC requirements. This information may also aid the staff in evaluating the estimates of radioactive waste that will be generated during decommissioning, the cost estimates for the decommissioning, and the ALARA evaluations developed by the licensee to support the decommissioning.

ACCEPTANCE CRITERIA

Regulatory Requirements

10 CFR 30.36(g), 40.42(g), 70.38(g), and 72.54(g)

Information to be Submitted

The information supplied by the licensee should be sufficient to allow the staff to fully understand what methods, procedures, and techniques the licensee intends to use to remediate the contaminated structure. In addition, the information should be sufficient to allow the staff to determine if the licensee's radiation safety procedures are appropriate, given the level of contamination and proposed method(s) for remediation. The staff's review should verify that all of the following information is included in the authorized activities section of the facility DP:

- A summary of the remediation tasks planned for each room or area in the contaminated structure in the order in which they will occur, including which activities will be conducted by licensee staff and which will be performed by a contractor.

- A description of the remediation techniques (such as scabbling, hydrolazing or grit blasting) that will be employed in each room or area of the contaminated structure. Licensees may generically describe these techniques once at the beginning of the "Contaminated Structures" section and refer to them in the descriptions of the remediation of the individual rooms or areas.

- A summary of the radiation protection methods (such as PPE, step-off pads and exit monitoring) and control procedures (such as scabbler shrouds, HEPA vented enclosures or superfine water misting) that will be employed in each room or area. (The staff's technical review of the adequacy of the licensee's radiation safety procedures should be performed pursuant to the criteria in Section 17.3 of this NUREG. In this section, the staff should make a qualitative assessment of the adequacy of the radiation protection and control methods proposed by the licensee to determine if the procedures described in the "Radiation Safety and Health" section of the DP have been followed.)

- A summary of the procedures already authorized under the existing license and those for which approval is being requested in the DP.

- A commitment to conduct decommissioning activities in accordance with written, approved procedures.

- A summary of any unique safety or remediation issues associated with remediating the room or area.

- For Part 70 licensees, a summary of how the licensee will ensure that the risks addressed in the facility's Integrated Safety Analysis will be addressed during decommissioning.

If the licensee intends to dismantle structures with contamination present in excess of the unrestricted use limits, the DP should provide a separate summary of the information listed above for the areas containing contamination in excess of the unrestricted use limits. In addition, the licensee should provide a description of the techniques and procedures that will be used to dismantle the building or structure and the licensee's procedures for evaluating the areas prior to dismantlement.

EVALUATION FINDINGS

Evaluation Criteria

The staff's review should verify that the licensee has described the remediation activities and associated safety precautions in sufficient detail to allow the staff to make a qualitative assessment of the adequacy of the proposed activities with respect to safety in compliance with NRC requirements. The staff should verify that the information summarized under "Information to be Submitted," above, is included in the licensee's description of the decommissioning activities portion of the DP. The staff should make a qualitative assessment of the adequacy of the licensee's proposed remediation methods and procedures to accomplish the remediation objectives in a manner that is protective of workers and the public and in compliance with NRC requirements. Detailed technical review of the safety precautions and procedures should be conducted pursuant to the criteria in Appendix D of this volume.

Sample Evaluation Findings

The staff may combine the evaluation finding for the licensee's description of the planned decommissioning activities with the findings for the remaining areas in this section of this volume as follows:

> "The NRC staff has reviewed the decommissioning activities described in the Decommissioning Plan for the [insert name and license number of facility] located at [insert location of facility] according to the Consolidated Decommissioning Guidance, Volume 1, Section 17.1 ("Planned Decommissioning Activities"). Based on this review, the NRC staff has determined that the licensee, [insert name], has provided sufficient

information to allow NRC staff to evaluate the licensee's planned decommissioning activities to ensure that the decommissioning can be conducted in accordance with NRC requirements."

17.1.2 CONTAMINATED SYSTEMS AND EQUIPMENT

The purpose of the review of the description of the planned decommissioning activities for contaminated systems and equipment is to allow the staff to fully understand what methods and procedures the licensee will undertake to remediate the contaminated systems or equipment at its facility. This will allow the staff to evaluate the licensee's methods and procedures to qualitatively assess if they can be performed safely and in compliance with NRC requirements. This information may also aid the staff in evaluating the estimates of radioactive waste that will be generated during decommissioning, the cost estimates for the decommissioning, and the ALARA evaluations developed by the licensee to support the decommissioning.

ACCEPTANCE CRITERIA

Regulatory Requirements

10 CFR 30.36(g), 40.42(g), and 70.38(g)

Information to be Submitted

The information supplied by the licensee should be sufficient to allow the staff to fully understand what methods, procedures, and techniques the licensee intends to use to remediate the contaminated systems and equipment. In addition, the information should be sufficient to allow the staff to determine if the licensee's radiation safety procedures are appropriate, given the level of contamination and proposed method(s) for remediation. The staff's review should verify that the following information is included in the authorized activities section of the facility DP:

- A summary of the remediation tasks planned for each system in the order in which they will occur, including which activities will be conducted by licensee staff and which will be performed by a contractor.

- A description of the techniques (such as scabbling, hydrolazing or grit blasting) that will be employed to remediate each system in the facility or site. Licensees may generically describe these techniques once at the beginning of the "Contaminated Systems" section and refer to them in the descriptions of the remediation of the individual systems.

- A description of the radiation protection methods (such as personal protective equipment (PPE), step-off pads and exit monitoring) and control procedures (such as scabbler shrouds, HEPA vented enclosures or superfine water misting) that will be employed while remediating each system. (The staff's technical review of the adequacy of the licensee's radiation safety procedures should be performed pursuant to the criteria in Section 17.3 of this NUREG. In this section, the staff should make a qualitative assessment of the adequacy of the radiation protection and control methods proposed by the licensee to determine if the procedures described in the Radiation Safety and Health section of the DP have been followed.)

- A summary of the equipment that will be removed or decontaminated and how the decontamination will be accomplished.

- A summary of the procedures already authorized under the existing license and those for which approval is being requested in the DP.

- A commitment to conduct decommissioning activities in accordance with written, approved procedures.

- A summary of any unique safety or remediation issues associated with remediating any system or piece of equipment.

- For Part 70 licensees, a summary of how the licensee will ensure that the risks addressed in the facility's Integrated Safety Analysis will be addressed during decommissioning.

EVALUATION FINDINGS

Evaluation Criteria

The staff's review should verify that the licensee has described the remediation activities and associated safety precautions in sufficient detail to allow the staff to determine if the proposed activities can be conducted safely and in compliance with NRC requirements. The staff should verify that the information summarized under "Information to be Submitted," above, is included in the licensee's description of the decommissioning activities portion of the DP. The staff should make a qualitative assessment of the adequacy of the licensee's proposed remediation methods and procedures to accomplish the remediation objectives in a manner that is protective of workers and the public and in compliance with NRC requirements. Detailed technical review of the safety precautions and procedures should be conducted pursuant to the criteria in Section 17.3 of this volume.

Sample Evaluation Findings

None required. The staff should combine the evaluation finding for the licensee's description of decommissioning activities for contaminated systems and equipment with the findings for the remaining areas in this section of this volume (see Section 17.1.1).

17.1.3 SOIL

The purpose of the review of the description of the planned decommissioning activities for soil is to allow the staff to fully understand what methods and procedures the licensee will undertake to remove or remediate the surface and subsurface soil at the site. This will allow the staff to evaluate the licensee's methods and procedures to qualitatively assess if they can be performed safely and in compliance with NRC requirements. This information may also aid the staff in evaluating the estimates of radioactive waste that will be generated during decommissioning, the cost estimates for the decommissioning, and the ALARA evaluations developed by the licensee to support the decommissioning. Additional guidance on the use of intentional mixing of soil to remediate surface and subsurface soil at the site is provided in Section 15.13.

ACCEPTANCE CRITERIA

Regulatory Requirements

10 CFR 30.36(g), 40.42(g), and 70.38(g)

Information to be Submitted

The information supplied by the licensee should be sufficient to allow the staff to fully understand what methods, procedures, and techniques the licensee intends to use to remove or remediate contaminated soil at the site. In addition, the information should be sufficient to allow the staff to determine if the licensee's radiation safety procedures are appropriate, given the level of contamination in the soil and proposed method(s) for removal or remediation. The staff's review should verify that the following information is included in the description of soil decommissioning activities in the facility DP:

- a summary of the removal/remediation tasks planned for surface and subsurface soil at the site in the order in which they will occur, including which activities will be conducted by licensee staff and which will be performed by a contractor;

- a description of the techniques that will be employed to remove or remediate surface and subsurface soil at the site;

- a description of the radiation protection methods (such as PPE, or area exit monitoring) and control procedures (such as the use of HEPA vented enclosures during excavation or covering soil piles to prevent wind dispersion) that will be employed during soil removal/remediation (The staff's technical review of the adequacy of the licensee's radiation safety procedures should be performed pursuant to the criteria in Section 17.3 of this NUREG. In this section, the staff should make a qualitative assessment of the adequacy of the radiation protection and control methods proposed by the licensee to determine if the procedures described in the Radiation Safety and Health section of the DP have been followed.);

- a summary of the procedures already authorized under the existing license and those for which approval is being requested in the DP;

- a commitment to conduct decommissioning activities in accordance with written, approved procedures;

- a summary of any unique safety or removal/remediation issues associated with remediating the soil; and

- for Part 70 licensees, a summary of how the licensee will ensure that the risks addressed in the facility's Integrated Safety Analysis will be addressed during decommissioning.

EVALUATION FINDINGS

Evaluation Criteria

The staff's review should verify that the licensee has described the remediation activities and associated safety precautions in sufficient detail to allow the staff to determine if the proposed activities can be conducted safely and in compliance with NRC requirements. The staff should verify that the information summarized under "Information to be Submitted," above, is included in the licensee's description of the decommissioning activities portion of the DP. The staff should make a qualitative assessment of the adequacy of the licensee's proposed remediation methods and procedures to accomplish the remediation objectives in a manner that is protective of workers and the public and in compliance with NRC requirements. Detailed technical review of the safety precautions and procedures should be conducted pursuant to the criteria in Section 17.3 of this volume.

Sample Evaluation Findings

None required. The staff should combine the evaluation finding for the licensee's description of decommissioning activities for soil with the findings for the remaining areas in this NUREG volume (see Section 17.1.1).

17.1.4   SURFACE AND GROUND WATER

The purpose of the review of the description of the planned decommissioning activities for surface and ground water is to allow the staff to fully understand what methods and procedures the licensee will undertake to remediate the contaminated water. This will allow the staff to evaluate the licensee's methods and procedures to qualitatively assess if they can be performed safely and in compliance with NRC requirements. This information may also aid the staff in evaluating the estimates of radioactive waste that will be generated during decommissioning, the cost estimates for the decommissioning, and the ALARA evaluations developed by the licensee to support the decommissioning.

ACCEPTANCE CRITERIA

Regulatory Requirements

10 CFR 30.36(g), 40.42(g), 70.38(g)

Information to be Submitted

The information supplied by the licensee should be sufficient to allow the staff to fully understand what methods, procedures, and techniques the licensee intends to use to remediate the contaminated ground or surface water. In addition, the information should be sufficient to allow the staff to determine if the licensee's radiation safety procedures are appropriate, given the level of contamination and proposed method(s) for remediation. The staff's review should verify that the following information is included in the authorized activities section of the facility DP:

- a summary of the remediation tasks planned for ground and surface water in the order in which they will occur, including which activities will be conducted by licensee staff and which will be performed by a contractor;

- a description the remediation techniques that will be employed to remediate the ground or surface water;

- a description of the radiation protection methods and control procedures that will be employed during ground or surface water remediation;

- a summary of the procedures already authorized under the existing license and those for which approval is being requested in the DP;

- a commitment to conduct decommissioning activities in accordance with written, approved procedures; and

- a summary of any unique safety or remediation issues associated with remediating the ground or surface water.

EVALUATION FINDINGS

Evaluation Criteria

The staff's review should verify that the licensee has described the remediation activities and associated safety precautions in sufficient detail to allow the staff to determine if the proposed activities can be conducted safely and in compliance with NRC requirements. The staff should verify that the information summarized under "Information to be Submitted," above, is included in the licensee's description of the decommissioning activities portion of the DP. The staff should make a qualitative assessment of the adequacy of the licensee's proposed remediation methods and procedures to accomplish the remediation objectives in a manner that is protective of workers and the public and in compliance with NRC requirements. Detailed technical review of the safety precautions and procedures should be conducted pursuant to the criteria in Section 17.3 of this volume.

Sample Evaluation Findings

None required. The staff should combine the evaluation finding for the licensee's description of decommissioning activities for surface and ground water with the findings for the remaining areas in this section of this NUREG volume (see Section 17.1.1).

17.1.5   SCHEDULES

The purpose of the review of the licensee's schedule is to determine whether it complies with NRC requirements for the completion of decommissioning activities.

ACCEPTANCE CRITERIA

Regulatory Requirements

10 CFR 30.36(h), 10 CFR 40.42(h), 70.38(h), and 72.54(j)

Information to be Submitted

The schedule supplied by the licensee should be sufficient to allow the staff to fully understand what activities will be performed to complete the decommissioning, the amount of time required to perform the activity, and the timeframe for performing the activities. The staff's review should verify that the licensee has included the all of the following:

- A Gantt or PERT chart is included that details the proposed remediation tasks in the order in which they will occur and including the amount of time required to perform each decommissioning activity and the initiation and completion dates for the activities.

- A statement is included that acknowledges that the dates in the schedule are contingent on NRC approval of the DP.

- A statement is included that acknowledges that circumstances can change during decommissioning, and, if the licensee determines that the decommissioning cannot be completed as outlined in the schedule, the licensee will provide an updated schedule to NRC.

- If the decommissioning is not expected to be completed within the timeframes outlined in NRC regulations at 10 CFR 30.36(h)(1), 10 CFR 40.42(h)(1), 70.38(h)(1), or 72.54(j)(1), the staff should verify that the licensee has requested an alternative schedule for completing the decommissioning and has addressed the criteria in NRC regulations at 10 CFR 30.36(h)(2)(i)(1–5), 10 CFR 40.42(h)(2)(i) (1–5), 70.38(h)(2)(i)(1–5), or 72.54(k)(1–5).

EVALUATION FINDINGS

Evaluation Criteria

The staff's review should verify that the licensee's schedule for decommissioning its facility is in compliance with NRC requirements. The staff should verify that the information summarized under "Information to be Submitted," above, is included in the licensee's description of the decommissioning activities portion of the DP.

Sample Evaluation Findings

None required. The staff should combine the evaluation finding for the licensee's description of decommissioning activities for soil with the findings for the remaining areas in this section of this NUREG volume (see Section 17.1.1).

17.2 DECOMMISSIONING PLAN: PROJECT MANAGEMENT AND ORGANIZATION

OVERVIEW

The staff should review the information supplied by the licensee to determine if the description of the licensee's decommissioning project organization and management structure is sufficient to allow the staff to fully understand how the licensee will ensure that it will exercise adequate control over the decommissioning project. This information should include a description of the management structure for the project, including individual organizational unit reporting responsibilities and lines of authority; a description of how radioactive material work procedures/practices (such as Radiation Work Permits) are developed reviewed, implemented, and managed; a description of the qualifications necessary for individuals performing the various project management and safety functions; a description of the relationship between the various organizational units within the decommissioning organization (such as remedial activities and health and safety units), including the responsibilities and authority to revise or stop work; a description of the licensee's training program; and a description of how contractors performing work at the facility will be managed during the decommissioning project.

REVIEW PROCEDURES

Safety Evaluation

The material to be reviewed is informational in nature, and no specific detailed technical analysis is required. The staff should make a qualitative assessment as to whether the licensee's descriptions of the proposed decommissioning project management and organization are adequate to serve as the basis for concluding that the licensee's management program will ensure that the appropriate control will be exercised during decommissioning operations.

17.2.1   DECOMMISSIONING MANAGEMENT ORGANIZATION

The purpose of the review of the description of the decommissioning project management organization is to verify that the licensee has a management organization and the personnel resources to ensure that the decommissioning of the facility can be completed safely and in accordance with NRC requirements.

ACCEPTANCE CRITERIA

Regulatory Requirements

10 CFR 30.36(g)(4)(ii), 40.42(g)(4)(ii), 70.38(g)(4)(ii) and 72.54(g)(2)

Information to be Submitted

The information supplied by the licensee should be sufficient to allow the staff to fully understand the structure and functions of the decommissioning project management organization. The staff's review should verify that the following information is included in the description of the decommissioning project management organization:

- a description of the decommissioning organization, including descriptions of the individual decommissioning project units within the decommissioning project; organization, such as project management, health and safety, and remedial activities;

- a description of the responsibilities of each of these decommissioning project units;

- a description of the reporting hierarchy within the decommissioning project management organization, including a chart or diagram showing the relationship of each decommissioning project unit to other project units and decommissioning project management; and

- a description of the responsibility and authority of each unit to ensure that decommissioning activities are conducted in a safe manner and in accordance with approved written procedures, including both stop-work authority of each unit and the manner in which concerns about safety issues are managed within the overall decommissioning project.

EVALUATION FINDINGS

Evaluation Criteria

The staff should verify that the information summarized under "Information to be Submitted," above, is included in the licensee's description of the decommissioning project management organization. NRC staff should verify that the descriptions of the decommissioning project management organization and individual project unit responsibilities are sufficiently detailed to allow the staff to understand the manner in which the organization will ensure that decommissioning will be conducted safely. The staff should verify that the individual project unit reporting hierarchy and lines of authority within the decommissioning project do not create conflicts that could compromise safety during decommissioning and that, as appropriate, individual units report directly to the unit responsible for overall decommissioning project management. The staff should verify that the individual project units, and individuals within each unit, have the responsibility and authority to bring safety concerns to decommissioning project management and that stop-work authority is provided to the unit responsible for safety and health. The staff should make a qualitative assessment of the adequacy of the licensee's proposed decommissioning management organization to accomplish the remediation objectives in a manner that is protective of workers and the public and in compliance with NRC requirements.

Sample Evaluation Findings

The NRC staff has reviewed the description of the decommissioning project management organization, position descriptions, management and safety position qualification requirements and the manner in which the licensee, [insert name and license number of licensee], will use contractors during the decommissioning of its facility located at [insert location of facility] according to the Consolidated Decommissioning Guidance, Section 17.2 ("Decommissioning Plan: Project Management and Organization"). Based on this review, the NRC staff has determined that the licensee, [insert name], has provided sufficient information to allow the NRC staff to evaluate the licensee's decommissioning project management organization and structure to determine if the decommissioning can be conducted safely and in accordance with NRC requirements. (Note that this finding incorporates the results of the staff's assessment under Sections 17.2.2–17.2.5.)

17.2.2   DECOMMISSIONING TASK MANAGEMENT

The purpose of the review of the description management of decommissioning tasks is to verify that all decommissioning activities will be conducted in accordance with written, approved procedures and that the licensee has a methodology in place to manage the development of, review, and maintenance of the procedures.

ACCEPTANCE CRITERIA

Regulatory Requirements

10 CFR 30.36(g)(4)(ii), 40.42(g)(4)(ii), 70.38(g)(4)(ii) and 72.54(g)(2)

Information to be Submitted

The information supplied by the licensee should be sufficient to allow the staff to fully understand the manner in which the licensee will evaluate decommissioning tasks and develop and manage the procedures necessary for conducting the tasks. The staff's review should verify that the following information is included in the description of decommissioning task management:

- A description of the manner in which the decommissioning tasks are managed, such as through the use of Radiation Work Permits (RWPs). The term "RWP" will be used throughout this section to refer to the written procedure used to manage individual decommissioning tasks.

- A description of how individual decommissioning tasks are evaluated and how the RWPs are developed for each task.

- A description of how the RWPs are reviewed and approved by the decommissioning project management organization.

- A description of how RWPs are managed throughout the decommissioning project (i.e., how they are issued, maintained, revised, and terminated).

- A description of how individuals performing the decommissioning tasks are informed of the procedures in the RWP, including how they are initially informed and how they are informed when an RWP is revised or terminated.

EVALUATION FINDINGS

Evaluation Criteria

The staff should verify that the information summarized under "Information to be Submitted," above, is included in the licensee's description of the manner in which decommissioning tasks will be managed. The staff should verify that the licensee will control decommissioning tasks through the use of written procedures. These procedures should be developed by individuals/units familiar with the physical and safety requirements necessary to complete the tasks safely. The procedures should be reviewed and approved by units responsible for physical, radiological, chemical, and occupational safety, as well as decommissioning project management. Note that NRC staff is not responsible for ensuring that physical, chemical or occupational safety procedures are adequate. Rather, the intent is to ensure that the licensee has an integrated approach for reviewing and approving procedures that could impact radiological safety. Procedures should also undergo separate review by a group charged with ensuring that activities are conducted safely and in a manner that ensures that exposures to radiation are ALARA. Staff should verify that the licensee has a methodology to issue, modify (after appropriate review and approval), and terminate radiation work permits (RWPs), as well as a program for ensuring that individuals performing the tasks are informed or trained in the procedures. The staff should make a qualitative assessment of the adequacy of the licensee's proposed decommissioning task management procedures to accomplish the decommissioning in a manner that is protective of workers and the public and in compliance with NRC requirements.

Sample Evaluation Findings

None required. The staff should combine the assessment of this section of the DP with Section 17.2.1.

17.2.3 DECOMMISSIONING MANAGEMENT POSITIONS AND QUALIFICATIONS

The purpose of the review of the licensee's decommissioning management positions and qualifications is to ensure that the licensee has the personnel resources to safely conduct and manage the decommissioning of its facility.

ACCEPTANCE CRITERIA

Regulatory Requirements

- 10 CFR 30.33(3), 40.32(b), 70.22(a)(6), 72.28(a–d)
- 10 CFR 30.36(g)(4)(ii), 40.42(g)(4)(ii), 70.38(g)(4)(ii) and 72.54(g)(2)

Information to be Submitted

The information supplied by the licensee should be sufficient to allow the staff to fully understand the responsibilities and minimum qualifications required for each of the management and safety-related positions within the licensee's decommissioning project organization. The staff's review should verify that the following information is included in the description of decommissioning positions and qualifications:

- a description of the duties and responsibilities of each management position in the decommissioning organization and the reporting responsibility of the position;

- a description of the duties and responsibilities of each chemical, radiological, physical and occupational safety-related position in the decommissioning organization, and the reporting responsibility of the position;

- a description of the duties and responsibilities of each engineering, quality assurance, and waste management position in the decommissioning organization and the reporting responsibilities of their respective positions;

- the minimum qualifications for each of the positions described above, and the qualifications of the individuals currently occupying the positions (the licensee should also commit to providing the staff with the qualifications of any newly hired employees or replacements for these positions); and

- a description of all decommissioning and safety committees, including the membership of the committees, the duties and responsibilities of each committee, and the authority of each committee.

EVALUATION FINDINGS

Evaluation Criteria

The staff should verify that the information summarized under "Information to be Submitted," above, is included in the licensee's description of the previous decommissioning activities carried out under the license. The staff should make a qualitative assessment of the adequacy of the licensee's decommissioning position and qualification requirements to ensure that the decommissioning can be conducted in a manner that is protective of workers and the public and in compliance with NRC requirements.

Sample Evaluation Findings

None required. The staff should combine its assessment of this section of the DP with Section 17.2.1.

Minimum qualifications should be summarized in tabular form, and the licensee should submit the *curricula vitae* of the individuals currently occupying the positions.

17.2.3.1   Radiation Safety Officer

The purpose of the review of the Radiation Safety Officer (RSO) position is to ensure that a qualified individual is designated and empowered to oversee the licensee's radiation protection program. The RSO must be qualified by training and experience for the types and quantities of radionuclides that will be encountered during decommissioning operations, as well as the operations that will be undertaken to decommission the facility. In addition, the RSO must be empowered by the licensee and be responsible for the implementation of the radiation protection program.

ACCEPTANCE CRITERIA

Regulatory Requirements

10 CFR 33.13(c)(2), 33.14(b)(1), 34.42, 35.900, and 36.13(d)

Information to be Submitted

The information supplied by the licensee should be sufficient to allow the staff to fully evaluate the qualifications, authority and responsibilities of the RSO. The staff's review should verify that the following information is included in the description of the RSO's qualifications, duties, and responsibilities:

- a description of the health physics and radiation safety education and experience required for individuals acting as the licensee's RSO;

- a description of the responsibilities and duties of the RSO; and

- a description of the specific authority of the RSO to implement and manage the licensee's radiation protection program, including the RSO's access and "stop-work" authority for all activities involving radioactive material at the site.

EVALUATION FINDINGS

Evaluation Criteria

The staff should verify that the information summarized under "Information to be Submitted," above, is included in the licensee's description of the duties and responsibilities of the RSO. The staff should verify that the description of the RSO's duties and responsibilities are sufficiently detailed to allow the staff to determine whether the RSO can, and will be able to, oversee the site radiation protection program effectively. The staff should verify that the RSO has clearly defined authority and responsibility to oversee the radiation protection program, such that if conflicts arise regarding the appropriate manner in which to conduct the decommissioning, the RSO can ensure that the decommissioning will be conducted safely.

The RSO is adequately qualified if he/she meets the following criteria:

- Education: A Bachelors' degree in the physical sciences, industrial hygiene or engineering from an accredited college or university or an equivalent combination of training and relevant experience in radiological protection. Two years of relevant experience are generally considered equivalent to 1 year of academic study;

- Health physics experience: At least 1 year of work experience in applied health physics, industrial hygiene or similar work relevant to radiological hazards associated with site remediation. This experience should involve actually working with radiation detection and measurement equipment, not simply administrative or "desk" work; and

- Specialized knowledge: A thorough knowledge of the proper application and use of all health physics equipment used for the radionuclides present at the site, the chemical and analytical procedures used for radiological sampling and monitoring, and methodologies used to calculate personnel exposure to the radionuclides present at the site.

Note that if the RSO does not have the decommissioning experience indicated above, the RSO could be supported by a contractor or someone on his/her staff who does have the experience.

The description of the RSO's duties and responsibilities should include the responsibility and authority to review and approve all procedures involving the use of radioactive material at the facility; to review and approve individuals as radiation workers at the site; to conduct audits and inspections to ensure that activities involving the use of radioactive material are being conducted safely; to monitor materials use and storage areas at the site; to oversee the inventory, ordering, receipt and shipment of all radioactive material and radioactive waste at the site; to ensure that all personnel at the site are trained in site radiation safety procedures and practices; to ensure that sealed sources are leak-tested per NRC requirements; to respond to and investigate incidents and accidents involving radioactive material at the site; monitor and evaluate radiation worker exposures at the site; and to maintain all required records.

The RSO should have the authority and access to all areas involved in decommissioning or radioactive material usage at the site and the specific authority and responsibility to stop any operations that in the RSO's opinion are not being conducted safely.

Sample Evaluation Findings

None required. The staff should combine their assessment of this section of the DP with Section 17.2.1.

17.2.4   TRAINING

The purpose of the review of the licensee's training program is to provide the staff with sufficient information to determine if the licensee can provide its employees with the training necessary to complete the decommissioning safely and in accordance with NRC requirements. Note that training related to the Radiation Health and Safety Program will be evaluated under Section 17.3.1.2.

ACCEPTANCE CRITERIA

Regulatory Requirements

- 10 CFR 19, 30.33(3), 40.32(b), 70.22(a)(6), 72.28(a), (b) and (d)
- 10 CFR 30.36(g)(4)(ii), 40.42(g)(4)(ii), 70.38(g)(4)(ii) and 72.54(g)(2)

Information to be Submitted

The information supplied by the licensee should be sufficient to allow the staff to determine whether the licensee has an acceptable program to train employees in the remediation and safety procedures that will be used to decommission the facility. The staff's review should verify that the following information is included in the description of the training program for the facility:

- a description of the radiation safety training that the licensee will provide to each employee including pre-employment, annual/periodic training and specialized training to comply with 10 CFR Part 19;

- a description of any daily worker "jobside" or "tailgate" training that will be provided at the beginning of each workday or job task to familiarize workers with job-specific procedures or safety requirements; and

- a description of the documentation that will be maintained to demonstrate that training commitments are being met.

EVALUATION FINDINGS

Evaluation Criteria

The staff should verify that the information summarized under "Information to be Submitted," above, is included in the licensee's description of training at its facility. The staff should make a qualitative assessment of the adequacy of the licensee's training programs to ensure that workers are adequately informed of the hazards, preventative measures, and procedures associated with performing each decommissioning task.

Sample Evaluation Findings

None required. The staff should combine its assessment of this section of the DP with Section 17.2.1.

17.2.5   CONTRACTOR SUPPORT

The purpose of the review of the licensee's description of interaction between the licensee and contractors is to determine if the interactions will occur such that both licensee and contractor personnel are adequately protected and that the decommissioning can be conducted in accordance with NRC requirements.

ACCEPTANCE CRITERIA

Regulatory Requirements

10 CFR 30.36(g)(4)(ii), 40.42(g)(4)(ii), 70.38(g)(4)(ii) and 72.54(g)(2)

Information to be Submitted

The information supplied by the licensee should be sufficient to allow the staff to determine whether the licensee's radiation protection procedures are adequate to ensure the safety of contractor and licensee personnel. The staff's review should verify that the following information is included in the discussion of contractor support at the facility:

- a summary of decommissioning tasks that will be performed by contractors, including the areas at the site where they will perform these tasks;

- a description of the management interfaces that will be in place between the licensee's management and onsite supervisors, and contractor management and onsite supervisors;

- a description of the oversight responsibilities and authority that the licensee will exercise over contractor personnel;

- a description of the training that will be provided to contractor personnel by the licensee, and the training that will be provided by the contractor; and

- a commitment that the contractor will comply with all radiation safety and license requirements at the facility.

EVALUATION FINDINGS

Evaluation Criteria

The staff should verify that the information summarized under "Information to be Submitted," above, is included in the licensee's description of contractor support at the site. The staff should make a qualitative assessment of the adequacy of the licensee's planned management interface procedures with contractor management to ensure that both licensee and contractor personnel are adequately informed of the hazards, preventative measures, and procedures associated with performing each decommissioning task. The staff should verify that the licensee has the authority and responsibility to ensure that contractor personnel perform decommissioning activities in accordance with all license commitments and NRC requirements. The staff should verify that all contractor personnel will receive adequate training (as described in the training program in Section 17.2.4), either as part of the licensee's training program or as part of the contractor's training program.

Sample Evaluation Findings

None required. The staff should combine its assessment of this section of the DP with Section 17.2.1.

AREAS OF REVIEW

NRC staff should review the information supplied by the licensee to determine if the health and safety measures to be used to control and monitor the impacts of ionizing radiation on workers comply with the NRC regulations in 10 CFR Parts 19 and 20. NRC staff should review only those parts of the applicant's RH&SP that were not previously approved in the original submission for a licensing action. The information requested should address the following aspects of the RH&SP program: a description of the radiation safety controls and types of monitoring to be used to ensure that internal and external exposures to workers are ALARA (including administrative procedures); a commitment in the licensee's RH&SP program to written procedures (and changes to procedures); a commitment to perform periodic inspections and audits; and a commitment to a recordkeeping program.

REVIEW PROCEDURES

Safety Evaluation

The material to be reviewed is technical in nature. The staff should make a quantitative assessment as to whether the licensee's proposed health and safety program complies with the regulatory requirements in 10 CFR Parts 19 and 20 and is adequate to protect workers from ionizing radiation during decommissioning activities. The staff should assess whether the applicant's radiological safety measures for workers are commensurate with the risks associated with licensed activities as required by 10 CFR 20.1101.

17.3   DECOMMISSIONING PLAN: RADIATION SAFETY AND HEALTH PROGRAM DURING DECOMMISSIONING

AREAS OF REVIEW

NRC staff should review the information supplied by the licensee to determine if the health and safety measures to be used to control and monitor the impacts of ionizing radiation on workers comply with the NRC regulations in 10 CFR Parts 19 and 20. NRC staff should review only those parts of the applicant's Radiation Health and Safety Program (RH&SP) that were not previously approved in the original submission for a licensing action. The information requested should address the following aspects of the RH&SP program: a description of the radiation safety controls and types of monitoring to be used to ensure that internal and external exposures to workers are ALARA (including administrative procedures); a commitment in the licensee's RH&SP program to written procedures (and challenges to procedures); a commitment to perform periodic inspections and audits; and a commitment to a recordkeeping program.

REVIEW PROCEDURES

Safety Evaluation

The material to be reviewed is technical in nature. NRC staff should make a quantitative assessment as to whether the licensee's proposed health and safety program complies with the regulatory requirements in 10 CFR Parts 19 and 20 and is adequate to protect workers from ionizing radiation during decommissioning activities. NRC staff should assess whether the applicant's radiological safety measures for workers are commensurate with the risks associated with licensed activities as required by 10 CFR 20.1101.

17.3.1   RADIATION SAFETY CONTROLS AND MONITORING FOR WORKERS

17.3.1.1   Workplace Air Sampling Program

The purpose of the review of the description of the licensee's air sampling program is to verify that the licensee has a program adequate to demonstrate compliance with the dose assessment requirements of 10 CFR 20.1204, the survey requirements in 10 CFR 20.1501(a)–(b), and the requirements in 10 CFR 20.1703(a)(3)(i)–(ii), when respirators are worn.

ACCEPTANCE CRITERIA

Regulatory Requirements

10 CFR 20.1204, 20.1501(a)–(b), 20.1502 (b), and 20.1703(a)(3)(i)–(ii)

Regulatory Guidance

Regulatory Guide 8.25, Rev. 1, Air Sampling in the Workplace, June 1992

Information to be Submitted

The information supplied by the licensee should be sufficient to allow the staff to fully understand the licensee's air sampling program under routine and emergency conditions. The staff's review should verify that the following information is included in the description of the licensee's air sampling program:

- a demonstration that the air sampling program is representative of the workers' breathing zones and will be initiated whenever a worker's intake is likely to exceed the criteria in 20.1502(b);

- a description of the criteria used for selection of the placement of air samplers in work areas where potential for airborne hazards exists;

- a description of the criteria demonstrating that air samplers with appropriate sensitivities will be used; and that samples will be collected at appropriate frequencies;

- a description of the conditions under which constant air monitors (CAMs) (or similar equipment), general air and breathing zone samplers will be used, including a description of their readouts, annunciators, and alarm setpoints;

- a description of the criteria used to determine the frequency of calibration of the flow meters on the air samplers;

- a description of the action levels for air sampling results, including the actions to be taken when they are exceeded; and

- a description of how minimum detectable activities (MDAs) for each specific radionuclide that may be collected in air samples are determined.

EVALUATION FINDINGS

Evaluation Criteria

The staff's review should verify that the air sampling program proposed by the licensee will be in compliance with 10 CFR 20.1204, 20.1501(a)–(b), 20.1502(b) 20.1703(a)(3)(i)–(ii), and Regulatory Guide 8.25. The staff should verify that the licensee's air sampling program will:

- Require air samples when a worker's intake is likely to exceed the criteria in 20.1502(b) and will demonstrate that the air samples are representative of the air inhaled in any work areas in which a potential exists for airborne radioactive materials, as indicated in Regulatory Position 3 of Regulatory Guide 8.25.

- Provide the bases for selection of the locations of air samplers in all work areas in which a potential exists for airborne radioactivity, as indicated in Regulatory Position 2 of Regulatory Guide 8.25.

- Measure air concentrations with sufficient sensitivity over the ranges of concentrations encountered in the various work areas, and with frequencies of sampling, as indicated in Regulatory Position 1 of Regulatory Guide 8.25.

- Specify the conditions under which CAMs will be used, and provide a description of their readouts, annunciators, and alarm setpoints, as indicated in Regulatory Position 1.6 of Regulatory Guide 8.25.

- Ensure that the frequency of calibration of the flow meters on the air samplers is as indicated in Regulatory Position 5 of Regulatory Guide 8.25.

- Provide action levels for air sampling results, actions to be taken when they are exceeded, and their technical bases, as indicated in Regulatory Position 6.1 of Regulatory Guide 8.25.

- Provide the MDA for each specific radionuclide that may be collected in air samples, as indicated in Regulatory Position 6.3 of Regulatory Guide 8.25.

Sample Evaluation Findings

The NRC staff has reviewed the information in the Decommissioning Plan for the [insert name and license number of facility] located at [insert location of facility] according to the Decommissioning Consolidated Guidance, Volume 1, Section 17.3.1.1 ("Workplace Air Sampling Program"). Based on this review, the NRC staff has determined that the licensee, [insert name], has provided sufficient information on when air samples will be taken in work areas, the types of air sample equipment to be used and where they will be located in work areas, calibration of flow meters, minimum detectable activities (MDA) of equipment to be used for analyses of radionuclides collected during air sampling, action levels for airborne radioactivity (and corrective actions to be taken when these levels are exceeded), to allow the NRC staff to conclude that the licensee's air sampling program will comply with 10 CFR 20.1204, 20.1501(a)–(b), 20.1502(b), 20.1703(a)(3)(i)–(ii), and Regulatory Guide 8.25.

17.3.1.2   Respiratory Protection Program

The purpose of the review of the description of the respiratory protection program is to verify that the measures used by the licensee in its respiratory protection program adequately limit intakes of airborne radioactive materials for workers in restricted areas and to keep the total effective dose equivalent ALARA.

ACCEPTANCE CRITERIA

Regulatory Requirements

10 CFR 20.1101(b), 20.1701, 20.1702, 20.1703, and 20.1704

Regulatory Guidance

- Draft Regulatory Guide DG–8022, "Acceptable Programs for Respiratory Protection"
- NUREG–0041, Rev. 1, "Manual of Respiratory Protection Against Airborne Radioactive Material"

Information to be Submitted

The staff's review should verify that the licensee's program description for respiratory protection will meet the requirements of 10 CFR 20.1101(b), 20.1701–20.1704, Appendix A of 10 CFR Part 20, and of the guidance in Draft Regulatory Guide DG–8022. The staff's review should verify that the following information is included in the description of the licensee's respiratory protection program:

- a description of the process controls, engineering controls, or procedures to control concentrations of radioactive materials in air;
- a description of the evaluation that will be performed when it is not practical to apply engineering controls or procedures, that demonstrates that the use of respiratory protection equipment is ALARA;
- a description of the considerations used to demonstrate that respiratory protection equipment is appropriate for a specific task, based on the guidance on assigned protection factors (APF);
- a description of the medical screening and fit testing required before workers will use any respirator that is assigned a protection factor;

- a description of the written procedures maintained to address all the elements of the respiratory protection program;

- a description of the use, maintenance, and storage of respiratory protection devices in such a manner that they are not modified and are in like-new condition at the time of issue;

- a description of the respiratory equipment users' training program; and

- a description of the considerations made when selecting respiratory protection equipment to mitigate existing chemical or other respiratory hazards instead of (or in addition to) radioactive hazards.

EVALUATION FINDINGS

Evaluation Criteria

The staff's review should verify that the licensee's respiratory protection program will be in compliance with the requirements of 20.1101(b), 20.1701–20.1704, Appendix A of 10 CFR Part 20, and of Draft Regulatory Guide DG–8022. The staff should verify that the licensee's program for respiratory protection for workers in restricted areas will:

- Apply process controls, engineering controls or procedures to control concentrations of radioactive materials in air as required by 10 CFR 20.1702 when practical.

- When it is not practical to apply engineering controls or procedures, perform an evaluation to show the use of respiratory equipment is ALARA, as indicated in Regulatory Positions C2.2 and C2.3 of Draft Regulatory Guide DG–8022.

- Consider which respiratory protection equipment is appropriate for a specific task based on the guidance on APF in Regulatory Position C2.3 of Draft Regulatory Guide DG–8022.

- Require medical screening and fit testing before workers will use any respirator that is assigned a protection factor, as indicated in Regulatory Position C5 of Regulatory Guide DG–8022.

- Maintain written procedures to address all the elements of the respiratory protection program as required by 10 CFR 20.1703 and as identified in Regulatory Position C3 of Regulatory Guide DG–8022.

- Use, maintain, and store respiratory protection devices in such a manner that they are not modified and are in like-new condition at the time of issue, as indicated in Regulatory Position C4 of Regulatory Guide DG–8022.

- Establish and implement a program to train respirator users, as indicated in Regulatory Position C5.2 of Regulatory Guide DG–8022.

- Comply with the safety concerns as indicated in Regulatory Position C6 of Regulatory Guide DG–8022.

- Consult the Occupational Safety and Health regulations of the Department of Labor when selecting respiratory protection equipment to mitigate existing chemical or other respiratory hazards instead of (or in addition to) radioactive hazards, as required by footnote a of Appendix A of 10 CFR Part 20.

Sample Evaluation Findings

The NRC staff has reviewed the information in the Decommissioning Plan for the [insert name and license number of facility] located at [insert location of facility] according to the Decommissioning Consolidated Guidance, Volume 1, Section 17.3.1.2 ("Respiratory Protection Program"). Based on this review, the NRC staff has determined that the licensee, [insert name], has provided sufficient information to implement an acceptable respiratory protection program so as to allow the NRC staff to conclude that the licensee's program will comply with 10 CFR 20.1101(b), and 10 CFR 20.1701 to 20.1704 and Appendix A of 10 CFR Part 20.

17.3.1.3   Internal Exposure Determination

ACCEPTANCE CRITERIA

The purpose of the review of the description of the Internal Exposure Determination Program is to verify that the measures used by the licensee to determine a worker's internal exposure complies with 10 CFR Part 20 and NRC guidance documents, focusing on techniques used to estimate intake of radionuclides by workers and the calculations necessary for the conversion of an intake either to a committed effective dose equivalent or to a total organ dose equivalent.

Regulatory Requirements

10 CFR 20.1101(b), 20.1201(a)(1), 20.1201 (d) and (e), 20.1204, and 20.1502(b)

Regulatory Guidance

- Regulatory Guide 8.9, Rev 1, "Acceptable Concepts, Models Equations, and Assumptions For A Bioassay Program"

- Regulatory Guide 8.25, "Air Sampling in the Workplace"

- Regulatory Guide 8.34, "Monitoring Criteria and Methods to Calculate Occupational Radiation Doses"

- Regulatory Guide 8.36, "Radiation Dose to the Embryo/Fetus"

Information to be Submitted

The information supplied by the licensee should be sufficient to allow the staff to fully understand what methods, procedures, and techniques the licensee intends to use to determine a worker's internal exposure. The staff's review should verify that the following information is included in the description of the licensee's program:

- A description of the monitoring to be performed to determine worker exposure during routine operations, special operations, maintenance, and clean-up activities.

- A description of how worker intakes are determined using measurements of quantities of radionuclides excreted from, or retained in the human body. The licensee should include in its description the following:

  — how frequencies for bioassay measurements for baseline, periodic, special, and termination assays are assigned;

  — how radioactivity measured in the human body by bioassay techniques are converted into worker intake; and

  — action levels for bioassay samples, actions to be taken when they are exceeded, and their technical bases.

- A description of how worker intakes are determined by measurements of the concentrations of airborne radioactive materials in the workplace. To determine worker intake by measurements of the concentrations of airborne radioactive materials in the workplace, the licensee should include the following:

  — how airborne concentrations of radioactivity are measured;

  — how airborne concentrations are converted to determine intakes;

  — action levels for a worker's intake based on dose, and actions to be taken when they are exceeded; and

  — action levels for a worker's intake based on chemical toxicity if soluble uranium is present in the work area.

- A description of how worker intakes, for an adult, a minor, and a declared-pregnant woman (DPW) are determined using any combination of the measurements above.

- A description of how worker intakes are converted into committed effective dose equivalent (and organ-specific committed dose equivalent), including how the intake of radioactivity by a DPW will be converted into a dose to the embryo/fetus.

EVALUATION FINDINGS

Evaluation Criteria

The staff's review should verify that the measures used to determine a worker's internal exposure will be in compliance with 10 CFR 20.1101(b), 20.1201(a)(1), (d) and (e), 20.1204 and 20.1502(b). The staff should verify that the licensee's program to determine internal exposure will:

- Monitor workers who meet the criteria in 10 CFR 20.1502(b)(1) and (2) for potential internal exposures during routine operations, special operations, maintenance, and clean-up activities.

- Determine worker intake by measurements of quantities of radionuclides excreted from, or retained in the human body by:

    — assigning frequencies for bioassay measurements for baseline, periodic, special, and termination assays, as indicated in Regulatory Position 2 in Regulatory Guide 8.9, Rev. 1;

    — converting radioactivity measured in the human body by bioassay techniques into worker intake, as indicated in Regulatory Position 4 of Regulatory Guide 8.9, Rev. 1; and

    — providing action levels for bioassay samples, actions to be taken when they are exceeded, and their technical bases as indicated in Regulatory Position 2.3 of Regulatory Guide 8.9, Rev. 1.

- Licensees may also determine worker intake by measurements of the concentrations of airborne radioactive materials in the workplace by:

    — measuring airborne concentrations of radioactivity, as indicated in Section 17.3.1.1 of this volume;

    — converting airborne concentrations to intakes, as indicated in Regulatory Position 3.3 of Regulatory Guide 8.34;

    — providing action levels for a worker's intake based on dose, and actions to be taken when they are exceeded (these will be found in Section 17.3.1.1 of this guidance); and

    — providing action levels for a worker's intake based on chemical toxicity, if soluble uranium is present in the work area, as indicated in 10 CFR 20.1201(e).

- Determine worker intake for an adult, a minor, and a DPW by any combination of the measurements above as may be necessary, as required by 10 CFR 20.1204(a)(1)–(4).

- Convert worker intakes into committed effective dose equivalent (and organ-specific committed dose equivalent) as indicated in Regulatory Positions 4, 5, and 6 of Regulatory Guide 8.34. The intake of radioactivity by a DPW should be converted into a dose to the embryo/fetus, as identified in Regulatory Position 2 (or 3) of Regulatory Guide 8.36.

- Maintain worker internal exposures ALARA, as required by 10 CFR 20.1101(b) and as described in Section 17.3.1.3 of this NUREG volume.

Sample Evaluation Findings

The NRC staff has reviewed the information in the Decommissioning Plan for the [insert name and license number of facility] located at [insert location of facility] according to the Consolidated Decommissioning Guidance, Volume 1, Section 17.3.1.3 ("Internal Exposure Determination"). Based on this review, the NRC staff has determined that the licensee, [insert name], has provided sufficient information on methods to calculate internal dose of a worker based upon measurements from air samples or bioassay samples to allow the NRC staff to conclude that the licensee's program to determine internal exposure will comply with 10 CFR 20.1101(b), 20.1201(a)(1), (d) and (e), 20.1204, and 20.1502(b).

17.3.1.4   External Exposure Determination

The purpose of the review of the description of the licensee's external exposure determination program is to verify if the licensee has a program adequate to demonstrate that the workers' external exposure program complies with 10 CFR Part 20 and NRC Guidance Documents. External exposure can be measured with dosimeters worn on the human body or calculated from measurements with appropriate instruments during surveys in areas where decommissioning activities are carried out.

ACCEPTANCE CRITERIA

Regulatory Requirements

10 CFR 20.1101(b), 20.1201, 20.1203, 20.1501(a)(2)(i), and (c), 20.1502(a), and 20.1601

Regulatory Guidance

- Regulatory Guide 8.4, "Direct-reading and Indirect-reading Pocket Dosimeters"

- Regulatory Guide 8.28, "Audible-Alarm Dosimeters"

- Regulatory Guide 8.34, "Monitoring Criteria and Methods to Calculate Occupational Radiation Doses"

Information to be Submitted

The information supplied by the licensee should be sufficient to allow the staff to fully understand what methods, procedures, and techniques the licensee intends to use to determine a worker's external exposure. The staff's review should verify that the following information is included in the description of the licensee's program:

- a description of the individual-monitoring devices that will be provided to workers who meet the criteria in 10 CFR 20.1502(a) and 20.1601 for external exposures;

- a description of the type, range, sensitivity, and accuracy of each individual-monitoring device;

- a description of the use of extremity and whole body monitors when the external radiation field is non-uniform;

- a description of when audible-alarm dosimeters and pocket dosimeters will be provided, and a description of their performance specifications;

- a description of how external dose from airborne radioactive material is determined;

- a description of the procedure to insure that surveys necessary to supplement personnel monitoring are performed; and

- a description of the action levels for workers' external exposure, including the technical bases and actions to be taken when they are exceeded.

EVALUATION FINDINGS

Evaluation Criteria

The staff's review should verify that the measures used to determine a worker's external exposure will be in compliance with the requirements of 10 CFR 20.1101(b), 20.1201(c), 20.1203, 20.1501(a)(2)(i) and (c), 20.1502(a), and 20.1601, and the guidance in Regulatory Guides 8.4, 8.28 and 8.34. The staff should verify that the licensee's program to determine external exposure will:

- Provide individual-monitoring devices to workers who meet the criteria in 10 CFR 20.1502(a) and 20.1601 for external exposures.

- Provide a description of the type, range, sensitivity, and accuracy of each individual-monitoring device.

- Require that individual monitoring devices be worn near the location on the human body that is expected to receive the highest dose, as required by 10 CFR 20.1201(c), and as indicated in Regulatory Positions C2.1 and C2.2 of Regulatory Guide 8.34.

- Require that all personnel dosimeters, which require processing to determine radiation dose, be processed and evaluated by a dosimetry processor that meets the criteria in 10 CFR 20.1501(c).

- Use extremity monitors when the external radiation field is non-uniform, as indicated in Regulatory Position C2.3 of Regulatory Guide 8.34.

- Use only audible-alarm dosimeters and pocket dosimeters that meet the performance specifications identified in Regulatory Guide 8.28 and Regulatory Guide 8.4; respectively.

- Determine external dose from airborne radioactive material, as required by 10 CFR 20.1203.

- Conduct a reasonable number of surveys to supplement personnel monitoring, as required by Section 20.1501(a)(2)(i).

- Provide action levels for workers' external exposure, including actions to be taken when they are exceeded.

Sample Evaluation Findings

The NRC staff has reviewed the information in the Decommissioning Plan for the [insert name and license number of facility] located at [insert location of facility] according the Decommissioning Consolidated Guidance, Volume 1, Section 17.3.1.4 ("External Exposure Determination"). Based upon this review, the NRC staff has determined that the licensee, [insert name], has provided sufficient information on methods to measure or calculate the external dose of a worker to allow the NRC staff to conclude that the licensee's program to determine external exposure will comply with the requirements of 10 CFR 20.1101(b), 20.1201(c), 20.1203, 20.1501(a)(2)(i) and (c), 20.1502(a), and 20.1601.

17.3.1.5 Summation of Internal and External Exposures

The purpose of the review of the licensee's description of its radiation monitoring program is to verify that the calculations and procedures used to sum external and internal doses satisfy the provisions of 10 CFR Part 20.

ACCEPTANCE CRITERIA

Regulatory Requirements

10 CFR 20.1202, 20.1208(c)(1) and (2), 20.2106

Regulatory Guidance

- Regulatory Guide 8.7, "Instructions for Recording and Reporting Occupational Radiation Exposure Data"

- Regulatory Guide 8.34, "Monitoring Criteria and Methods to Calculate Occupational Radiation Doses"

- Regulatory Guide 8.36, "Radiation Dose to the Embryo/Fetus"

Information to be Submitted

The information supplied by the licensee should be sufficient to allow the staff to fully understand the calculations and procedures used in summing external and internal doses. The staff's review should verify that the following information is included in the licensee's program to sum internal and external doses:

- a description of how the internal and external monitoring results are used to calculate Total Organ Dose Equivalent (TODE) and TEDE doses to occupational workers;

- a description of how internal doses to the embryo/fetus, which is based on the intake of an occupationally-exposed, declared-pregnant woman, will be determined;

- a description of the monitoring of the intake of a declared-pregnant woman if determined to be necessary; and

- a description of the program for the preparation, retention and reporting of records for occupational radiation exposures.

EVALUATION FINDINGS

Evaluation Criteria

The staff's review should verify that the method used to sum internal and external exposures will be in compliance with 10 CFR 20.1202, 20.1208(c)(1) and (2), and 20.2106. The staff should verify that the licensee's calculations to sum internal and external exposures will:

- Use the results of internal and external monitoring to calculate TODE and TEDE to occupational workers as indicated in Regulatory Positions 7.1–C7.3 of Regulatory Guide 8.34 (a sample calculation is can be found in the Appendix to Regulatory Guide 8.34).

- Sum the internal exposure to the embryo/fetus, which is based on the intake of an occupationally-exposed DPW, as indicated in Regulatory Positions C1 to C3 of Regulatory Guide 8.36, with external dose to the DPW to obtain the "dose equivalent" to the embryo/fetus.

- Monitor the intake of a DPW if her internal exposure is likely to exceed the intake criteria indicated in Regulatory Position C1.1 of Regulatory Guide 8.36.

- Follow the program for the preparation, retention, and reporting of records for occupational radiation exposures, as indicated in Regulatory Guide 8.7.

Sample Evaluation Findings

The NRC staff has reviewed the information in the Decommissioning Plan for the [insert name and license number of facility at [insert location of facility] according to the Decommissioning Consolidated Guidance, Volume 1, Section 17.3.1.5 ("Summation of Internal and External Exposures"). Based on this review, the NRC staff has determined that the licensee, [insert name], has provided sufficient information to conclude that the licensee's program for summation of internal and external exposures will comply with 10 CFR 20.1202, 20.1208(c)(1) and (2), and 20.2106.

17.3.1.6   Contamination Control Program

The purpose of the staff's review of the licensee's description of its program to monitor and control contamination during decommissioning activities is to verify that it complies with the requirements of 10 CFR Part 20. This section focuses on surveys of skin, protective and personal clothing, fixed and removable surface contamination, transport vehicles, equipment (including ventilation surveys), and packages.

NRC requires testing to determine whether there is any radioactive leakage from sealed sources. The NRC NUREG–1556 series lists guidance documents specific to the many license applications for sealed sources and sealed sources used in devices.

ACCEPTANCE CRITERIA

Regulatory Requirements

10 CFR 20.1501, 20.1702, 20.1906 (b), (d), and (f), 20.2103, 30.53

Regulatory Guidance

- Information Notice 97–55, "Calculation of Surface Activity for Contaminated Equipment and Materials"

- Regulatory Guide 8.21, "Health Physics Surveys for Byproduct Material at NRC–Licensed Processing and Manufacturing Plants"

- Regulatory Guide 8.23, "Radiation Surveys at Medical Institutions"

- Regulatory Guide 8.24, "Health Physics Surveys During Enriched Uranium-235 Processing and Fuel Fabrication"

- Regulatory Guide 8.25, "Air Sampling in the Workplace"

- NUREG–1660, "Specific Schedules of Requirements for Transport of Specified Types of Radioactive Material Consignments"

- Branch Technical Position, "License Condition for Leak Testing Sealed Sources"

Information to be Submitted

The information supplied by the licensee should be sufficient to allow the staff to fully understand how the licensee will implement and modify its contamination control program throughout the schedule phases of the decommissioning activities.

The staff's review should verify that the following information is included in the description of the licensee's contamination control program:

- a description of the written procedures to control both access to and stay time in contaminated areas by workers, if they are needed;

- a description of surveys to supplement personnel monitoring for workers during routine operations, maintenance, clean-up activities, and special operations;

- a description of the surveys that will be performed to determine the baseline of background radiation levels and radioactivity from natural sources for areas where decommissioning activities will take place;

- a description in matrix or tabular form that describes contamination action limits (i.e., actions taken either to decontaminate a person, place or area, or to restrict access, or to modify the type or frequency of radiological monitoring);

- a description (included in the matrix or table mentioned above) of proposed radiological contamination guidelines for specifying and modifying the frequency for each type of survey used to assess the reduction of total contamination; and

- a description of the procedures used to test sealed sources and to insure that sealed sources are leak tested at appropriate intervals.

EVALUATION FINDINGS

Evaluation Criteria

The staff's review should verify that the measures used to control contamination will be in compliance with the requirements of 10 CFR 20.1501(a); 20.1702, 20.1906 (b), (d) and (f); the guidance in Regulatory Guides 8.21, 8.23, 8.24, Rev. 1, and 8.25; and, for Part 70 licensees, the Fuel Cycle Branch Technical Positions for leak testing sealed sources. The staff should verify that the licensee's contamination control program during decommissioning operations (prior to the final status survey) will perform all of the following:

- Establish a program and written procedures to control both access to and stay time in contaminated areas by workers, as required by 10 CFR 20.1702.

- Require surveys to supplement personnel monitoring for workers during routine operations, maintenance, clean-up activities, and special operations.

- Require surveys to determine the baseline of background radiation levels and radioactivity from natural sources for areas where decommissioning activities will take place.

- Require surveys of air quality based on Regulatory Guide 8.25, as described in Section 17.3.1.1 of this volume.

- Follow the procedures for surveys as indicated in Regulatory Position C.1, Types of Surveys, in Regulatory Guide 8.21, 8.23, or 8.24, Rev. 1 (depending on the kind of nuclear facility being decommissioned).

- Propose and justify administrative limits for removable surface contamination that will be allowed for restricted and unrestricted areas before decontamination will be performed. Refer to Regulatory Position C.1 of the appropriate Regulatory Guide 8.21, 8.23 or 8.24, for an illustration of generic administrative limits for contamination of surfaces, and of generic limits for contamination of clothing to be worn inside and outside restricted areas. Refer to Regulatory Guide 1.86 and FC 83–23 for an illustration of administrative limits for the uncontrolled release of equipment for sites with DPs approved before August 20, 1999. Refer to Table 1 in 63 FR 64132, November 18, 1998 for acceptable license termination screening values of common radionuclides for building surface contamination. Refer to NUREG–1660 for Limits of Contamination established by the Department of Transportation.

- Calculate the surface activity of contaminated materials with a 4-pi surface-efficiency factor for gamma emitters, and 2-pi surface-efficiency factor for beta emitters as required by NRC Information Notice No. 97–55.

- Propose and justify administrative guidelines for the frequency for each type of survey used to assess trends in the reduction of total contamination during decontamination of each work area, as indicated in Regulatory Position C.2 in the appropriate Regulatory Guide 8.21, 8.23 or 8.24, Rev. 1.

- Leak-test sealed sources on a regular basis in accord with the guidance in Annex A.2.1 of ANSI/HPS N43.6–1997 (or for Part 70 licenses, as indicated in NRC's Branch Technical Positions for Leak Testing, April 1993).

Sample Evaluation Findings

The NRC staff has reviewed the information in the Decommissioning Plan for the [insert name and license number of facility] located at [insert location of facility] according to the Decommissioning Consolidated Guidance, Volume 1, Section 17.3.1.6 ("Contamination Control Program"). Based on this review, the NRC staff has determined that the licensee, [insert name], has provided sufficient information to control contamination on skin, on protective and personal clothing, on fixed and removable contamination on work surfaces, on transport vehicles, on equipment (including ventilation hoods), and on packages to allow the NRC staff to conclude that the licensee's contamination control program will comply with 20.1501(a), 20.1702, 20.1906(b), (d); and (f) of 10 CFR Part 20. The staff has verified that the information summarized under "Evaluation Criteria" above is included in the licensee's description of the methodology used to control contamination at the facility.

ACCEPTANCE CRITERIA

17.3.1.7   Instrumentation Program

The purpose of the staff's review is to verify that the licensee's description of its instruments and equipment used to make quantitative radiation measurements during surveys are calibrated periodically and have sufficient sensitivity to detect the types and magnitudes of ionizing radiation. Instrumentation will be used to: conduct radiation and contamination surveys, sample airborne radioactivity, monitor radiation levels in work areas, monitor airborne radionuclides in effluents, monitor personal dose, and analyze environmental air, water, soil and vegetation samples.

Regulatory Requirements

10 CFR 20.1501(b) and (c)

Regulatory Guidance

- NUREG–1506, "Measurement Methods for Radiological Surveys in Support of New Decommissioning Criteria"

- NUREG–1507, "Minimum Detectable Concentrations with Typical Radiation Survey Instruments for Various Contaminants and Field Conditions"

- NUREG–1549, "Decision methods for Dose Assessment to Comply With Radiological Criteria for License Termination"

- NUREG–1575, "Multi-Agency Radiation Survey and Site Investigation Manual," (MARSSIM)

- Table 10.1 of NCRP Report 127 "Operational Radiation Safety Program," 1998

Information to be Submitted

The information supplied by the licensee should be sufficient to allow the staff to fully understand how the licensee will implement and maintain its radiological instrumentation program. The staff's review should verify that the following information is included in the licensee's instrumentation program:

- a description of the instruments to be used to support the health and safety program including the manufacturer's name, the intended use of the instrument, the number of units available for the intended use, the ranges on each scale, the counting mode and the alarm set-points;

- a description of instrumentation storage, calibration and maintenance facilities for instruments used in field surveys, including onsite facilities used for laboratory analyses of samples collected during surveys;

- a description of the method used to estimate the Minimum Detectable Concentration (MDC) or Minimum Detectable Activity (MDA) (at the 95 percent confidence level) for each type of radiation to be detected;

- a description of the instrument calibration and quality assurance procedures;

- a description of the methods used to estimate uncertainty bounds for each type of instrumental measurement; and

- a description of air sampling calibration procedures or a statement that the instruments will be calibrated by an accredited laboratory.

EVALUATION FINDINGS

Evaluation Criteria

The staff's review should verify that the licensee's instrumentation program will meet the requirements of 10 CFR 20.1501(b) and (c) and the guidance in NUREG–1506, NUREG–1507 and NUREG–1575. The selection of the instruments to be used for each type of field survey or laboratory analysis should comply with the general guidance on selection of instruments during decommissioning activities, as recommended in Sections 6.1–6.5.3 and Appendix H of NUREG–1575. The method used to estimate the MDC or MDA (at the 95 percent confidence level) for each type of radiation to be detected should comply with the methods recommended in Section 6.7 of NUREG–1575. Chapters 4 and 5 of NUREG–1507 provide additional information on the extent to which the ideal MDC and MDA values may be affected when a contaminated surface is covered by paint, dust, oil, or moisture. The description of the instrument calibration and quality assurance procedures should comply with Table 10.1 of NCRP Report 127; the description of the methods used to estimate uncertainty bounds for each type of instrumental measurement should comply with recommendations indicated in Section 6.8 of NUREG–1575.

Sample Evaluation Findings

The NRC staff has reviewed the information in the Decommissioning Plan for the [insert name and license number of facility], located at [insert location of facility] according to the Decommissioning Consolidated Guidance, Volume 1, Section 17.3.1.7 ("Instrumentation Program"). Based on this review, the NRC staff has determined that the licensee, [insert name], has provided sufficient information on the sensitivity and the calibration of instruments and equipment to be used to make quantitative measurements of ionizing radiation during surveys to allow the NRC staff to conclude that the licensee's instrumentation program will comply with 10 CFR 20.1501(b) and (c).

17.3.2 NUCLEAR CRITICALITY SAFETY

The purpose of the review of the licensee's nuclear criticality safety program description is to verify that the licensee has an adequate program to maintain the criticality safety basis established in the facility's existing safety analyses.

It is essential that all operations and personnel involved in decommissioning maintain the safety basis as established in the facility's existing safety analyses. In principle, the criticality safety requirements and other Items Relied on for Safety (IROFS) resulting from Nuclear Criticality Safety Analysis (NCSA) or Integrated Safety Analysis (ISA) of plant processes will have covered all credible operations involving that process, including shutting the process down and rendering it safe by removal of all fissile material. However, decommissioning challenges this existing safety basis in two ways:

- Certain unique operations may not be covered by the existing safety analysis because decommissioning involves actions differing from normal shutdown, such as dismantlement or special decontamination; and

- Decommissioning may involve the use of different personnel than normal operations.

Therefore, in selected cases, new or updated safety analyses may be required. This is not a new provision, but is simply the existing fundamental Nuclear Criticality Safety standard from consensus standard ANSI/ANS 8.1 that "Before a new operation with fissionable materials is begun or before an existing operation is changed, it shall be determined that the entire process will be subcritical under both normal and credible abnormal conditions."

This provision, although not usually present verbatim in the license, is normally implemented by specific commitments stated in the NCS section of the license application. To the extent that decommissioning operations are new or involve changes to existing operations, compliance with the above fundamental standard means that re-analysis to assure subcriticality would be needed. Therefore, before decommissioning operations involving new steps are begun on processes that may contain fissionable material, a review of the NCSA or ISA for that operation must be conducted. It is expected that a summary of this review be submitted as part of the DP. NRC staff should review this summary to assure completeness and adequacy of items relied on for safety during decommissioning.

ACCEPTANCE CRITERIA

Regulatory Requirements

10 CFR Parts 70 and 76

Regulatory Guidance

Regulatory Guide 3.71 and endorsed standards of ANSI/ANS Series 8

Information to be Submitted

The staff's review should verify that the following information (at a minimum) is included in the licensee's NCS information:

- A description of how the NCS functions, including management responsibilities and technical qualifications of safety personnel, will be maintained when needed throughout the decommissioning process.

- A description of how an awareness of procedures and other items relied on for safety will be maintained throughout decommissioning among all personnel with access to systems that may contain fissionable material in sufficient amounts for criticality.

- A summary of the review of NCSAs or the ISA indicating either that the process needs no new safety procedures or requirements, or that new requirements or analysis have been performed.

- A summary of any generic NCS requirements to be applied to general decommissioning, decontamination, or dismantlement operations, including those dealing with systems that may unexpectedly contain fissionable material.

The description of NCS functions for decommissioning is acceptable if its implementation would reasonably assure the continuance of necessary NCS functions where and when needed throughout the decommissioning process.

The description of how an awareness of procedures and other items relied on for safety will be maintained is acceptable if it provides for measures that would reasonably assure that all personnel with access to systems that might contain fissionable material will conform to necessary NCS requirements. To be acceptable, the general methods for informing or training of personnel involved in decommissioning but who are not qualified operators of processes with fissionable materials should be sufficient to assure that such personnel do not inadvertently violate safety requirements. It is not necessary that all such personnel be trained in the details of all NCS requirements of systems, but they should be aware that operations involving such systems where fissile material may be present are subject to NCS requirements. For instance, certain operations may need to be conducted under the supervision of appropriately trained personnel.

The summary of the review of NCSAs or the ISA is acceptable if it indicates, for each process that may contain fissionable material in amounts of concern, whether the analysis is already adequate to cover all operations needed for decommissioning, or if new analysis or requirements were developed to address decommissioning tasks. In addition, the reviewer should make a selection of individual processes that is representative of the whole facility but based on risk. These selected safety analyses should then be reviewed for adequacy. The analyses are acceptable if they comply with the same criteria and commitments as for NCSAs applied during normal operations; namely, those specified in the license and plant procedures in conformance with the regulations and guidance. The guidance on acceptance NCS criteria includes the

ANSI/ANS Series 8 standards endorsed by Regulatory Guide 3.71, as well as more detailed criteria in this NUREG series applicable to the licensee.

The summary of generic NCS requirements for decommissioning is acceptable if they provide reasonable assurance that existing specific NCS requirements will be complied with despite the general dismantlement and decontamination operations involved in decommissioning. Specifically, these requirements are acceptable if they provide, as necessary, reasonable assurance that potentially critical masses of fissionable material in unexpected but credible locations will be detected and safely dispositioned. The potential for mobilizing or moderating such material by introduction of fluids should be addressed, as well as changes in any other parameters affecting criticality.

EVALUATION FINDINGS

The results of the NRC staff's review of the licensee's submittal should be stated in the form of findings of fact and acceptability for compliance with the regulations as guided by this volume. In particular, the evaluation should make findings as to the acceptability and adequacy of the items addressed by this volume to provide reasonable assurance of protection of public health and safety from the risk of nuclear criticality during decommissioning.

17.3.3 HEALTH PHYSICS AUDITS, INSPECTIONS, AND RECORDKEEPING PROGRAM

The staff should review the applicant's proposed audit, internal inspection, and recordkeeping procedures. The program should identify the scope of the audit and inspections, their frequency, the responsibilities of all participants in these programs, and any corrective actions to be taken if deficiencies are found.

ACCEPTANCE CRITERIA

Regulatory Requirements

Broad Scope Licensees:

- 10 CFR 33.13(c); 33.14(b); and 33.15(c)

All Licensees:

- 10 CFR 20.1101; and 20.2102

Regulatory Guidance

- Information Notice 96–28, "Suggested Guidance Relating to Development and Implementation of Corrective Action," dated May 1, 1996

- NUREG–1460, "Guide to NRC Reporting and Recordkeeping Requirements," Rev. 1, July 1994

Information to be Submitted

The information supplied by the licensee should be sufficient to allow the staff to fully evaluate the applicants' executive management and RSO audit program established to insure compliance with license conditions, commitments and regulatory requirements. The staff review should verify that all of the following information is included in the description of the audit program:

- A general description of the annual program review conducted by executive management is included.

- A description of the records to be maintained of the annual program review and executive audits is included.

- A description of the types and frequencies of surveys and audits to be performed by the RSO and RSO staff is included. These surveys and audits should be frequent enough to ensure close communications and proper surveillance of individual radiation workers. Applicants should consider developing survey and audit schedules based on activity and use (e.g., highly contaminated areas or facilities involving volatile radioactive materials may be audited weekly or biweekly, moderately contaminated areas or facilities may be audited monthly, and slightly contaminated facilities may be audited quarterly). The audit program should include routine unannounced inspections).

- A description of the process used in evaluating and dealing with violations of NRC requirements or license commitments identified during audits is included.

- A description of the records maintained of RSO audits is included (e.g., the date of each audit, name of person(s) who conducted the audit, persons contacted by the auditor(s), areas audited, audit findings, corrective actions, and follow-up).

EVALUATION FINDINGS

The staff's review should verify that the licensee's audit and recordkeeping program implemented to evaluate, control, and monitor health and safety procedures is appropriate and consistent with the guidance in this volume The proposed audit program should insure timely identification and correction of health and safety issues, such that compliance with NRC requirements for the protection of the public health and safety and the environment is insured.

Sample Evaluation Findings

The NRC staff has reviewed the description of the licensee's, [insert name and license number of licensee], audit and recordkeeping program, which the licensee will use during the decommissioning of its facility located at [insert location of facility] according to the Decommissioning Consolidated Guidance, Volume 1, Section 17.3.3 ("Health Physics Audits, Inspections, and Recordkeeping Program"). Based on this review, the NRC staff has determined that the licensee, [insert name], has provided sufficient information to allow the NRC staff to evaluate the licensee's executive management and RSO audit and recordkeeping program to determine if the decommissioning can be conducted safely and in accordance with NRC requirements.

17.4 DECOMMISSIONING PLAN: ENVIRONMENTAL MONITORING AND CONTROL PROGRAM

OVERVIEW

The NRC staff should review the information submitted by the licensee to determine if the environmental monitoring and control program complies with the regulatory requirements in 10 CFR Part 20 and if it is adequate to protect workers, the public, and the environment from ionizing radiation during decommissioning activities. The staff should verify that the licensee's radiological effluent management practices are adequate to ensure that radiological effluent levels are maintained within applicable standards and are ALARA. The environmental monitoring and control program should include descriptions of (1) the environmental exposure evaluations to be performed during decommissioning; (2) the effluent monitoring for radioactive material at potential points of release to the environment; and (3) the controls that the licensee will use to ensure that radioactive material in effluents does not exceed applicable NRC, State, or local requirements.

REVIEW PROCEDURES

Safety Evaluation

The material to be reviewed is technical in nature. The staff should make a quantitative assessment as to whether the licensee's proposed effluent monitoring and control program complies with the regulatory requirements in 10 CFR Part 20 and is adequate to protect workers, the public and the environment from ionizing radiation during decommissioning activities. The staff should assess whether the applicant's environmental monitoring and control measures are commensurate with the risks associated with the proposed decommissioning activities.

17.4.1 ENVIRONMENTAL ALARA EVALUATION PROGRAM

The purpose of the review of the licensee's environmental ALARA evaluation program description is to verify if the licensee has a program adequate to demonstrate compliance with the requirements of 10 CFR Part 20 to maintain releases of radioactive material to the environment ALARA.

ACCEPTANCE CRITERIA

Regulatory Requirements

10 CFR Part 20.1101(b) and (d)

Regulatory Guidance

- Regulatory Guide 8.37, "ALARA Levels for Effluents from Materials Facilities," July 1993

- Regulatory Guide 4.20, "Constraint on Releases of Airborne Radioactive Materials to the Environment for Licensees Other Than Power Reactors," December 1998

Information to be Submitted

The information supplied by the licensee should be sufficient to allow the staff to fully understand the licensee's environmental evaluation activities and procedures. The staff's review should verify that the following information is included in the description of the licensee's environmental ALARA evaluation program:

- a description of ALARA goals for effluent control;

- a description of the procedures, engineering controls, and process controls to maintain doses ALARA (see Section 17.4); and

- a description of the ALARA reviews and reports to management.

EVALUATION FINDINGS

Evaluation Criteria

The staff should verify that the information summarized under "Evaluation Criteria" is included in the licensee's environmental ALARA evaluation program description. The staff should verify that the licensee's program for the management of radiological materials released to the environment complies with NRC requirements at 10 CFR Part 20, and that the program uses appropriate methods and procedures based upon recognized NRC and other professional health physics organizations' guidance documents.

The staff should verify that the licensee's ALARA goals are a fraction (10 to 20 percent) of the values in (a) Table 2 (Columns 1 and 2) and Table 3 in Appendix B of 10 CFR Part 20, (b) the external exposure limit in 10 CFR 20.1302(b)(2)(ii), and (c) the applicable dose limit for members of the public. An approach is acceptable if it is consistent with guidance found in Regulatory Guide 4.20 and if the description of the approach provides sufficient detail to demonstrate specific application of the guidance to the proposed operations. The licensee should use sound, commonly accepted, and well-established procedures, engineering controls, and process controls to achieve ALARA goals for effluent minimization. These include filtration, encapsulation, adsorption, containment, recycling, leakage reduction, and the storage of materials for radioactive decay. Practices for large, diffuse sources such as contaminated soils or surfaces include covers, wetting during operations, and the application of stabilizers. In addition, the licensee must demonstrate a commitment to reducing unnecessary exposure to members of the public and releases to the environment.

ALARA program management should include a commitment to perform annual reviews of the content and implementation of the environmental radiation protection program. This review includes an analysis of trends in release concentrations, environmental monitoring data, and radionuclide usage, a determination of whether operational changes are needed to achieve the ALARA effluent goals, and an evaluation of all designs for system installations or modifications.

The description should also include a commitment to report the results to senior management along with recommendations for changes in facilities or procedures that are necessary to achieve ALARA goals.

Sample Evaluation Findings

The NRC staff has reviewed the information in the Decommissioning Plan for the [insert name and license number of facility] located at [insert location of facility] according to the Decommissioning Consolidated Guidance, Section 17.4 ("Decommissioning Plan: Environmental Monitoring and Control Program"). Based on this review, the NRC staff has determined that the licensee, [insert name], has provided sufficient information on the staff to conclude that the licensee's program will comply with 10 CFR Part 20.

Note that the results from the staff's evaluation of the Environmental ALARA, Environmental Monitoring, and Effluent Control programs should be combined in this finding.

17.4.2   EFFLUENT MONITORING PROGRAM

The purpose of the review of the description of the licensee's effluent monitoring program is to determine if the licensee has an adequate program for the collection and analysis of airborne and liquid effluents, for assessing radiation exposures to members of the public, and for demonstrating compliance with applicable regulations.

ACCEPTANCE CRITERIA

Regulatory Requirements

- 10 CFR 20.1301(a) and (d), 20.1302(a) and (b), 20.1501, 2001(a), 20.2003(a), 20.2103 (b), 20.2107(a), 20.2202(a), 20.2203(a), and 70.59.

Regulatory Guidance

- ANSI N13.1–1982, "Guide to Sampling Airborne Radioactive Materials in Nuclear Facilities"

- ANSI N42.18–1980, "Specification and Performance of On-site Instrumentation for Continuously Monitoring Radioactive Effluents"

- NCRP Report No. 123, "Screening Models for Releases of Radionuclides to Atmosphere, Surface Water, and Ground," January 1996

- NRC Information Notice 94–07, "Solubility Criteria for Liquid Effluent Releases to Sanitary Sewerage Under the Revised 10 CFR Part 20," January 28, 1994

- NRC Regulatory Guide 4.15, "Quality Assurance for Radiological Monitoring Programs (Normal Operations)–Effluent Streams and the Environment"

- NRC Regulatory Guide 4.16, "Monitoring and Reporting Radioactivity in Releases of Radioactive Materials in Liquid and Gaseous Effluents from Nuclear Fuel Processing and Fabrication Plants and Uranium Hexafluoride Production Plants"

Information to be Submitted

The information supplied by the licensee should be sufficient to allow the staff to fully understand how the licensee will implement and conduct its effluent monitoring program. The staff's review should verify that the following information is included in the licensee's effluent monitoring program:

- a demonstration that background and baseline concentrations of radionuclides in environmental media have been established through appropriate sampling and analysis;

- a description of the known or expected concentrations of radionuclides in effluents;

- a description of the physical and chemical characteristics of radionuclides in effluents;

- a summary or diagram of all effluent discharge locations;

- a demonstration that samples will be representative of actual releases;

- a summary of the sample collection and analysis procedures, including the minimum detectable concentrations of radionuclides (if this information is not already described pursuant to Section 17.4 of this volume);

- a summary of the sample collection frequencies;

- a description of the environmental monitoring recording and reporting procedures; and

- a description of the quality assurance program to be established and implemented for the effluent monitoring program (if this is not already described under Section 17.6 of this volume).

EVALUATION FINDINGS

Evaluation Criteria

The staff should verify that the information summarized under "Evaluation Criteria," above, is included in the licensee's description of its effluent monitoring program. The staff should verify that the licensee's program complies with NRC requirements at 10 CFR Part 20 and that the program uses appropriate methods and procedures based upon recognized NRC and other professional health physics organizations' guidance documents. Concentrations of radioactive materials in airborne and liquid effluents as well as physical and chemical characteristics should be estimated based on operational data for the facility.

Releases shall be maintained below the limits in Table 2 of Appendix B to 10 CFR Part 20 or below site-specific limits established in accordance with 20.1302(c) and should be ALARA. NRC regulations require that licensees demonstrate that releases are maintained below the limits in 10 CFR Part 20 by calculation or measurement. If a licensee elects to make this demonstration by calculation, the estimate should be based on the total volume of effluents (air or liquid) released from the facility during a year and the total activity of radioactive material possessed by the licensee during the year. The total activity of radioactive material may be adjusted to reflect the actual activity that could have been released in effluents, as long as the licensee can justify the adjustment through materials inventory and balance records.

If the licensee elects to demonstrate compliance with NRC requirements by sampling, all liquid and airborne effluent discharge locations should be described, with a description of how each location is monitored such that the samples collected are representative of the concentration and quantity of radiological material released to the environment. A description of the effluents that are continuously sampled from radiological operations associated with the plant, such as laboratories, experimental areas, and storage areas, should also be included.

For liquid effluents, representative samples should be taken at each release point for the determination of concentrations and quantities of radionuclides released to an unrestricted area, including discharges to sewage systems. For continuous releases, samples should be continuously collected at each release point. For batch releases, a representative sample of each batch should be collected. If periodic sampling is used in lieu of continuous sampling, the description should demonstrate that the samples are representative of actual releases. Sample collection frequencies are appropriate for the effluent medium and the radionuclide(s) being sampled if they are performed during activities that could generate effluents in the medium being sampled and the samples collected can be shown to be representative of the concentrations of radionuclides in the medium. Reporting procedures are adequate if they comply with the guidance specified in Regulatory Guide 4.16. Reports of the concentrations of principal radionuclides released to unrestricted areas in liquid and gaseous effluents should be provided and include the MDC for the analysis and the error for each data point.

If the licensee believes that radioactivity in effluents is insignificant and will remain so during decommissioning and after license termination, a justification for this assertion should be included. For the purposes of this NUREG series, an effluent is significant if the concentration averaged over a calendar quarter is equal to 10 percent or more of the applicable concentration listed in Table 2 of Appendix B to 10 CFR Part 20.

17.4.3 EFFLUENT CONTROL PROGRAM

The purpose of the review of the licensee's effluent control program description is to verify that the licensee has a program to control radioactive material in effluents and to comply with all applicable standards and permit requirements related to the release of radioactive material in effluents.

ACCEPTANCE CRITERIA

Regulatory Requirements

10 CFR 20.1301(a) and (d), 20.1302(a) and (b), 20.1501, 2001(a), 20.2003(a), 20.2103 (b), 20.2107(a), 20.2202(a), and 20.2203(a)

Regulatory Guidance

- Regulatory Guide 4.20, "Constraints on Releases of Airborne Radioactive Materials to the Environment for Licensees other than Power Reactors," December 1996.

- NRC Information Notice 94–23: "Guidance to Hazardous, Radioactive and Mixed Waste Generators on the Elements of a Waste Minimization Program," March 25, 1994.

- IAEA, No. 16, "Manual on Environmental Monitoring in Normal Operations," Vienna, 1996.

- IAEA, No. 18, "Environmental Monitoring in Emergency Situations", Vienna, 1966.

- IAEA, Safety Series No. 41, "Objectives and Design of Environmental Monitoring Programs for Radioactive Contaminants," Vienna, 1975.

- NCRP Report No. 50, "Environmental Radiation Measurements," December 1976.

- NCRP Report No. 123, "Screening Models for Releases of Radionuclides to Atmosphere, Surface Water, and Ground," January 1996.

Information to be Submitted

The information supplied by the licensee should be sufficient to allow the staff to fully understand how the licensee will implement and conduct its effluent control program. The staff's review should verify that the following information is included in the licensee's effluent control program:

- a description of the controls that will be used to minimize releases of radioactive material to the environment;

- a summary of the action levels and description of the actions to be taken, should a limit be exceeded;

- a description of the leak detection systems for ponds, lagoons, and tanks;

- a description of the procedures to ensure that releases to sewer systems are controlled and maintained to meet the requirements of 10 CFR 20.2003; and

- a summary of the estimates of doses to the public from effluents and a description of the method used to estimate public dose.

EVALUATION FINDINGS

Evaluation Criteria

The staff should verify that the information summarized under "Evaluation Criteria," above, is included in the description of the licensee's effluent control program. The staff should verify that the licensee's program for the control of radiological materials released to the environment complies with NRC requirements at 10 CFR Part 20, and that the program uses appropriate methods and procedures, based upon recognized NRC and other professional health physics organizations' guidance documents. The staff should verify that the licensee has identified all possible effluent pathways, based on current and expected future site conditions, and evaluated the likelihood of releases via these pathways. The controls proposed by the licensee to minimize releases of radioactive material to the environment should be based on well-recognized industry practices and procedures.

Proposed action levels should be a fraction (10 to 20 percent) of limits and should be justified. Action levels should be incremental, such that each increasing action level results in a more aggressive action to assure and control effluents. A slightly higher than normal concentration of a radionuclide in effluent triggers an investigation into the cause of the increase. In addition, an action level should be specified that will result in the shutdown of an operation if this level is exceeded. These action levels should be selected on the likelihood that a measured increase in concentration could indicate potential violation of the effluent limits. Actions to be taken if the levels are exceeded should be described in sufficient detail to allow the staff to fully understand the scope and results of the actions.

The description of the system(s) for the detection of leakage from ponds, lagoons, and tanks are adequate if they are based on well-recognized engineering practices and allow for the intervention and response to leaks before radioactive material enters unrestricted areas.

Controls for releases to sewer systems shall meet the requirements of 10 CFR 20.2003, including (i) the material is water soluble; (ii) known or expected discharges meet the effluent limits of 10 CFR 20 Appendix B, Table 3; and (iii) the known or expected total quantity of radioactive material released into the sewer system in a year does not exceed 5 Ci (185 GBq) of H-3, 1 Ci (37 GBq) of C-14, and 1 Ci (37 GBq) of all other radioactive materials combined. Solubility is determined in accordance with the procedure described in NRC Information Notice 94–07. If the licensee proposes to demonstrate compliance with 10 CFR 20.1301 through a calculation of the TEDE to the individual likely to receive the highest dose in accordance with 20.1302(b)(1), calculation of the TEDE by pathways analyses uses appropriate models and codes and assumptions that accurately represent the facility, the site, and the surrounding area. It is also required that assumptions are reasonable, input data are accurate, all applicable pathways are considered, and the results are interpreted correctly. NCRP Report No. 123, "Screening Models for Releases of Radionuclides to Atmosphere, Surface Water, and Ground," January 1996, provides acceptable methods for calculating the dose from radioactive effluents. Computer codes are acceptable tools for pathways analysis if the applicant is able to demonstrate that the code has undergone validation and verification to demonstrate the validity of estimates developed using the code for established input sets. Dose conversion factors used in the pathways analyses are acceptable if they are based on the methodology described in ICRP 30, "Limits for Intakes of Radionuclides by Workers," as reflected in Federal Guidance Report 11.

Sample Evaluation Findings

None. The staff should combine the findings from the review of the Effluent Control Program with the findings from Section 17.4.1.

17.5 DECOMMISSIONING PLAN: RADIOACTIVE WASTE MANAGEMENT PROGRAM

OVERVIEW

The NRC staff should review the information supplied by the licensee to determine if the description of the program for the management of radioactive waste generated as part of the decommissioning of the facility is adequate to allow the staff to fully understand the types of radioactive waste that will be generated by decommissioning operations and the manner in which the licensee will manage these wastes. This information will be used by the staff to ensure that the waste will be managed in accordance with NRC requirements, to support the staff's evaluation of the licensee's health and safety program, the evaluation of potential accidents, and the licensee's cost estimates for decommissioning. This information should include descriptions of the types, volumes, and activities of radioactive waste generated by the decommissioning operations, a description of how the wastes will be stored, treated (if onsite treatment is anticipated), and packaged for transport and disposal, and the name and location of the facility where the licensee intends to treat and/or dispose of the waste.

REVIEW PROCEDURES

Safety Evaluation

The material to be reviewed is informational in nature, and no specific detailed technical analysis is required. The staff should verify that the manner in which the licensee intends to package the waste for transport and disposal is acceptable by comparing the descriptions of the waste and the packaging procedures with the relevant NRC regulations. The staff should verify that the waste disposal locations are appropriate for the wastes generated during decommissioning by comparing the waste generated by the decommissioning operations with publicly available information on the types of wastes that are accepted by the disposal facility. The staff should make a qualitative assessment as to whether the licensee's descriptions of the types, volumes, and activities of radioactive waste generated by the decommissioning operations appear accurate (given the information presented in the facility radiological status section of the DP) and if the descriptions of how the wastes will be stored and treated are appropriate for the types and volumes of wastes, as well as being protective of worker and public health and safety.

17.5.1   SOLID RADIOACTIVE WASTE

The purpose of the review of the description of the management of solid radioactive waste generated during decommissioning operations is to ensure that the manner in which the licensee proposes to manage the waste will be protective of the public health and safety and that the waste will be treated and disposed of in accordance with NRC requirements. The information will also be used to support the staff's evaluation of potential accidents and the licensee's cost estimates for decommissioning.

ACCEPTANCE CRITERIA

Information to be Submitted

The information supplied by the licensee should be sufficient to allow the staff to fully understand the types, volumes, and activities of solid radioactive waste generated during decommissioning operations and the manner in which the licensee intends to manage and dispose of the wastes. The staff's review should verify that the following information is included in the solid radioactive waste section of the facility DP:

- a summary of the types of solid radioactive waste that are expected to be generated during decommissioning operations, including (but not limited to) soil, structural and component metal, concrete, activated components, contaminated piping, wood, and plastic;

- a summary of the estimated volume, in cubic feet, of each solid radioactive waste type summarized under bullet 1, above;

- a summary of the radionuclides (including the estimated activity of each radionuclide) in each estimated solid radioactive waste type summarized under bullet 1, above;

- a summary of the volumes of Classes A, B, C, and Greater-than-Class-C solid radioactive waste that will be generated by decommissioning operations;

- a description of how and where each of the solid radioactive wastes summarized under bullet 1, above, will be stored onsite prior to shipment for disposal;

- a description of how the each of the solid radioactive wastes summarized under the first bullet above, will be treated and packaged to meet disposal site acceptance criteria prior to shipment for disposal;

- if appropriate, a description of how the licensee intends to manage volumetrically contaminated material;

- a description of how the licensee will prevent contaminated soil, or other loose solid radioactive waste, from being re-disbursed after exhumation and collection; and

- the name and location of the disposal facility that the licensee intends to use for each solid radioactive waste type summarized under the first bullet, above.

EVALUATION FINDINGS

Evaluation Criteria

The staff should verify that the information summarized under "Information to be Submitted," above, is included in the licensee's description of the solid radioactive waste management program. The staff should verify that the licensee's program for the management of solid radioactive waste complies with NRC requirements at 10 CFR Part 20, Subpart K, 10 CFR 61.55, 61.56, 61.57 and 71.5. The staff should make a qualitative assessment of the accuracy of the licensee's descriptions of the types, volumes, and activities of the solid radioactive waste by comparing them with the information presented in the facility description, planned decommissioning activities, and radiological status portions of the DP. The staff should make a qualitative assessment of the licensee's proposed methods to store solid radioactive waste prior to disposal, including the manner in which volumetrically contaminated waste will be managed. The staff should verify that the waste disposal locations are appropriate for the solid wastes generated during decommissioning by comparing the solid waste generated by the decommissioning operations with publicly available information on the types of solid wastes that are accepted by the disposal facility.

Sample Evaluation Findings

The staff may combine the evaluation finding for the licensee's description of solid radioactive waste management programs with the findings for the remaining areas in this section of this guidance, as follows:

> The NRC staff has reviewed the licensee's descriptions of the radioactive waste management program for the [insert name and license number of facility] located at [insert location of facility] according to the Decommissioning Consolidated Guidance, Volume 1, Section 17.5 ("Decommissioning Plan: Radioactive Waste Management Program"). Based on this review, the NRC staff has determined that the licensee's [insert name] programs for the management of radioactive waste generated during decommissioning operations ensure that the waste will be managed in accordance with NRC requirements and in a manner that is protective of the public health and safety.

17.5.2   LIQUID RADIOACTIVE WASTE

The purpose of the review of the description of the management of liquid radioactive waste generated during decommissioning operations is to ensure that the manner in which the licensee proposes to manage the waste will be protective of the public health and safety and that the waste will be treated and disposed of in accordance with NRC requirements. The information will also be used to support the staff's evaluation of potential accidents and the licensee's cost estimates for decommissioning.

ACCEPTANCE CRITERIA

Information to be Submitted

The information supplied by the licensee should be sufficient to allow the staff to fully understand the types, volumes, and activities of liquid radioactive waste generated during decommissioning operations and the manner in which the licensee intends to manage and dispose of the wastes. The staff's review should verify that the following information is included in the liquid radioactive waste section of the facility DP:

- a summary of the types of liquid radioactive waste that are expected to be generated during decommissioning operations;
- a summary of the estimated volume, in liters, of each liquid radioactive waste type summarized under the first bullet above;
- a summary of the radionuclides (including the estimated activity of each radionuclide) in each liquid radioactive waste type summarized under the first bullet above;
- a summary of the estimated volumes of Class A, B, C and Greater-than-Class-C liquid radioactive waste that will be generated by decommissioning operations;

- a description of how and where each of the liquid radioactive wastes summarized under the first bullet above, will be stored onsite prior to shipment for disposal;

- a description of how the each of the liquid radioactive wastes summarized under the first bullet above, will be treated and packaged to meet disposal site acceptance criteria prior to shipment for disposal; and

- the name and location of the disposal facility that the licensee intends to use for each liquid radioactive waste type summarized under the first bullet, above.

EVALUATION FINDINGS

Evaluation Criteria

The staff should verify that the information summarized under "Information to be Submitted," above, is included in the licensee's description of the liquid radioactive waste management program. The staff should verify that the licensee's program for the management of liquid radioactive waste complies with NRC requirements at 10 CFR Part 20, Subpart K, 61.55, 61.56, 61.57 and 71.5. The staff should make a qualitative assessment of the accuracy of the licensee's descriptions of the types, volumes, and activities of liquid radioactive waste by comparing them with the information presented in the facility description, planned decommissioning activities, and radiological status portions of the DP. The staff should make a qualitative assessment of the licensee's proposed methods to store liquid radioactive waste prior to disposal. The staff should verify that the waste disposal locations are appropriate for the liquid wastes generated during decommissioning by comparing the liquid waste generated by the decommissioning operations with publicly available information on the types of liquid wastes that are accepted by the disposal facility.

Sample Evaluation Findings

None. The staff should combine the evaluation finding for the licensee's description of liquid radioactive waste management programs with the findings for the remaining areas in this section of this guidance (see Section 17.5.1).

17.5.3 MIXED WASTE

The purpose of the review of the description of the management of mixed waste generated during decommissioning operations is to ensure that the manner in which the licensee proposes to manage the mixed waste will be protective of the public health and safety and that the waste will be managed, treated and disposed of in accordance with NRC and Environmental Protection Agency (EPA) or EPA–authorized State requirements. The information will also be used to support the staff's evaluation of potential accidents and the licensee's cost estimates for decommissioning.

ACCEPTANCE CRITERIA

Information to be Submitted

The information supplied by the licensee should be sufficient to allow the staff to fully understand the types, volumes, and activities of mixed waste generated during decommissioning operations and the manner in which the licensee intends to manage and dispose of the wastes. The staff's review should verify that the following information is included in the mixed waste section of the facility DP:

- a summary of the types of solid and liquid mixed waste that are expected to be generated during decommissioning operations;

- a summary of the estimated volumes, in cubic feet, of each solid mixed waste type summarized under bullet 1 above and in liters for each liquid mixed waste;

- a summary of the radionuclides (including the estimated activity of each radionuclide) in each type of mixed waste type summarized under bullet 1 above;

- a summary of the estimated volumes of Class A, B, C and Greater-than-Class-C mixed waste that will be generated by decommissioning operations;

- a description of how and where each of the mixed wastes summarized under bullet 1 above, will be stored onsite prior to shipment for disposal;

- a description of how the each of the mixed wastes summarized under bullet 1 above, will be treated and packaged to meet disposal site acceptance criteria prior to shipment for disposal;

- the name and location of the disposal facility that the licensee intends to use for each mixed waste type summarized under bullet 1 above;

- a discussion of the requirements of all other regulatory agencies having jurisdiction over the mixed waste; and

- a demonstration that the licensee possesses the appropriate EPA or State permits to generate, store and/or treat the mixed wastes.

EVALUATION FINDINGS

Evaluation Criteria

The staff should verify that the information summarized under "Information to be Submitted," above, is included in the licensee's description of the liquid radioactive waste management program. The staff should verify that the licensee's program for the management of mixed waste complies with NRC requirements at 10 CFR Part 20, Subpart K, 61.55, 61.56, 61.57 and 71.5. The staff should make a qualitative assessment of the accuracy of the licensee's descriptions of the types, volumes, and activities of mixed waste by comparing it to the information presented in the facility description, planned decommissioning activities, and radiological status portions of the DP. The staff should make a qualitative assessment of the licensee's proposed methods to store mixed waste prior to disposal. The staff should verify that the waste disposal locations are appropriate for the mixed wastes generated during decommissioning by comparing the mixed waste generated by the decommissioning operations to publicly available information on the types of mixed wastes that are accepted by the disposal facility.

Note that the NRC staff is NOT responsible for ensuring that the licensee's program complies with the requirements of 40 CFR 260–270 or the Department of Transportation regulations pertaining to the transportation of the hazardous component of the mixed waste. The staff should make a qualitative assessment of the acceptability of the licensee's descriptions of the methods they will employ to comply with the requirements of other agencies with regulatory responsibility for the mixed waste.

Sample Evaluation Findings

None. The staff should combine the evaluation finding for the licensee's description of mixed waste management programs with the findings for the remaining areas in this section of the guidance (see Section 17.5.1).

17.6   DECOMMISSIONING PLAN: QUALITY ASSURANCE PROGRAM

OVERVIEW

The staff should review the information supplied by the licensee to determine if the description of the quality assurance (QA) program is adequate to allow the staff to conclude that the licensee has adequate controls in place to support the decommissioning. Further, if the licensee effectively implements the QA program described, the data collected should be accurate and of sufficient quality to justify the conclusions drawn from the information. This information should include the following:

- a description of the organization responsible for implementing the QA program;

- a description of the QA program, including descriptions of the manner in which QA activities are controlled; a description of the manner in which QA program documents are controlled;

- a description of how measuring and test equipment is controlled;

- a description of how conditions adverse to quality are corrected; a description of the QA records that will be maintained; and

- a description of the audits and surveillances that are performed as part of the QA program.

REVIEW PROCEDURES

Safety Evaluation

The material to be reviewed is informational in nature, and no specific detailed technical analysis is required. The staff should make a qualitative assessment as to whether the licensee's QA program is adequate to ensure that accurate, high-quality information is developed to support the decommissioning of the facility.

17.6.1   ORGANIZATION

The purpose of the review of the QA organization is to verify that the licensee has an adequate organization, sound management philosophy, and the resources necessary to ensure that the information submitted to support the decommissioning is accurate and of sufficient quality to justify the conclusions drawn from the information.

ACCEPTANCE CRITERIA

Information to be Submitted

The staff should review the licensee's description of its organizational structure to ensure that persons and organizations performing quality affecting activities have sufficient authority and freedom to identify quality problems, provide solutions, and verify that solutions have been implemented. The staff's review should verify that the following information is included in the description of the QA program organization:

- a description of the QA program management organization;

- a description of the duties and responsibilities of each unit within the organization and how delegation of responsibilities is managed within the decommissioning program;

- a description of how work performance is evaluated;

- a description of the authority of each unit within the QA program; and

- an organization chart of the QA program organization.

EVALUATION FINDINGS

Evaluation Criteria

The staff should verify that the information summarized under "Information to be Submitted," above, is included in the licensee's description of the QA program. The staff should verify that the organization or individual responsible for submitting the license application exercises and retains the responsibility for the establishment and execution of the overall program. The staff should verify that major delegations of work are fully described and that in each case, organizational responsibilities and methods for control of the work by the applicant are described, including how responsibility for delegated work is to be retained and exercised. The staff should verify that the licensee and its prime contractors describe how responsibility is exercised for the overall QA program and that the extent of management responsibility and authority are addressed. The staff should verify that policies regarding the implementation of the QA program are documented and made mandatory.

The staff should verify that the licensee and its contractors will evaluate the performance of work delegated to other organizations, including audits/surveillances of the contractor's QA programs and audits/surveillances of subcontractors, consultants, and vendors furnishing equipment or services to the applicant or its contractors. The frequency and method of this evaluation should be specified.

The staff should verify that the licensee and prime contractors identify a management position that retains overall authority and responsibility for the QA program (normally, this position is filled by the QA Manager). The staff should verify that the QA Manager position is at the same or a higher organization level than the position of the highest line manager directly responsible for performing activities affecting quality (such as engineering, procurement, construction, and operation) and is sufficiently independent from cost and schedule restraints (this does not mean that the QA position must report outside of the project or program). The staff should verify that the authority and duties of persons and organizations performing functions related to meeting the performance objectives are clearly established and delineated in writing, including both the performing functions of attaining the requisite quality of work (quality achieving) and the assurance functions of verifying the attainment of quality (quality assuring). The staff should verify that designated QA personnel, sufficiently free from direct pressures resulting from cost and schedule, have the responsibility, delineated in writing, to stop unsatisfactory work and control further processing or delivery of nonconforming material.

The staff should verify that persons and organizations performing quality assurance functions have sufficient authority and organizational freedom (1) to identify quality problems, (2) to initiate, recommend, or provide solutions through designated channels, and (3) to verify implementation of solutions. The staff should verify that persons and organizations with the above authority are identified and a description of how those actions are carried out is provided.

The staff should verify that provisions are established for the resolution of disputes involving quality arising from a difference of opinion between QA personnel and other department personnel. The staff should verify that the position description ensures that the individual directly responsible for the definition, direction, and effectiveness of the overall QA program has sufficient authority to implement responsibilities effectively. This position is to be sufficiently free from cost and schedule responsibilities.

The staff should verify that the person responsible for the onsite QA program is identified by position and has the appropriate organizational position, responsibilities, and authority to exercise proper control over the QA program.

The staff should verify that organization charts clearly identify all the onsite and offsite organizational elements that function under the cognizance of the QA program.

17.6.2   QUALITY ASSURANCE PROGRAM

The purpose of the review of the QA program is to verify that the licensee's QA program and activities affecting quality will be controlled by written policies, procedures and instruction, which, if effectively implemented, should ensure that the information submitted to support the decommissioning is accurate and of sufficient quality to justify the conclusions drawn from the information.

ACCEPTANCE CRITERIA

Information to be Submitted

The staff should review the licensee's QA program to determine if activities affecting quality will be conducted in accordance with written policies, procedures, and instructions, and that activities affecting quality are accomplished by suitably trained and qualified individuals. The staff should review the licensee's QA program to ensure that quality affecting activities are prescribed by documented procedures, drawings, or instructions. The staff should verify that the following information is included in the description of the QA program:

- a commitment that activities affecting the quality of site decommissioning will be subject to the applicable controls of the QA program and activities covered by the QA program are identified on program defining documents;

- a brief summary of the company's corporate QA policies;

- a description of provisions to ensure that technical and quality assurance procedures required to implement the QA program are consistent with regulatory, licensing, and QA program requirements and are properly documented and controlled;

- a description of the management reviews, including the documentation of concurrence in these quality-affecting procedures;

- a description of the quality-affecting procedural controls of the principal contractors, including documentation of the acceptance of the controls before the initiation of activities affected by the program;

- a description of how NRC will be notified of changes (a) for review and acceptance in the accepted description of the QA program as presented or referenced in the DP before implementation and (b) in organizational elements within 30 days after the announcement of the changes (note that editorial changes or personnel reassignments of a nonsubstantive nature do not require NRC notification);

- a description is provided of how management (above or outside the QA organization) regularly assesses the scope, status, adequacy, and compliance of the QA program;

- a description of the instruction provided to personnel responsible for performing activities affecting quality pertaining to the purpose, scope, and implementation of the quality-related manuals, instructions, and procedures;

- a description of the training and qualifications of personnel verifying activities affecting quality in the principles, techniques, and requirements of the activity being performed;

- for formal training and qualification programs, documentation includes attendees, date of attendance, and the objectives and content of the program;

- a description of the self-assessment program to confirm that activities affecting quality comply with the QA program;

- a commitment that persons performing self-assessment activities are not to have direct responsibilities in the area they are assessing;

- a description of the organizational responsibilities for ensuring that activities affecting quality are (a) prescribed by documented instructions, procedures, and drawings; and (b) accomplished through implementation of these documents; and

- a description of the procedures to ensure that instructions, procedures, and drawings include quantitative acceptance criteria (such as those pertaining to dimensions, tolerances, and operating limits) and qualitative acceptance criteria (such as workmanship samples) for determining that important activities have been satisfactorily performed.

EVALUATION FINDINGS

Evaluation Criteria

The staff should verify that the information summarized under "Information to be Submitted," above, is included in the description of the QA program. Licensees are encouraged to submit the information in electronic format.

17.6.3 DOCUMENT CONTROL

The purpose of the review of the licensee's description of how QA program documents are issued and amended is to ensure that adequate control is exercised over the development, issuance and revision of the documents.

ACCEPTANCE CRITERIA

Regulatory Requirements

10 CFR 30.36(g)(4)(ii), 40.42(g)(4)(ii),40.28(b)(3), 70.22(f), 70.38(g)(4)(ii), and 72.54(g)(6)

Information to be Submitted

The information supplied by the licensee should be sufficient to allow the staff to understand how the licensee will develop, issue and revise documents associated with the QA program. The staff's review should verify that the following information is included in the description of the QA document control program:

- a summary of the types of QA documents included in the program; and

- a description of how the licensee develops, issues, revises and retires QA documents.

EVALUATION FINDINGS

Evaluation Criteria

The staff should verify that the information summarized under "Information to be Submitted," above, is included in the licensee's description of the QA document control program. The staff should verify that the scope of the document control program is described, and the types of controlled documents are identified. As a minimum, controlled documents include quality assurance, quality control manuals, quality-affecting procedures, and technical reports. The staff should verify that procedures for the review, approval, and issuance of documents and changes will be established and described to ensure technical adequacy and inclusion of appropriate quality requirements before implementation. The staff should verify that procedures will be established to ensure that changes to documents are reviewed and approved by the same organizations as those that performed the initial review and approval or by other qualified responsible organizations delegated by the applicant. The staff should verify that procedures will be established to ensure that documents are available at the location where the activity will be performed prior to commencing work. The staff should verify that procedures will be established to ensure that obsolete or superseded documents are removed and replaced by applicable revisions in work areas in a timely manner. Licensees are encouraged to submit the information in electronic format.

17.6.4 CONTROL OF MEASURING AND TEST EQUIPMENT

The purpose of the review of the description of the test and measurement equipment calibration program is to verify that the licensee has a program to ensure that equipment used to support decommissioning activities is properly controlled, calibrated, and maintained.

ACCEPTANCE CRITERIA

Regulatory Requirements

10 CFR 30.36(g)(4)(ii), 40.42(g)(4)(ii), 40.28(b)(3), 70.22(f), 70.38(g)(4)(ii), and 72.54(g)(6)

Information to be Submitted

The information supplied by the licensee should be sufficient to allow the staff to fully understand the methods and procedures that the licensee will use to ensure that only accurate and calibrated test and measurement equipment will be used during the decommissioning project. The staff's review should verify that the following information is included in the description of the test and measurement equipment QA program:

- a summary of the test and measurement equipment used in the program;

- a description of how and at what frequency the equipment will be calibrated;

- a description of the daily calibration checks that will be performed on each piece of test or measurement equipment; and

- a description of the documentation that will be maintained to demonstrate that only properly calibrated and maintained equipment was used during the decommissioning.

EVALUATION FINDINGS

Evaluation Criteria

The staff should verify that the information summarized under "Information to be Submitted," above, is included in the licensee's description of the test and measurement equipment program. The staff should verify that the scope of the program for the control of measuring and test equipment is described and the types of equipment to be controlled are established. The staff should verify that QA and other organizations' responsibilities are described for establishing, implementing, and ensuring effectiveness of the calibration and adjustment program. The staff should verify that procedures will be established for calibration (technique and frequency), maintenance, and control of the measuring and test equipment. The staff should also verify that the review of and documented concurrence in these procedures are described, and the organization responsible for these functions is identified. The staff should further verify that measuring and test equipment are identified and traceable to the calibration test data. The staff should verify that measuring and test equipment will be labeled or tagged or "otherwise controlled" to indicate due date of the next calibration. The method to "otherwise control" equipment should be described. The staff should verify that measuring and test equipment will be calibrated at specified intervals on the basis of the required accuracy, purpose, degree of usage, stability characteristics, and other conditions affecting the measurement.

17.6.5 CORRECTIVE ACTION

The staff should review the licensee's QA program to ensure that measures have been established to assure that conditions adverse to quality are promptly identified and corrected.

ACCEPTANCE CRITERIA

Regulatory Requirements

10 CFR 30.36(g)(4)(ii), 40.42(g)(4)(ii),40.28(b)(3), 70.22(f), 70.38(g)(4)(ii), and 72.54(g)(6)

Information to be Submitted

The information supplied by the licensee should be sufficient to allow the staff to determine whether adequate procedures and controls are in place to identify and correct conditions that will adversely affect quality. The staff's review should verify that the following information is included in the description of the corrective action program portion of the QA program:

- a description of the corrective action procedures for the facility, including a description of how the corrective action is determined to be adequate; and

- a description of the documentation maintained for each corrective action and any follow-up activities by the QA organization, after the corrective action is implemented.

EVALUATION FINDINGS

Evaluation Criteria

NRC staff should verify that the information summarized under "Information to be Submitted," above, is included in the licensee's description of the corrective action. The staff should verify that procedures will be established for a corrective action program and that the QA organization reviews and documents concurrence in the procedures. The staff should verify that corrective action will be documented and initiated following the determination of a condition adverse to quality (such as nonconformance, failure, malfunction, deficiency, deviation, and defective material and equipment) to preclude recurrence. The staff should verify that the QA organization will be included in the concurrence chain regarding the adequacy of the corrective action. The staff should verify that follow-up action will be taken by the QA organization to verify the proper implementation of corrective action and to close out the corrective action in a timely manner. The staff should verify that significant conditions adverse to quality, the cause of the conditions, and the corrective action taken to preclude repetition will be documented and reported to immediate management and upper levels of management for review and assessment.

17.6.6 QUALITY ASSURANCE RECORDS

The purpose of the review of the QA records program is to verify that the licensee has procedures and facilities in place to adequately maintain and store the QA program records.

ACCEPTANCE CRITERIA

Regulatory Requirements

10 CFR 30.36(g)(4)(ii), 40.42(g)(4)(ii),40.28(b)(3), 70.22(f), 70.38(g)(4)(ii), and 72.54(g)(6)

Information to be Submitted

The information supplied by the licensee should be sufficient to allow the staff to fully understand the types of procedures that will be in place to manage the QA program records. The staff should verify that the following information is included in the description of the QA records program:

- a description of the manner in which the QA records will be managed;

- a description of the responsibilities of the QA organization as well as all other units involved in the decommissioning to implement and maintain QA records; and

- a description of the QA records storage facility.

EVALUATION FINDINGS

Evaluation Criteria

The staff should verify that the information summarized under "Information to be Submitted," above, is included in the licensee's description of the QA records program. The staff should verify that the QA records program is described, and includes results of reviews, inspections, tests, audits, and material analyses; monitoring records of work performance; and records on the qualification of personnel, procedures, and equipment. The staff should verify that QA and other organizations are identified and their responsibilities are described for the definition and implementation of activities related to QA records. The staff should verify that suitable facilities for the storage of records are described and satisfy the requirements of ANSI/ASME NQA–1. Alternatives to the fire protection rating provisions are acceptable if record storage facilities conform to National Fire Protection Association Standard NFPA 232, Class 1, for permanent records and if the 2-hour fire-rating requirement contained in proposed ANSI N45.2.9 is met by the applicant in any one of the following three ways: (1) a 2-hour-rated vault meeting NFPA 232, (2) 2-hour-rated file containers meeting NFPA 232 (Class B), or (3) a 2-hour-rated fire resistant file room meeting NFPA 232.

17.6.7 AUDITS AND SURVEILLANCE

The purpose of the staff's review of the licensee's description of audits and surveillances is to ensure that the licensee has a comprehensive system of audits planned to verify compliance with all aspects of the QA program, and to determine the effectiveness of the QA program.

ACCEPTANCE CRITERIA

Regulatory Requirements

10 CFR 30.36(g)(4)(ii), 40.42(g)(4)(ii),40.28(b)(3), 70.22(f), 70.38(g)(4)(ii), and 72.54(g)(6)

Information Criteria

The information supplied by the licensee's should be sufficient to allow the staff to determine if the audit and surveillance program is adequate to ensure that a comprehensive system of audits planned to verify compliance with all aspects of the QA program is in place to determine the effectiveness of the QA program. The following information should be included in the description of the audit program:

- a description of the audit program, including the procedures for conducting the audits or surveillances;

- a description of the records and documentation generated during the audits and the manner in which the documents are managed;

- a description of all followup activities associated with audits or surveillances; and

- a description of the trending/tracking that will be performed on the results of audits and surveillances.

EVALUATION FINDINGS

Evaluation Criteria

The staff should verify that the information summarized under "Information to be Submitted," above, is included in the licensee's description of the audits program for the facility. The staff should verify that audits and surveillances will be performed in accordance with pre-established written procedures or checklists and conducted by trained personnel not having direct responsibilities for the achievement of quality in the areas being audited. The staff should verify that audit and surveillance results will be documented and then reviewed with management having responsibility in the area audited. The staff should verify that provisions exist such that appropriate follow-up corrective action to audit and surveillance reports will be undertaken by responsible management and that auditing organizations schedule and conduct appropriate follow-up to assure that the corrective action is effectively accomplished. The staff should verify that both technical and QA programmatic audits and surveillances will be performed to provide a comprehensive independent verification and evaluation of procedures and activities affecting quality. The staff should verify that audits and surveillances objectively assess the effectiveness and proper implementation of the QA program and address the technical adequacy of the activities being conducted. The staff should verify that provisions will be provided such that audits and surveillances are required to be performed in all areas where the requirements of the QA program are applicable. The staff should verify that audit and surveillance deficiency data are analyzed and trended. The staff should verify that reports that indicating quality trends and the effectiveness of the QA programs will be given to management for review, assessment, corrective action, and follow up.

17.7   RESTRICTED USE

17.7.1   OVERVIEW

NRC staff should review the information supplied by the licensee to determine if the description of the activities undertaken by the licensee is adequate to allow the staff to conclude that the licensee has complied with the requirements of 10 CFR 20.1403 for those licensees who intend to request termination of their radioactive materials licenses using the restricted use provisions of 10 CFR Part 20, Subpart E.

If the licensee is requesting license termination under restricted use in 10 CFR 20.1403, this information should include: a demonstration that the licensee qualifies for license termination under 10 CFR 20.1403(a); a description of the institutional controls the licensee has instituted or plans to institute at the site; a description of the activities undertaken by the licensee to obtain advice from the public on the proposed institutional controls and the results of these activities; a demonstration that the potential doses from residual radioactive material at the site will not exceed the limits in 10 CFR 20.1403 and are ALARA; and a description of the amount and mechanism for financial assurance required under 10 CFR 20.1403(c).

The LTR established a system of controls to sustain protection at restricted use or alternate criteria sites. This approach is described in Appendix M. The total system includes the following six elements: (1) legally enforceable institutional controls; (2) engineered barriers; (3) monitoring and maintenance; (4) independent third party oversight; (5) sufficient funding; and (6) maximum limits on dose (i.e., "dose caps") if institutional controls fail. While elements 1, 3, 4, 5, and 6 are required by the LTR, element 2 (engineered barriers) is not required but could be used to mitigate adverse processes (e.g., infiltration or erosion) so that the dose criteria of the LTR can be met (see Section 3.5 of Volume 2 of this NUREG report). The licensee should describe how it proposes to apply the total system approach to its specific site.

The licensee should describe how it has used the risk-informed graded approach (described in Appendix M of this Volume and Section 3.5 of Volume 2 of this NUREG report) to select the appropriate institutional controls and engineered barriers for decommissioning under restricted use or alternate criteria, so that restrictions and engineered barriers are most effectively targeted and are based on duration and magnitude of the hazard. This approach is flexible and uses risk insights from dose assessments to tailor site-specific restrictions and engineered barriers that would prevent potential disruptive land uses or natural processes important to compliance with the dose criteria. Appendix M also describes how institutional controls combine with other elements, such as engineered barriers, to form a total system to sustain protection.

If a licensee cannot establish acceptable institutional controls or independent third party arrangements, the licensee may propose one of the two new options involving NRC: an NRC long-term control (LTC) license or an NRC legal agreement and restrictive covenant (LA/RC). Both of these options are described in Appendix M of this Volume and are summarized below. These options are new types of legally enforceable and durable institutional controls established by Commission policy (see SECY-03-0069). These options are not for the purpose of storage of radioactive materials; they are to serve as an institutional control mechanism for restricted use decommissioning. These options should not be considered a guaranteed option for decommissioning, but would be used as a last resort for those sites that could not decommission to unrestricted use levels and could not arrange for other institutional controls. Therefore, these options should not encourage or lead to the proliferation of restricted use sites. In addition, for both of these options, all the restricted use requirements of the LTR must be met, to ensure protection of the public health and safety. Furthermore, NRC is taking measures to prevent future decommissioning problem sites (including reducing the number of future restricted use sites) by considering changes to financial assurance requirements and licensee operations, as described in SECY-03-0069 and RIS-2004-08.

The LTC license option is a possession-only license that would be used to satisfy the LTR requirement for legally enforceable and durable (if needed) institutional controls. The conditions of the LTC license would require the licensee to maintain restrictions on site use and any necessary monitoring, maintenance, and reporting. NRC would use inspections and enforcement, if needed, to assure that the licensee's controls and other activities are effective.

The LA/RC option is a combination of a legal agreement and restrictive covenant that provides a legally enforceable and durable institutional control, with the NRC having an oversight role. Under the LA/RC option, the current licensee or site owner and NRC enter into a legal agreement on the restrictions and controls needed for license termination under restricted conditions. The legal agreement includes using a restrictive covenant, which outlines the restrictions on site use and any necessary maintenance, monitoring, or reporting. In accordance with the legal agreement, the licensee or site owner is required to record the restrictive covenant with the appropriate recordation body in the jurisdiction where the site is located, before the site is released under restricted conditions.

It is noted that the LA/RC option has not been implemented by the NRC or legally tested, and NRC's ability to enforce the LA/RC depends on the laws of the jurisdiction where the site is located. Therefore, the licensee must demonstrate that the LA/RC is a legally enforceable institutional control in the jurisdiction where the site is located.

If complex monitoring or maintenance activities are needed at a restricted use site, the LTC license could be an appropriate institutional control option (compared to the LA/RC). Under the LTC license option, NRC would need to review and approve the transfer of site ownership (including determining whether the new owner had the technical capability or means to conduct any complex monitoring or maintenance activities), as the new owner(s) would become the LTC licensee to provide the necessary controls outlined in the LTC license. Complex monitoring or maintenance activities could include maintenance of an engineered barrier or groundwater or radiological monitoring activities, which require the site owner to have necessary knowledge, expertise, or technical abilities to carry out these activities and comply with all provisions of the LTC license. If the restrictions on site use or monitoring and reporting activities are simpler, the LA/RC may be an appropriate institutional control option. Simpler restrictions or activities related to the restrictions could include the site owner responding to an annual NRC inquiry as to how the site is being used or allowing the NRC to conduct a periodic inspection of the site.

Figure 17.1 illustrates the process for selecting appropriate institutional controls (including institutional control options involving NRC) for restricted use or alternate criteria decommissioning. Refer to Section 17.8 of this volume for guidance on decommissioning using the alternate criteria provisions in 10 CFR 20.1404. The steps in this process are described below.

1. Determine if the site is a lower or higher risk site, based on the following criteria. (See also Table M.1 and Section M.2 of this volume.)

    Lower risk sites (legally enforceable institutional controls):

    > Shorter hazard duration: shorter dose persistence or shorter radionuclide half-life (less than 100 years).

    > Lower hazard level: calculated dose is less than 1.0 mSv/y (100 mrem/y) assuming institutional controls are not in place.

    Higher risk sites (legally enforceable and durable institutional controls):

    > Longer hazard duration: longer dose persistence or longer radionuclide half-life (more than 100 years).

    > Higher hazard level: calculated dose is 1.0–5.0 mSv/y (100–500 mrem/y) assuming institutional controls are not in place.

2. For lower risk sites, determine if legally enforceable institutional controls and independent third party arrangements can be established.

    Yes: Licensee can arrange for appropriate legally enforceable institutional controls and independent third party arrangements. The licensee must comply with all LTR requirements in 10 CFR 20.1403 or 10 CFR 20.1404, for license termination under restricted conditions or alternate criteria, respectively.

    No: Licensee should document that it could not arrange for legally enforceable institutional controls or independent third party arrangmements and may go to step 4 to select institutional control options involving NRC. The institutional control options involving NRC, which provide legally enforceable and durable institutional controls, may be used, even though durable controls are not necessary for lower risk sites.

3. For higher risk sites, determine if legally enforceable and durable institutional controls, via government (Agency other than NRC) ownership or control, and independent third party arrangements can be established.

   Yes: Licensee can arrange for legally enforceable and durable institutional controls and independent third party arrangements. The licensee must comply with all LTR requirements in 10 CFR 20.1403 or 10 CFR 20.1404, for license termination under restricted conditions or alternate criteria, respectively.

   No: Licensee should submit a letter from Federal, State, or local government declining to take ownership or control responsibility and may go to step 4 to select institutional control options involving NRC.

4. Determine the appropriate NRC option for providing an institutional control.

   LTC license option could be used if:

   — Substantial restrictions, monitoring, or maintenance require special expertise (e.g., groundwater monitoring or maintaining an engineered barrier); or

   — Maintaining single ownership of a site with both restricted use and unrestricted use areas is desirable to sustain future ownership to ensure long-term protection of public health and safety.

   LA/RC option could be used if:

   — Current licensee or formerly licensed site owner demonstrates that the LA/RC option would be effective and legally enforceable by NRC in the jurisdiction where the site is located; and

   — Restrictions, monitoring, or maintenance activities are simple and do not require special expertise (e.g., annual letter certifying restrictions are in place, fence repair, or sign replacement).

   Both the LTC license and LA/RC institutional control options provide legally enforceable and durable (if needed) institutional controls. The licensee must comply with all LTR requirements in 10 CFR 20.1403 or 10 CFR 20.1404, for license termination under restricted conditions or alternate criteria, respectively.

Demonstrate LTR compliance in the DP: After the above steps are completed and an institutional control option is selected, the licensee would incorporate the option into its DP and demonstrate compliance with all other applicable LTR requirements in 10 CFR 20.1403 or 10 CFR 20.1404, for license termination under restricted conditions or alternate criteria, respectively.

Before preparation and submittal of the DP to NRC, licensees are encouraged to contact NRC or the appropriate Agreement State authority to discuss the selection of appropriate institutional controls (including institutional control options involving NRC) for restricted use or alternate criteria decommissioning.

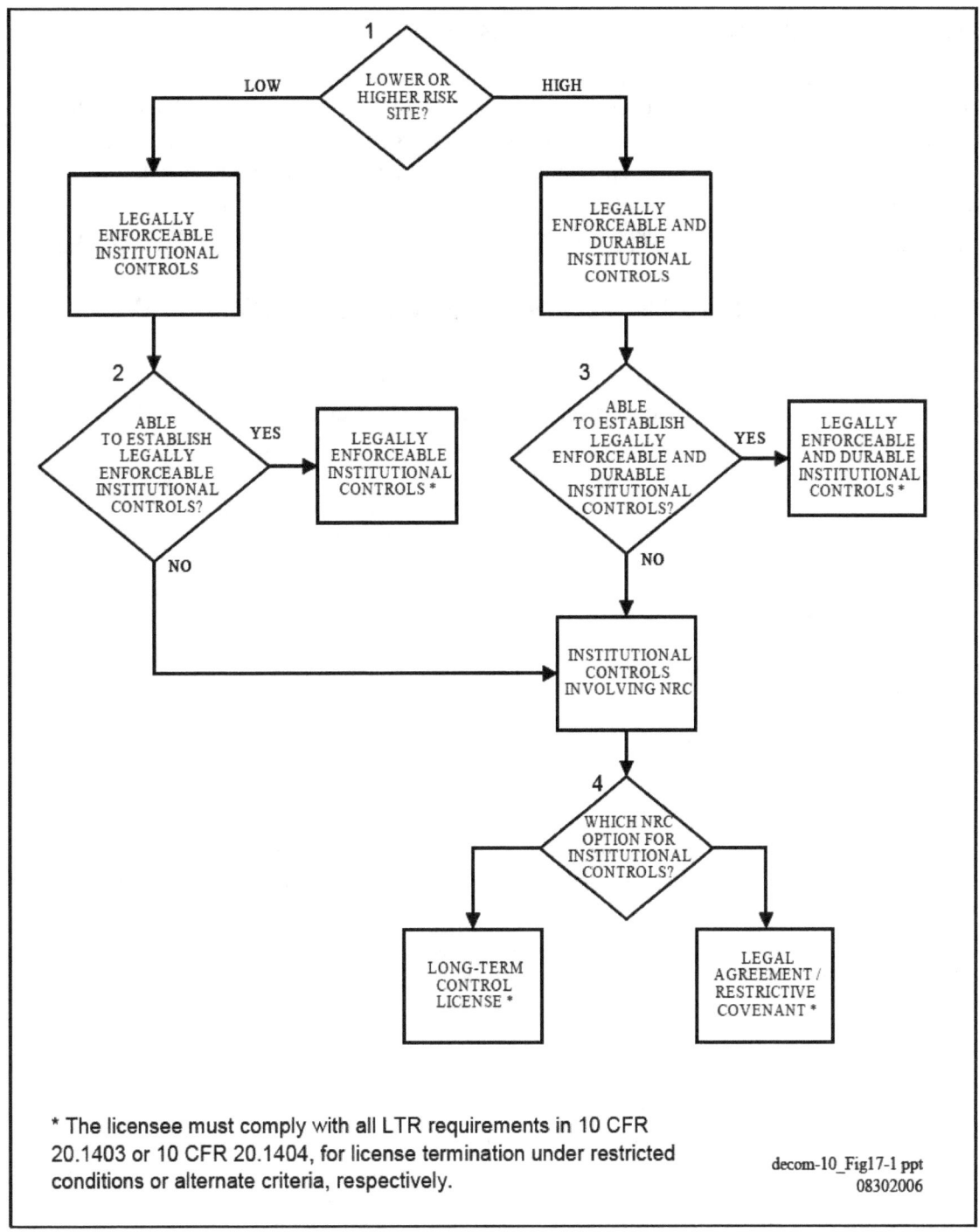

Figure 17.1   Process for Selecting Institutional Controls.

17.7.2   INITIAL ELIGIBILITY DEMONSTRATION

> The purpose of the review of the licensee's demonstration that it is initially eligible to further evaluate release of the site, under the provisions of 10 CFR 20.1403, is to verify that the licensee has demonstrated that further reductions in residual radioactivity at the site to meet the unrestricted release criteria in 10 CFR 20.1402 would: (1) result in net public or environmental harm; or (2) are not being undertaken because the residual radioactivity levels are ALARA.

ACCEPTANCE CRITERIA: INFORMATION TO BE SUBMITTED

The information supplied by the licensee should be sufficient to allow the staff to fully understand how the licensee has concluded that reducing radioactivity to the unrestricted use levels in 10 CFR 20.1402 would result in net public or environmental harm or are not being undertaken because the residual radioactivity levels are ALARA. The staff's review should verify that the following information is included in the licensee's demonstration that it is eligible for requesting license termination under the provisions of 10 CFR 20.1403:

- A demonstration that the benefits of dose reduction are less than the cost of doses, injuries, and fatalities (see Volume 2 of this NUREG series); or

- A demonstration that the proposed residual radioactivity levels at the site are ALARA.

EVALUATION FINDINGS: EVALUATION CRITERIA

If the licensee has concluded that further reductions in residual radioactivity levels would result in net public or environmental harm, the staff should verify that the licensee has accurately calculated the benefits versus costs of further remediation using the guidance in Chapter 6 and Appendix N of Volume 2 of this NUREG series. In considering the net public and environmental harm, a licensee's evaluation should consider the radiological and nonradiological impacts of decommissioning on a person that may be impacted, as well as the potential impact on ecological systems from decommissioning activities. (See Section B.3.2 of the "Statements of Consideration" for the License Termination Rule, 62 FR 39069.)

If the licensee has concluded that further reductions in residual radioactivity levels are not required because they are ALARA, the staff should verify that the licensee has considered all of the applicable benefits and costs of further reduction of residual radioactivity and accurately calculated the benefits and costs using the methodology described in Chapter 6 and Appendix N of Volume 2 of this NUREG series.

17.7.3   INSTITUTIONAL CONTROLS AND ENGINEERED BARRIERS

The purpose of the review of the description of the institutional controls and engineered barriers the licensee has provided for the site is to determine if the licensee has made provisions for legally enforceable institutional controls that will limit the dose to the average member of the critical group to less than 0.25 mSv/y (25 mrem/y).

ACCEPTANCE CRITERIA: INFORMATION TO BE SUBMITTED

The information supplied by the licensee should be sufficient to allow the staff to fully understand what institutional controls and engineered barriers the licensee plans to use or has provided for the site and the manner in which these institutional controls will limit doses to the average member of the critical group to 0.25 mSv/y (25 mrem/y).

In the NRC's view, engineered barriers are distinct and separate from institutional controls (NRC 2002). Used in the general sense, an engineered barrier could be one of a broad range of barriers with varying degrees of durability, robustness, and isolation capability. Generally, engineered barriers are passive, man-made structures or devices intended to enhance a facility's ability to meet the dose criteria in the LTR. Engineered barriers are usually designed to inhibit water from contacting waste, limit releases of radionuclides (e.g., through groundwater, biointrusion, erosion), or to mitigate doses for inadvertent intruders. Institutional controls are used to limit inadvertent intruder access to, and/or use of, the site to ensure that the exposure from the residual radioactivity does not exceed the established criteria. Institutional controls include legal mechanisms (e.g., land use restrictions) and may include, but are not limited to, physical controls (e.g., signs, markers, landscaping, and fences) to control access to the site and minimize disturbances to engineered barriers.

NRC reviewers and licensees should refer to Section 3.5 of Volume 2 of this NUREG series for guidance on engineered barriers. Section 3.5 provides guidance on the engineered barrier analysis process, including analysis of: (a) contribution of engineered barriers towards compliance, with institutional controls in place, including monitoring and maintenance; and (b) contribution of engineered barriers toward compliance, assuming loss of institutional controls (including monitoring and maintenance) and degradation of barriers. The guidance also discusses the main elements that support the assessment of engineered barrier performance, including: (a) design and functionality of engineered barriers, including interactive effects (both positive and negative) from the implementation of multiple barriers; (b) technical basis for design and functionality of engineered barriers; (c) degradation mechanisms and sensitivity analysis; (d) uncertainty in design and functionality of engineered barriers; (e) suitability of numerical models; (f) model support; and (g) quality assurance.

The licensee should summarize the total system of controls used to provide protection and include a general description of each system element and how it contributes to protection. Elements might include institutional controls, engineered barriers, monitoring and maintenance; independent third party oversight; trust fund; and maximum limits on dose (i.e., "dose caps") assuming institutional controls fail. Refer to Appendix M of this Volume, which describes this total system approach and apply the approach for the specific site. The staff's review should verify that the following information is included in the description of institutional controls that the licensee plans to use or has provided for the site:

Area and Type of Institutional Controls

- Area and description of the general type of institutional controls and the basis for selection, using NRC's risk-informed graded approach in Appendix M of this Volume. Using this approach, determine if the restricted area of the site is a lower or higher risk area and the general type of institutional controls that are needed. Consider both hazard duration, based on the dose persistance and the half-life of radionuclides, as well as hazard level [i.e., less than or greater than 1.0 mSv/y (100 mrem/y)] based on dose assessments assuming controls are no longer in effect. For a simpler site, the complete area of the current site would be restricted use. For a more complex site, this approach might result in identifying unrestricted use areas where no institutional controls are required, and restricted use areas using either legally enforceable institutional controls or durable and legally enforceable institutional controls.

- A demonstration that the size of the restricted use area has been minimized. The staff considers that minimizing the size of the restricted use area would contribute to demonstrating ALARA for sites that are considering subdividing the site into unrestricted and restricted use portions. It would also result in a smaller area to control, which may make access limitations like fencing and surveillance simpler and thus more effective.

  When determining what portion of the site needs to be restricted (or how a site could be subdivided between restricted and unrestricted portions), the licensee should consider and balance the goal of minimizing the restricted area of the site (and minimizing burdens associated with restricted use) with defining the area that will ensure long-term protection. Risk insights from dose assessments (extent of residual radioactivity and how it migrates/travels through the environment) will help determine what areas need to be monitored and the location and size of the restricted area.

  For a licensee considering an LTC license that has both restricted and unrestricted use areas of its site, the licensee should evaluate and choose whether to: (a) keep both restricted use and unrestricted use areas together under single ownership and LTC license; or (b) release the unrestricted use areas of the site while maintaining the restricted use areas under the LTC license. The licensee should consider and demonstrate how its choice will enhance long-term protection of public health and safety, through maintenance of the site controls and restrictions. For example, for a privately owned site, where the restricted use area has little or no resale value but the unrestricted use area does have resale value, keeping both areas together under an LTC license could maintain the value for the entire site and thus sustain future ownership. However, keeping both areas together could result in undue burdens at some sites. Thus, both benefits and burdens should be evaluated, considering the views of affected parties. See Appendix M of this volume for further discussion of this choice and the flexibility for partial restricted release (i.e., release of unrestricted use areas while maintaining restricted use areas under an LTC license). The licensee should show the boundaries of both the restricted use area(s) and the unrestricted use area(s) in its DP.

- A description of the specific type of legally enforceable institutional control(s) and an explanation of how the institutional control is a legally enforceable mechanism.

  If such controls cannot be arranged, provide justification for appropriateness of using either the LTC license or LA/RC options:

  — durable institutional controls are required; and

  — licensee was unable to establish other types of legally enforceable institutional controls or independent third party arrangements (e.g., letter from the State declining responsibility).

  For the LTC license, state that two specific types of institutional controls would be used. First, describe that the NRC LTC license is considered to be a specific type of legally enforceable and durable institutional control. Second, describe the licensee's responsibility to put in place and maintain a deed notice that notifies potential landowners of the LTC license requirement and the conditions of the LTC license.

  Use the following criteria to decide when the LTC license or LA/RC would be appropriate:

  LTC license could be used if:

  — Substantial restrictions, monitoring, or maintenance require special expertise (e.g., groundwater monitoring, use of an engineered barrier);

  — Maintaining single ownership of a site with both restricted use and unrestricted use areas is desirable to sustain future ownership to ensure long-term protection of public health and safety.

  LA/RC option could be used if:

  — Current licensee or formerly licensed site owner demonstrates that the LA/RC option would be effective and legally enforceable by NRC in the jurisdiction where the site is located; and

  — Restrictions, monitoring, or maintenance activities are simple and do not require maintaining special expertise (e.g., annual letter certifying restrictions are in place, fence repair, sign replacement).

Restrictions and Controls Implemented by Licensee

- A description of the restrictions on present and future landowners;

  Describe the access and land use restrictions, based on the dose assessments assuming no controls. Identify specific access and land use scenarios that would lead to non-compliance with the dose criteria of the LTR and therefore should be prohibited (e.g., farming, construction of a residence, excavation into the cell for any purpose, or groundwater use).

Indicate what access and land uses might be permitted (e.g., industrial, recreational, or wildlife conservation area).

Describe what restrictions on land use would be needed to maintain effective engineered barrier performance (e.g., prohibit excavation of the cell cap and removal of cell cap material or contaminated material), as well as permitted access and land use.

Describe the licensee's activities to restrict/control access and land use, including use of fences, signs, monuments, and periodic surveillance (e.g., annual site surveillance and adverse event surveillance).

If the LTC license or LA/RC is used, all of the above should be conditions in the LTC license or LA/RC. For the LTC license option, the licensee should prepare a Long-Term Control Plan that describes the details of how the licensee will implement the LTC license conditions.

- A discussion of the durability of the institutional control(s);

  Explain how the controls selected are durable based on the risk-informed graded approach.

  Note that NRC considers both the LTC license and LA/RC to be durable institutional controls.

Duration of the Institutional Controls

- A description of the duration of the institutional control(s), the basis for the duration, the conditions that will end the institutional control(s) and the activities that will be undertaken to end the institutional control(s).

  For the LTC license and LA/RC, discuss that the duration of these controls is as long as needed, but could be permanent for a site with long half-life radionuclides, such as uranium and thorium. However, the LTC license would be renewed at least every 5 years for the purpose of evaluating continued effectiveness and maintaining institutional and public awareness and information transfer. The appropriate renewal time period would be determined when establishing the LTC license for a specific site and could be adjusted in the future.

  Under the LTC license, further flexibility is provided for a licensee to request, in the future, approval for removing the residual radioactivity, terminating the license, and releasing the site for unrestricted use. For this approach, a licensee would submit a decommissioning plan for NRC review, as is currently done, and decommission the site in accordance with NRC's decommissioning regulations. NRC would assure that the site was properly decommissioned and suitable for unrestricted release, before terminating the LTC license. NRC would also allow a request to terminate the LTC license and release the site with restrictions using another acceptable type of legally enforceable or, if needed, durable institutional control and independent third party arrangement, if approved by NRC.

Records Retention and Availability

- A description of the records pertaining to the institutional controls, how and where will they will be maintained, and how the public will have access to the records.

    For the LTC license, identify both historical and new records to be retained by the licensee that are necessary for the licensee to provide effective long-term protection. This includes the decommissioning plan, final status survey report, LTC license, long-term control plan, and all correspondence under the LTC license. Identify the location and methods used for retention of records by the licensee. Note that NRC will retain all licensing records as part of its agency recordkeeping system and that these records will be available to the public in the future, as they are today.

    For the LA/RC, identify both historical and new records to be retained by the site owner that are necessary for the site owner to provide effective long-term protection, including the decommissioning plan, final status survey report, legal agreement (site owner at time of license termination), restrictive covenant, and correspondence between NRC and the site owner. Note that NRC has the primary responsibility for maintaining records and making those available to the public, as part of its Agency recordkeeping system. In accordance with the legal agreement, the licensee or site owner would be required to record the restrictive covenant with the appropriate recordation body responsible for maintaining records related to land ownership (e.g., Registrar of Deeds) in the jurisdiction where the site is located. These recordkeeping responsibilities should be outlined in the legal agreement and restrictive covenant.

Detriments Associated With Institutional Controls

- A description of any detriments or potential drawbacks associated with the maintenance of the institutional control(s). Detriments could result from restricted use of the land, independent of the type of legal instrument used for institutional controls. Include any applicable stakeholder inputs or advice, if provided.

    For the LTC license, describe any detriments to using the LTC license option. For example, describe potential impacts on sale of property or value of property due to the NRC license or perceptions that NRC could potentially require further cleanup in the future (i.e., lack of finality).

EVALUATION FINDINGS: EVALUATION CRITERIA

The staff should determine whether the information summarized in "Information to be Submitted," above satisfies the criteria summarized below. The application of the criteria below is dependent on the circumstances of the case. In each case, the staff should consult with the Office of the General Counsel on the application of the criteria and the sufficiency of the licensee's proposal.

A. For legally enforceable institutional controls on privately owned land:

Proprietary institutional controls on privately owned land, including LA/RC, should:

- Be enforceable against any owner of the affected property and any person that subsequently acquires the property or acquires any rights to use the property.

- Be enforceable by entities, other than the landowner, that have the legal authority to enforce the restriction. For LA/RC, the legal authority to enforce the restriction would be with the NRC.

- Be developed based on considerations of how durable the controls need to be.

- Include provisions to replace the entity with authority to enforce the restriction.

- Indicate actions the entity with authority to enforce the restrictions may take.

- Remain in place for the duration of the time they are needed.

- Have appropriate funds set aside.

- Be appropriately recorded, including in the deed and in land records, as appropriate.

- Include a legal opinion by an attorney specializing in real estate law, who is knowledgeable in the particular State and local land use laws, that demonstrates:

  — The property law of the particular State and locality in which the land is located ensures that the particular instrument selected will accomplish its intended purpose.

  — The restrictions have been reviewed and their validity affirmed for the locality.

  — The owner of the affected property (i.e., the possessor of the land) can be compelled to abide by the terms of the land use restriction.

  — The restrictions are binding on future owners (possessors) of the land (i.e., they should "run with the land").

- Include a legal opinion that the entity with the right to restrict the land's use and the responsibility to enforce the restriction has the legal authority to do so and is someone other than the owner or possessor of the land in question.

- Include a demonstration that the entity (or entities) with authority to enforce the restrictions have the knowledge, capability, and willingness to do so, and are appropriate for the specific situation.

- Include a demonstration that the institutional control is durable enough to provide an adequate level of protection of public health and safety and the environment for the amount of residual radioactivity remaining on the site. Use the risk-informed graded approach described in Appendix M of this Volume.

- Include a provision to replace the entity with authority to enforce the restriction if that entity is no longer willing or able to enforce the restriction.

- Clearly state the actions that the parties with authority to enforce the restrictions may take to keep the restrictions functioning (e.g., monitoring of deed compliance, control and maintenance of physical barriers).

- Include a demonstration that the restrictions will remain in place for the duration that they are needed, including periodic re-recording of the restrictions.

- If restrictions will end, the conditions that would end the restriction must be clearly stated, and the procedures for canceling or amending the restriction should be readily available. There should be no provisions in the restriction or in the land use law of the local jurisdiction that would cause the restrictions to end while they are still needed to protect the public.

- Identify corrective actions to be taken in case the restrictions need to be broken. For example, a no-excavation restriction may need to be broken if a water main under the site bursts and must be repaired.

- Include a demonstration that the information about restrictions is recorded in the deed and in land records and will contain:

  — a legal description of the property affected;

  — the name or names of the current owner or owners of the property as reflected in public land records;

  — identification of the parties that can enforce the restriction (i.e., own the rights to restrict use of the land);

  — the reason for the restriction, the nature of the radiation hazard, including the estimated dose if institutional controls fail, and that this restriction is established as a condition of license termination by NRC pursuant to 10 CFR 20.1403;

  — a statement describing the nature of the restriction, limitation, or control created by the restriction;

  — the duration of the restriction;

  — permission to install and maintain physical controls, if any are used; and,

  — the location of copies of the important records related to the decommissioning of the site and license termination under restricted conditions.

- For LA/RC, identify the reasons that the LA/RC is an appropriate option for institutional controls, given the criteria in "Area and Type of Institutional Controls" in Section 17.7.3.

B. For legally enforceable institutional controls on government owned land:

NRC may accept government ownership of land as a method to enforce controls on land use and to meet the legally enforceable institutional control requirements in 10 CFR 20.1403(b) and (e). Government ownership will generally be acceptable when the dose to an average member of the critical group could exceed 1.0 mSv (100 mrem) per year (but be less than 5.0 mSv (500 mrem) per year) if the institutional controls were no longer in effect. In reviewing restrictions involving government ownership of land, NRC staff should ensure that the restriction will remain in place for the entire time they are needed and that the nature of the controls and restrictions on the land are clearly stated in a publicly available legal record. Depending on the government entity involved, consider as appropriate the items under part A, above.

C. For institutional controls based on sovereign or police powers:

Institutional controls that are based on sovereign or police powers generally consist of zoning or other restrictive requirements. The permissibility and effectiveness of governmental controls at a particular site will depend on the applicable State and local law.

Institutional controls based on sovereign or police powers should:

- Include a legal opinion by an attorney specializing in real estate law who is knowledgeable in the particular State and local land use laws that verifies the following:

  — Zoning and other restrictive requirements have been reviewed and their validity affirmed.

  — They are binding on present and future owners of the land.

- Include a demonstration that the government agency imposing the zoning or restriction will assume responsibility for enforcing the restriction.

- Include a demonstration that the restrictions will remain in place for the entire time that they are needed or the conditions that can cause them to be changed.

- Include a demonstration that the restrictions or zoning requirements are clear to current and future owners of the land, local and State governments, and others, as appropriate, through public documents, notification, placement in land records, and so forth. Such documentation should include an indication of the activities allowable and the residual radioactivity remaining onsite.

D. For institutional controls based on NRC LTC license:

For the LTC license, identify the reason that the LTC license is an appropriate option for durable institutional controls, given the criteria in Section 17.7.3.

17.7.4 SITE MAINTENANCE AND LONG-TERM MONITORING

The purpose of the review of the information about the licensee's long-term monitoring and site maintenance program is to ensure that: (1) adequate arrangements have been made to ensure that the site will be maintained in accordance with the institutional controls described above, and that (2) the licensee has an adequate arrangement to ensure that an independent third party can assume and carry out responsibilities for any necessary control, monitoring, and maintenance of the site after NRC has terminated the license. Criteria for evaluating the licensee's mechanism to ensure that sufficient funds are available to allow an independent third party to assume and carry out responsibilities for any necessary control, monitoring, and maintenance of the site after NRC has terminated the license, are addressed in Part II of Volume 3 of this NUREG series.

ACCEPTANCE CRITERIA: INFORMATION TO BE SUBMITTED

The information supplied by the licensee should be sufficient to allow the staff to fully understand what arrangements for long-term monitoring and site maintenance have been provided by the licensee. This should include descriptions of how the site maintenance arrangements will ensure that the site will be managed per the institutional controls described above and how an independent third party will assume and carry out responsibilities for any necessary control and maintenance of the site after NRC has terminated the license.

The licensee should describe the long-term monitoring and maintenance activities. Under the LTC license, these would be required by the license conditions. Note that a Long-Term Control Plan would be developed before license amendment for the LTC license option, which would include the detailed plans and procedures for restrictions on access and use, long-term monitoring, and maintenance. For the LA/RC option, any monitoring and maintenance activities would be required by the provisions of the LA/RC. Note that the LA/RC option should be used only if there are no complex monitoring and maintenance activities that would require expertise to carry out. The staff's review should verify that the following information is included in the discussion of the site maintenance program in the facility DP:

Long-term Monitoring

- A description and basis for the long-term monitoring program, using the risk-informed graded approach. This approach consists of combining the prohibited access and land uses that could lead to non-compliance (see Section 17.7.3) with the human and natural disruptive processes for engineered barriers (see Section 17.7.3 of this volume and Section 3.5.2 of Volume 2) to form one list of disruptive human and natural processes which could lead to non-compliance and should be the focus of monitoring and maintenance.

   For these disruptive processes, identify how each would be monitored, including type of monitoring (e.g., visual surveillance of fence integrity, visual surveillance for indicators of disruptive erosion such as gullies; radiological monitoring of groundwater or surface water; visual surveillance of disruptive vegetation intrusion into an engineered barrier/cover) and how each type would be used to detect indicators or precursors of the disruptive process, either in the environment surrounding the engineered barrier or the engineered barrier itself. Also, include and justify the location, frequency, and duration of monitoring. Duration of monitoring should be determined by considering several site-specific factors such as nature of disruptive processes; time needed to reduce uncertainty in barrier performance; and barrier degradation rates. Thus, duration and amount of monitoring should be risk-informed. For example, little or no long-term monitoring might be needed for lower risk sites, sites without disruptive processes important to compliance, or sites with low uncertainty in the engineered barrier performance over the time needed. In contrast, monitoring would be needed for higher risk sites, sites with disruptive processes, or if there is high uncertainty with engineered barrier performance.

Maintenance

- A description of the site maintenance program and the basis for concluding that the program is adequate to control and maintain the site.

    The following risk-informed approach should be used for determining the maintenance that is needed, which consists of identifying the disruptive process important to compliance, describing the maintenance that would provide corrective actions to mitigate the disruptive process, and how monitoring information would be used to identify the need and appropriate type of maintenance.

    The risk-informed approach for maintenance also would be applied to engineered barriers, if they are used. For higher risk sites where robust barriers are designed so their performance does not rely on active ongoing maintenance, monitoring and maintenance should still be planned, particularly, for disruptive events that could lead to non-compliance or where there is higher uncertainty. This approach provides added confidence through redundancy and defense-in-depth, as described in Appendix M for the total system.

    For the LTC license and LA/RC, the maintenance activities would be conditions of the LTC license or LA/RC. For the LTC license, the detailed plans and procedures to implement the conditions would be included in the Long-Term Control Plan.

- A demonstration that an appropriately qualified entity has been provided to control and maintain the site.

    Under the LTC license, the entity could be the licensee or a contractor to the licensee. Describe the qualifications of the personnel that are necessary to conduct the planned LTC activities.

- If the licensee plans on using a contractor, a description of the arrangement or contract with the entity charged with carrying out the actions necessary to maintain control at the site.

- If the licensee plans on using a contractor, a demonstration that the contract or arrangement will remain in effect for as long as feasible, and include provisions for renewing or replacing the contract.

- A description of the plans for corrective actions that may be undertaken in the event the institutional control(s) fail.

- A description of the plans for corrective actions that may be undertaken in the event the site maintenance and control program fails.

  Identify reasonably foreseeable events (e.g., forced entry through fences or disruption of cap material) that could cause a failure of access and land use controls. Describe the corrective actions the licensee would take and the requirement that NRC would be notified of the events and planned corrective actions.

- A description of licensee reporting to NRC and State and local officials, including an annual report and event corrective actions reports, as needed. The annual report should describe licensee surveillance and routine maintenance. Event corrective action reports would identify the adverse event that occurred and the licensee's planned corrective actions. Follow- up reports would include a summary of the results of the corrective actions taken, an analysis of lessons learned from the event, and plans to prevent similar future events from occurring.

Enforcing Institutional Controls

- A description of the entities enforcing, and their authority to enforce, the institutional control(s);

  For the LTC license, specify that NRC will have jurisdiction for oversight of licensee activities and can take enforcement actions, if needed, under its licensing authority from the AEA. NRC's general role under the LTC license is to assure that the controls are maintained and remain protective over time. Also note that NRC activities would include review, inspection, license renewal, and enforcement.

  For the LA/RC, specify that NRC is responsible for (1) assuring that the site owner is complying with the LA/RC, and (2) taking actions to enforce the LA and RC (e.g., legal action in the courts), if the conditions of these legal tools are not met. The LA/RC, when written for a specific site, would describe the methods and frequency in which NRC (as the enforcing party) would monitor the site to verify the effectiveness of controls. Outline how the NRC would enforce the restrictions, and if the LA or RC were breached, what steps NRC would take to restore these instruments (and the land use restrictions, monitoring, or reporting actions they contain).

- A description of the activities that the entity with the authority to enforce the institutional controls may undertake to enforce the institutional controls;

This is not applicable for the LTC license option. Under the LA/RC option, the legal agreement and restrictive covenant should outline the activities in which NRC may undertake to enforce the controls.

• A description of the manner in which independent oversight of the entity charged with maintaining the site will be conducted and what entity will conduct the oversight.

For the LTC license and LA/RC, the above item is not applicable, because NRC is the entity that will conduct the oversight.

• A description of the periodic site inspections that will be performed by the third party, including the frequency of the inspections.

This is not applicable for the LTC license option. Under the LA/RC option, the legal agreement and restrictive covenant should outline any necessary details of the periodic site inspections NRC will perform, including the frequency of the inspections.

• A description of the manner in which the entity with the authority to enforce the institutional control(s) will be replaced if that entity is no longer willing or able to enforce the institutional control(s) (this may not be needed for Federal or State entities);

For the LTC license and LA/RC, the above item is not applicable, because NRC is the enforcing party.

Sufficient Financial Assurance

For the purposes of a LTC license and LA/RC, "sufficient" financial assurance, pursuant to 10 CFR 20.1403(c), is an amount that will (1) enable an independent third party to assume and carry out responsibilities for any necessary control and maintenance of the site, (2) provide for trust fund expenses, (3) provide for NRC fees applicable to the site, and (4) provide a 25% contingency factor. The financial assurance instrument used will be a trust fund with sufficient capital to cover the cost estimate. The cost estimate, trust agreement, and the trustee must be approved by NRC.

To develop the cost estimate, refer to NUREG-1757, Volume 3, which contains guidance on developing the cost estimate for long-term site control and maintenance. Once the amount is estimated, the licensee must provide sufficient funds to produce an annual average income that covers the annual surveillance, control, and maintenance/repair costs, NRC fees, and trustee expenses. By analogy to uranium mill tailings funds, a 1% rate of return may be used by the licensee to determine the minimum funding level. This rate would contribute to the LTR requirement for sufficient funds for a site with long-lived radionuclides needing control over a long time period. It is also justified because the current licensee responsible for the contamination should fund the long-term control so that no additional costs will be passed on to future site owners/licensees.

The cost estimate should include costs for at least the following activities:

- site surveillance of access and land use restrictions;

- maintenance;

- radiological monitoring of surface and groundwater, if needed and non-radiological monitoring of processes

- reporting; and

- records retention.

For the LTC license, the cost estimate should also include NRC oversight fees. The fees given below are in 2005 dollars and should be adjusted for inflation. To adjust for inflation, use the ratio of the cost of professional staff hours found in 10 CFR 170.20. In 2005, the staff-hour cost for NMSS was $197 per hour. In 2005, NRC fees in 2005 dollars are as follows:

- a fee of $10,000 for one inspection and one report each year; and

- $20,000 every 5 years for 5 year license renewal, inspection, and report.

For the LA/RC option, the cost estimate should include the above NRC oversight fees for periodic inspections (at a frequency/interval based on site-specific considerations). The cost estimate should also include fees for NRC review of the property laws in the jurisdiction where the site is located at the time site ownership changes (or at least every five years), to assure that the local laws still support the enforceability of the restrictive covenant.

Finally, the estimate should include reasonable trustee fees and expenses.

NUREG-1757 Volume 3 provides for a contingency factor of 25% to be added to the cost estimate. This contingency should be retained to buffer against potential market losses and to provide for unexpected costs. If the contingency proves insufficient, the licensee should add funds to the trust. As a matter of fairness, particularly in light of the long term existence of the fund, if the balance substantially exceeds the amount needed to produce sufficient annual income, a provision, to return excess funds to the licensee with NRC's approval, should be included in the trust.

Under the LTC license, further flexibility is provided for a future licensee to request approval for removing the residual radioactivity, terminating the license, and releasing the site for unrestricted use. For this approach, a licensee would submit a decommissioning plan for NRC review, as is currently done, and decommission the site in accordance with NRC's decommissioning regulations. NRC would assure that the site was properly decommissioned and suitable for unrestricted release, before terminating the LTC license. The trust fund does not have to include sufficient funds to clean the site to unrestricted release; the future site owner would need to independently cover this cost. NRC would also allow a request to terminate the LTC license and release the site with restrictions using another acceptable type of legally enforceable and, if needed, durable institutional control and independent third party arrangement, if approved by NRC.

Independent Third Party

- a description of the authority granted to the third party (including NRC under the LTC license or LA/RC options) to perform, or have performed, any necessary maintenance activities;

- unless the entity is a government entity, a demonstration that the third party is not the entity holding the financial assurance mechanism (this is not applicable for the LTC license or LA/RC options as NRC is the beneficiary of the financial assurance mechanism);

- a demonstration that sufficient records evidencing to official actions and financial payments made by the third party (including NRC under the LTC license or LA/RC) are open to public inspection.

EVALUATION FINDINGS: EVALUATION CRITERIA

The staff should determine whether the information summarized under "Information to be Submitted," above satisfies the criteria summarized below. The application of the criteria below is dependent on the circumstances of the case. In each case, the staff should consult with the Office of the General Counsel on the application of the criteria and the sufficiency of the licensee's proposal.

The entity to control and maintain the site may be the former licensee, the landowner, a governmental agency, an organization, a corporation or company, or occasionally a private individual. Control and maintenance of a site does not necessarily have to be carried out by an independent third party. The entity should be capable of carrying out its responsibilities and should be appropriate given the nature of the restrictions in place. The entity could be a contractor to the entity that holds the rights to restrict use of the property. Note that government control and/or ownership is generally appropriate for higher risk sites involving large quantities of uranium and thorium contamination and for those sites where the potential dose to the public could exceed 1.0 mSv/y (100 mrem/y) if institutional controls fail. See Appendix M for the risk-informed graded approach.

The maintenance and control program includes detailed descriptions of: (1) the repair/replacement and maintenance program for the site; (2) if appropriate, an environmental monitoring program, including the duration of the monitoring, who will be informed of the results, action levels and what action will be taken if the action levels are exceeded; and (3) the mechanism to detect and mitigate the loss of site controls; the mechanism to, if necessary, inform local emergency responders of the loss of controls.

An arrangement or contract is in place to carry out any actions necessary to maintain the controls so that the annual dose to the average member of the critical group does not exceed 0.25 mSv (25 mrem). The arrangement or contract should be for as long a time as is feasible, and there should be provisions for renewing or replacing the contract to be consistent with the duration of the restrictions. The arrangement may include oversight of the entity by a government entity or the courts.

A mechanism is in place to replace the entity controlling/maintaining the site if that becomes necessary. Replacement may be specified in the agreement with the conditions under which a government, the courts, or other entity can replace the entity.

The entity is authorized to either perform the necessary work to maintain the controls or to contract for the performance of the work. The entity would need the authority to contract for the necessary work, review and approve the adequacy of the work performed, replace contractors if necessary, and authorize payment for the work.

The entity performing the site control and maintenance should not hold the funds itself; that is to say, the entity should not serve as the provider of financial assurance (e.g., escrow agent, trustee, issuer of letter of credit). However, if the entity is a government, the licensee may elect to allow the government to hold the funds.

A demonstration that sufficient records evidencing the official actions of and financial payments made by the entity are open to public inspection.

The entity has the responsibility to perform periodic rechecks of the site no less frequently than every 5 years [if required by 10 CFR 20.1403(e)(2)(iii)] to ensure that the institutional controls continue to function. The periodic rechecks should include an onsite inspection to verify that prohibited activities are not being conducted and that markers, notices, and other physical controls remain in place.

Under the LTC license option, NRC would review and renew the LTC license every five years. Under the LA/RC option, NRC would review the property laws in the jurisdiction where the site is located at the time site ownership changes (or at least every five years), to assure that the laws of the jurisdiction where the site is located still support the enforceability of the restrictive covenant.

17.7.5   OBTAINING PUBLIC ADVICE

The purpose of the review of the license's description of activities undertaken to obtain advice from the public on institutional controls is to determine if the advice of individuals and institutions in the community that may be affected by the decommissioning has been sought, evaluated, and as appropriate, incorporated into the licensee's decisions following an analysis of the advice.

ACCEPTANCE CRITERIA: INFORMATION TO BE SUBMITTED

The information supplied by the licensee should be sufficient to allow the staff to determine whether the licensee has adequately sought, managed, and, as appropriate, incorporated, advice from individuals and institutions that may be affected by the decommissioning alternative proposed by the licensee.

10 CFR 20.1403(d)(1) requires that licensees proposing to decommission a site by restricting use of the site shall seek advice from affected parties on whether:

- The provisions for institutional controls will provide reasonable assurance that the TEDE distinguishable from background radiation will not exceed 0.25 mSv/y (25 mrem/y).

- The provisions for institutional controls will be enforceable.

- The provisions for institutional controls will not impose an undue burden on the community or other affected parties.

- Sufficient financial assurance has been provided to allow an independent third party to carry out any necessary control and maintenance activities at the site after license termination.

The staff's review should verify that the following information is included in the discussion of how advice was sought, obtained, evaluated, and as appropriate, incorporated for each of the issues identified above:

- a description of how individuals and institutions that may be affected by the decommissioning were identified and informed of the opportunity to provide advice to the licensee;

- a description of the manner in which the licensee obtained advice from these individuals or institutions;

- a description of how the licensee provided for participation by a broad cross-section of community interests in obtaining the advice;

- a description of how the licensee provided for a comprehensive, collective discussion of the issues by the participants represented;

- a copy of the publicly available summary of the results of discussions, including individual viewpoints of the participants on the issues and the extent of agreement and disagreement among the participants;

- a description of how this summary has been made available to the public; and

- a description of how the licensee evaluated the advice, and the rationale for incorporating, or not incorporating, the advice from affected members of the community into the DP.

EVALUATION FINDINGS: EVALUATION CRITERIA

The staff should verify that the information summarized under "Information to be Submitted," above, is included in the licensee's description of how advice was solicited, obtained, evaluated and as appropriate, incorporated into the licensee's decisions and DP. The staff should verify that the manner in which advice was sought and obtained and the activities associated with obtaining this advice are consistent with the criteria in Sections 17.7.5. and M.6 of this volume.

17.7.6 DOSE MODELING AND ALARA DEMONSTRATION

The purpose of the review of the licensee's estimates of doses from the site after termination of the license to verify that the dose to the average member of the critical group will not exceed 0.25 mSv/y (25 mrem/y) with the institutional controls in place required by 10 CFR 20.1403(b) and that the doses are as low as reasonably achievable. The staff's review should also verify that, if institutional controls are no longer in place, there is reasonable assurance that the dose to the average member of the critical group from residual radioactive material at the site will not exceed either 1.0 mSv/y (100 mrem/y), or 5.0 mSv/y (500 mrem/y) required by 10 CFR 20.1403(e), provided that the licensee:

- demonstrates that further reductions in residual radioactivity necessary to comply with the 1.0 mSv/y (100 mrem/y) requirement are not technically achievable, would be prohibitively expensive, or would result in net public or environmental harm;

- makes provisions for durable institutional controls; and

- provides sufficient financial assurance to allow an independent third party to carry out rechecks at the site no less frequently than every five years and to assume and carry out responsibilities for any necessary control and maintenance of the controls at the site.

ACCEPTANCE CRITERIA: INFORMATION TO BE SUBMITTED

The information supplied by the licensee should be sufficient to allow the staff to determine whether the residual radioactive material at the site will not result in a TEDE that exceeds 0.25 mSv/y (25 mrem/y) with institutional controls in place and is ALARA, or that if institutional controls are no longer in place that there is reasonable assurance that the TEDE to the average member of the critical group will not exceed either 1.0 mSv/y (100 mrem/y) or 5.0 mSv/y (500 mrem/y), with conditions. The information should also demonstrate that the financial assurance mechanism(s) are adequate for the site (See Section 17.7.4). Finally, the information should be adequate to allow the staff to determine if the institutional controls and site maintenance activities are adequate.

In conducting dose assessments, the licensee should identify realistic exposure scenarios assuming past, present, and reasonably foreseeable (i.e., that are likely within the next 100 years) land uses (as described in Chapter 5 of Volume 2 of this NUREG series). Note that the 100 years described here is only a timeframe for estimating future land uses; the licensee must evaluate doses that could occur over the 1000-year time period specified in the LTR. Included in the assumption that the institutional controls are no longer in place, the licensee should assume that there is no maintenance and no repair of engineered barriers (if used), and as a result, should analyze how the engineered barrier might degrade over time, for example, due to erosion or biointrusion.

If a licensee proposes that a portion of its site be released for unrestricted use, then the total dose from all portions of the site must meet the applicable dose criteria. Therefore, dose assessments for both restricted and unrestricted use portions of the site are needed and also must take into consideration the impact of the other portion of the site — impacts of the restricted use portion on the unrestricted use portion (e.g., the potential for future contaminated groundwater to migrate into the unrestricted area) and impacts of the unrestricted portion on the restricted use portion.

The staff's review should verify that the following information is included in the dose modeling/ALARA demonstration subsection of the restricted use section of the DP:

- a summary of the dose to the average member of the critical group with institutional controls in place, as well as the estimated doses if they are no longer in place;

- a summary of the evaluation performed pursuant to Chapter 6 and Appendix N in Volume 2 of this NUREG series demonstrating that these doses are ALARA. ALARA analyses should also use the more realistic scenario approach to identify reasonably foreseeable land uses;

- if the estimated dose to the average member of the critical group could exceed 1.0 mSv/y (100 mrem/y), but would be less than 5.0 mSv/y (500 mrem/y):

  — a demonstration that further reductions in residual radioactivity necessary to comply with the 1.0 mSv/y (100 mrem/y) requirement are not technically achievable, would be prohibitively expensive, or would result in net public or environmental harm;

  — provisions for durable institutional controls are in place; and

  — sufficient financial assurance has been provided to allow an independent third party to carry out rechecks at the site no less frequently than every 5 years and to assume and carry out responsibilities for any necessary control and maintenance of the controls at the site.

EVALUATION FINDINGS: EVALUATION CRITERIA

The staff should verify that the information summarized under "Information to be Submitted," above, is included in the dose modeling/ALARA demonstration subsection of the restricted use section of the DP. The staff should verify that the dose to the average member of the critical group does not exceed 0.25 mSv/y (25 mrem/y) with institutional controls in place and that the licensee estimated the dose in accordance with Chapter 5 of Volume 2 of this NUREG series. The staff should verify that these doses are ALARA and that the licensee has made this evaluation in accordance with the criteria in Chapter 6 and Appendix N of Volume 2 of this NUREG series. The staff should verify that the dose to the average member of the critical group will not exceed

1.0 mSv/y (100 mrem/y), without institutional controls, and that the licensee has estimated the dose in accordance with Chapter 5 of Volume 2 of this NUREG series.

If the dose to the average member of the critical group could exceed 1.0 mSv/y (100 mrem/y), without institutional controls, the staff should verify that the dose will not exceed 5.0 mSv/y (500 mrem/y) and that the licensee has estimated the dose in accordance with Chapter 5 of Volume 2 of this NUREG series. The staff also should verify that the licensee has determined that further reductions in residual radioactivity necessary to comply with the 1.0 mSv/y (100 mrem/y) requirement are not technically achievable, would be prohibitively expensive or would result in net public or environmental harm in accordance with Chapter 6 and Appendix N of Volume 2 of this NUREG series. The staff should verify that the institutional controls provided by the licensee meet the criteria for a durable institutional controls (i.e., government ownership or control or responsibility as the third party). The staff should verify that the licensee has provided sufficient financial assurance to allow an independent third party to carry out rechecks at the site no less than every five years. The staff should verify that the amount of financial assurance is sufficient to assume and carry out responsibilities for any necessary control and maintenance of the controls at the site in accordance in Part II of Volume 3 of this NUREG series.

17.8   ALTERNATE CRITERIA

For certain difficult sites with unique decommissioning problems, 10 CFR 20.1404 includes a provision by which NRC may terminate a license using alternate dose criteria. NRC expects the use of alternate criteria to be limited to rare situations. This provision was included in 10 CFR 20.1404 because NRC believed that it was preferable to codify provisions for these difficult sites in the rule rather than require licensees to seek an exemption outside the rule. Under 10 CFR 20.1404, NRC may consider terminating a license under alternate criteria that are greater than 0.25 mSv/y (25 mrem/y) [but less than 1.0 mSv/y (100 mrem/y)] with restrictions in place, but NRC limits the conditions under which a licensee could apply to NRC for, or be granted use of, alternate criteria to unusual site-specific circumstances.

The guidance in Section 17.7 for restricted use sites, including the risk-informed graded approach and use of new institutional control options involving NRC (i.e., LTC license and LA/RC), also applies to selecting the appropriate institutional controls for sites proposing to decommission using alternate criteria in 10 CFR 20.1404.

The purpose of the review of the licensee's discussion of why it is requesting license termination under the alternate criteria provisions of 10 CFR 20.1404 is to determine if the licensee can demonstrate that the estimated doses to the public from all man-made sources other than medical will be less than 1.0 mSv/y (100 mrem/y) and are ALARA, that appropriate restrictions are in place at the site and that the licensee has sought, obtained, evaluated and, as appropriate addressed, advice from individuals and institutions that may be affected by the decommissioning, in accordance with the criteria in 10 CFR 20.1404.

ACCEPTANCE CRITERIA: INFORMATION TO BE SUBMITTED

The information supplied by the licensee should be sufficient to allow the staff to determine whether the residual radioactive material at the site will result in a dose that exceeds 0.25 mSv/y (25 mrem/y), but will not exceed 1.0 mSv/y (100 mrem/y) (considering all man-made sources other than medical) and is ALARA. The information should also demonstrate that the financial assurance mechanism(s) are adequate for the site. Finally, the information should be adequate to allow the staff to determine if the institutional controls, site maintenance activities and the manner in which advice from individuals or institutions that could be affected by the decommissioning was sought, obtained, evaluated, and, as appropriate, addressed is in accordance with NRC requirements. The staff should verify that the following information is included in the discussion of why the licensee is requesting license termination under the provisions of 10 CFR 20.1404:

- a summary of the dose to the average member of the critical group (considering all man-made sources other than medical);

- a summary of the evaluation performed pursuant to Chapter 6 and Appendix N of Volume 2 of this NUREG series demonstrating that these doses are ALARA;

- an analysis of all possible sources of exposure to radiation at the site and a discussion of why it is unlikely that the doses from all man-made sources, other than medical, will be more than 1.0 mSv/y (100 mrem/y);

- an analysis demonstrating that if institutional controls are not in effect, the dose will not exceed the dose "caps" in 10 CFR 20.1403(e);

- a description of the legally enforceable institutional control(s) and an explanation of how the institutional control is a legally enforceable mechanism;

- a description of any detriments associated with the maintenance of the institutional control(s);

- a description of the restrictions on present and future landowners;

- a description of the entities enforcing and their authority to enforce the institutional control(s);

- a discussion of the durability[23] of the institutional control(s);

- a description of the activities that the party with the authority to enforce the institutional controls will undertake to enforce the institutional control(s);

- a description of the manner in which the entity with the authority to enforce the institutional control(s) will be replaced if that entity is no longer willing or able to enforce the institutional control(s);

---

[23] The Commission has stated (see Section B.3.3 of the "Statements of Consideration" for 10 CFR Part 20, Subpart E, "Radiological Criteria for License Termination") that stringent institutional controls would be needed for sites involving large quantities of uranium and thorium contamination. Typically, these would involve legally enforceable deed restrictions and/or controls backed up by State and local government control or ownership, engineered barriers, and as appropriate, Federal ownership.

- a description of the duration of the institutional control(s), the basis for the duration, the conditions that will end the institutional control(s) and the activities that will be undertaken to end the institutional control(s);

- a description of the corrective actions that will be undertaken in the event the institutional control(s) fail;

- a description of the amount of and mechanism for financial assurance, required under 10 CFR 20.1403(c);

- a description of the records pertaining to the institutional controls, how and where they will be maintained, and how the public will have access to the records.

- a description of how individuals and institutions that may be affected by the decommissioning were identified and informed of the opportunity to provide advice to the licensee;

- a description of the manner in which the licensee obtained advice from affected individuals or institutions;

- a description of how the licensee provided for participation by a broad cross-section of community interests in obtaining the advice;

- a description of how the licensee provided for a comprehensive, collective discussion on the issues by the participants represented;

- a copy of the publicly available summary of the results of discussions, including individual viewpoints of the participants on the issues and the extent of agreement and disagreement among the participants;

- a description of how this summary has been made available to the public; and

- a description of how the licensee evaluated advice from individuals and institutions that could be affected by the decommissioning and the manner in which the advice was addressed.

EVALUATION FINDINGS: EVALUATION CRITERIA

NRC staff should review the information supplied by the licensee to determine if the description of the activities undertaken by the licensee is adequate to allow the staff to conclude that the licensee has complied with the requirements of 10 CFR 20.1404.

The staff should determine whether the information summarized under "Information to be Submitted," above, is included in the discussion of why the licensee is requesting license termination under the provisions of 10 CFR 20.1404. The application of the criteria is dependent on the circumstances of the case. In each case the staff should consult with the Office of the General Counsel on the application of the criteria and the sufficiency of the licensee's proposal.

Review of the manner in which doses to the public should be estimated is addressed in Chapter 5 of Volume 2 of this NUREG series, and the staff should refer to Chapter 5 of Volume 2 of this NUREG series to determine if the dose estimates developed by the licensee are acceptable. The evaluation of these doses to determine if they are ALARA is addressed in Chapter 6 and Appendix N of Volume 2 of this NUREG series, and the staff should refer to Chapter 6 and Appendix N of Volume 2 of this NUREG series to review the licensee's demonstration that the doses are ALARA. The evaluation of the licensee's financial assurance mechanism(s) is addressed above and in Part II of Volume 3 of this NUREG series and the staff should refer to these sections to review the financial assurance mechanisms. The evaluation of institutional controls, site maintenance activities, and obtaining advice from individual and institutions that could be affected by the decommissioning are addressed in Sections 17.7.3, 17.7.4, and 17.7.5 of this volume.

18 DECOMMISSIONING PLANS: MODIFICATIONS TO DECOMMISSIONING PROGRAMS AND PROCEDURES

As the radiological contamination at a facility is reduced, the potential doses to workers and the public from the residual radioactive material is also generally reduced. Therefore, in some cases, it may be appropriate to allow licensees to revise their decommissioning programs and procedures to address this reduced threat. If a licensee wishes to revise its program without prior NRC review and approval, the decommissioning program description needs to be a detailed description of how the licensee will review and re-evaluate its program as conditions at the facility change and, as appropriate, modify its procedures to meet the reduced risk. If the staff is satisfied with the licensee's methodology for changes to its programs and procedures, NRC may approve a DP that allows revisions to programs and procedures without prior NRC approval, subject to the following conditions:

- The change does not conflict with requirements specifically stated in the license or impair the licensee's ability to meet all applicable NRC regulations.

- There is no degradation in safety or environmental commitments addressed in the NRC–approved DP.

- There are no significant adverse effects on the quality of the work, the remediation objectives, or health and safety.

- The change is consistent with the conclusions of actions analyzed in the Environmental Assessment, Environmental Impact Statement and Safety Evaluation Report developed for the decommissioning project.

- Licensees may not change programs and procedures related to dose modeling, final radiological surveys or restricted use/alternative criteria without prior NRC approval.

The purpose of the staff's review of the licensee's procedures for modifying its decommissioning program is to evaluate the licensee's description of its methodology to modify its programs and procedures as decommissioning progresses with the removal of the residual radioactive material from the facility. In addition, the staff's review should determine if the licensee can demonstrate that it can adequately evaluate, revise, and monitor any future revisions in its programs so as to ensure that the level of protection afforded by the revisions are commensurate with the potential risk from residual radioactive material remaining at the site and with the provisions of 10 CFR Parts 19 and 20.

Because modifying decommissioning programs/procedures could be applicable to any of the previous sections on DP guidance, as well as the guidance in Volumes 2 and 3, a discussion of the minimum information that should be included in a DP for these modifications is included here in lieu of in each section. In some instances, additional information may be required to support the modification of specific programs or procedures. NRC staff should work with licensees to identify this information and include it in the DP for that licensee.

All of the following information should be included in the licensee's description of how modifications to decommissioning programs/procedures will be managed:

- A description of how the licensee will evaluate the radiological conditions, including surface and soil contamination and determination of the potential doses to workers performing decommissioning activities and how the licensee will determine that the existing requirements are no longer necessary.

- A description of the method by which the licensee will use this information to develop the revised modifications to its program and how the licensee will compare and evaluate any revised procedures with the radiological conditions at the site.

- A demonstration that the modification and approval review process is as rigorous as the review and approval process for RWPs, includes approval from the same level of licensee management as revisions to the RWP, as well as review by all appropriate internal decommissioning organizations (including, but not limited to, the health and safety organization and the remediation organization). The review process should include an assessment relative to items 1–5, above.

- A description of how the various decommissioning organizations will monitor the implementation of the modifications to ensure that all personnel are following the revised procedures.

- A description of the immediate and long-term actions that will be taken in the event the revised procedures are found not to provide the same level of protection afforded by the existing procedures.

- A description of the periodic review of the procedures to ensure that the revisions are current and continue to be appropriate.

- A description of how the licensee will document each change to the procedures, and where it will be stored onsite, so it will be available for periodic review by NRC inspectors. This documentation should include: a description of each change, the technical justification for each change, when it became effective, how it was implemented, and who in management approved the change.

- A commitment to report all changes to NRC within 30 days of the change.

NRC staff should ensure that the licensee's proposed methodology will include all of the following actions:

- evaluates the radiological conditions against the existing programs/procedures prior to developing the proposed programs/procedures;

- develops the proposed modifications to the programs/procedures such that the level of protection is commensurate with the risk from the residual radioactive material at the facility;

- obtains the appropriate level of review and approval within the individual decommissioning unit and overall decommissioning management organization, including an assessment of the change relative to items 1–5, above;

- monitors the implementation of the modifications to the programs/procedures;

- includes provisions to respond to situations where the revised procedures are found to be inadequate;

- reviews periodically the revised programs/procedures to ensure that the revisions are current and continue to be appropriate;

- documents properly the revisions to the programs/procedures and their implementation; and

- includes a commitment to report the changes to NRC within 30 days of the change. This report must include a description of the changes, a summary of the safety and environmental evaluations performed for each change, and the revised DP pages reflecting the changes.

A licensee may replace a Radiation Safety Officer (RSO) without prior approval from NRC, as long as (a) the new RSO meets the criteria listed in Section 17.2.1 of this guidance; (b) the licensee maintains the documentation that the individual meets the criteria listed in Section 17.2.1 of this guidance and makes it available during inspections; and (c) the licensee informs NRC, in writing, within 30 days of the date of the change. The procedure for replacing the RSO should be included in the licensee's description of how modifications to decommissioning programs/procedures will be managed.

Appendix A
U.S. Nuclear Regulatory Commission
Form 314

NRC FORM 314	U.S. NUCLEAR REGULATORY COMMISSION	APPROVED BY OMB: NO. 3150-0028 EXPIRES: MM/DD/YYYY
(MM-YYYY) 10 CFR 30.36(c)(1)(iv) 10 CFR 40.42(c)(1)(iv) 10 CFR 70.38(c)(1)(iv) **CERTIFICATE OF DISPOSITION OF MATERIALS** INSTRUCTIONS: ALL ITEMS MUST BE COMPLETED -- PRINT OR TYPE SEND THE COMPLETED CERTIFICATE TO THE NRC OFFICE SPECIFIED ON THE REVERSE		Estimated burden per response to comply with this mandatory information collection request: 30 minutes. This submittal is used by NRC as part of the basis for its determination that the facility has been cleared of radioactive material before the facility is released for unrestricted use. Forward comments regarding burden estimate to the Records Management Branch (T-6 F33), U.S. Nuclear Regulatory Commission, Washington, DC 20555-0001, and to the Paperwork Reduction Project (3150-0028), Office of Management and Budget, Washington, DC 20503. If an information collection does not display a currently valid OMB control number, the NRC may not conduct or sponsor, and a person is not required to respond to, the information collection.

LICENSEE NAME AND ADDRESS

LICENSE NUMBER

LICENSE EXPIRATION DATE

A. MATERIALS DATA *(Check one and complete as necessary)*

THE LICENSEE OR ANY INDIVIDUAL EXECUTING THIS CERTIFICATE ON BEHALF OF THE LICENSEE CERTIFIES THAT:
(Check and/or complete the appropriate item(s) below.)

☐ 1. NO MATERIALS HAVE EVER BEEN PROCURED OR POSSESSED BY THE LICENSEE UNDER THIS LICENSE.

OR

☐ 2. ALL ACTIVITIES AUTHORIZED BY THE LICENSE HAVE CEASED AND ALL MATERIALS PROCURED AND/OR POSSESSED BY THE LICENSEE UNDER THE LICENSE NUMBER CITED ABOVE HAVE BEEN DISPOSED OF IN THE FOLLOWING MANNER. *(If additional space is needed, use the reverse side or provide attachments.)*

Describe specific material transfer actions and, if there were radioactive wastes generated in terminating this license, the disposal actions including the disposition of low-level radioactive waste, mixed waste, Greater-than-Class-C waste, and sealed sources, if applicable.

For transfers, specify the date of the transfer, the name of the licensed recipient, and the recipient's NRC license number or Agreement State name and license number.

If materials were disposed of directly by the licensee rather than transferred to another licensee, licensed disposal site or waste contractor, describe the specific disposal procedures *(e.g., decay in storage).*

B. OTHER DATA

☐ 1. OUR LICENSE HAS NOT YET EXPIRED; PLEASE TERMINATE IT.

2. A RADIATION SURVEY WAS CONDUCTED BY THE LICENSEE TO CONFIRM THE ABSENCE OF LICENSED RADIOACTIVE MATERIALS AND TO DETERMINE WHETHER ANY CONTAMINATION REMAINS ON THE PREMISES COVERED BY THE LICENSE. *(Check one)*

☐ NO *(Attach explanation)*
☐ YES, THE RESULTS *(Check one)*
  ☐ ARE ATTACHED, or
  ☐ WERE FORWARDED TO NRC ON *(Date)*

3. THE PERSON TO BE CONTACTED REGARDING THE INFORMATION PROVIDED ON THIS FORM	NAME	TELEPHONE NUMBER *(Include Area Code)*

4. MAIL ALL FUTURE CORRESPONDENCE REGARDING THIS LICENSE TO

CERTIFYING OFFICIAL

I CERTIFY UNDER PENALTY OF PERJURY THAT THE FOREGOING IS TRUE AND CORRECT

PRINTED NAME AND TITLE	SIGNATURE	DATE

WARNING: FALSE STATEMENTS IN THIS CERTIFICATE MAY BE SUBJECT TO CIVIL AND/OR CRIMINAL PENALTIES. NRC REGULATIONS REQUIRE THAT SUBMISSIONS TO THE NRC BE COMPLETE AND ACCURATE IN ALL MATERIAL RESPECTS. 18 U.S.C. SECTION 1001 MAKES IT A CRIMINAL OFFENSE TO MAKE A WILLFULLY FALSE STATEMENT OR REPRESENTATION TO ANY DEPARTMENT OR AGENCY OF THE UNITED STATES AS TO ANY MATTER WITHIN ITS JURISDICTIONS.

PRINTED ON RECYCLED PAPER

NRC FORM 314 (MM-YYYY)

Appendix B
Screening Values

Table B.1 Acceptable License Termination Screening Values of Common Radionuclides for Building-Surface Contamination

Radionuclide	Symbol	Acceptable Screening Levels[a] for Unrestricted Release (dpm/100 cm^2)[b]
Hydrogen-3 (Tritium)	^{3}H	120000000
Carbon-14	^{14}C	3700000
Sodium-22	^{22}Na	9500
Sulfur-35	^{35}S	13000000
Chlorine-36	^{36}Cl	500000
Manganese-54	^{54}Mn	32000
Iron-55	^{55}Fe	4500000
Cobalt-60	^{60}Co	7100
Nickel-63	^{63}Ni	1800000
Strontium-90	^{90}Sr	8700
Technetium-99	^{99}Tc	1300000
Iodine-129	^{129}I	35000
Cesium-137	^{137}Cs	28000
Iridium-192	^{192}Ir	74000

Notes:

a Screening levels are based on the assumption that the fraction of removable surface contamination is equal to 0.1. For cases when the fraction of removable contamination is undetermined or higher than 0.1, users may assume, for screening purposes, that 100 percent of surface contamination is removable, and therefore the screening levels should be decreased by a factor of 10. Alternatively, users having site-specific data on the fraction of removable contamination, based on site-specific resuspension factors (e.g., within 10 percent to 100 percent range), may calculate site-specific screening levels using DandD, Version 2.

b Units are disintegrations per minute (dpm) per 100 square centimeters (dpm/100 cm^2). One dpm is equivalent to 0.0167 becquerel (Bq). Therefore, to convert to units of Bq/m^2, multiply each value by 1.67. The screening values represent surface concentrations of individual radionuclides that would be deemed in compliance with the 0.25 mSv/y (25 mrem/y) unrestricted release dose limit in 10 CFR 20.1402. For radionuclides in a mixture, the "sum of fractions" rule applies; see Part 20, Appendix B, Note 4.

APPENDIX B

Table B.2 Screening Values (pCi/g) of Common Radionuclides for Soil Surface Contamination Levels

Radionuclide	Symbol	Surface Soil Screening Values[a]
Hydrogen-3	^{3}H	1.1 E+02
Carbon-14	^{14}C	1.2 E+01
Sodium-22	^{22}Na	4.3 E+00
Sulfur-35	^{35}S	2.7 E+02
Chlorine-36	^{36}Cl	3.6 E-01
Calcium-45	^{45}Ca	5.7 E+01
Scandium-46	^{46}Sc	1.5 E+01
Manganese-54	^{54}Mn	1.5 E+01
Iron-55	^{55}Fe	1.0 E+04
Cobalt-57	^{57}Co	1.5 E+02
Cobalt-60	^{60}Co	3.8 E+00
Nickel-59	^{59}Ni	5.5 E+03
Nickel-63	^{63}Ni	2.1 E+03
Strontium-90	^{90}Sr	1.7 E+00
Niobium-94	^{94}Nb	5.8 E+00
Technetium-99	^{99}Tc	1.9 E+01
Iodine-129	^{129}I	5.0 E-01
Cesium-134	^{134}Cs	5.7 E+00
Cesium-137	^{137}Cs	1.1 E+01
Europium-152	^{152}Eu	8.7 E+00
Europium-154	^{154}Eu	8.0 E+00
Iridium-192	^{192}Ir	4.1 E+01
Lead-210	^{210}Pb	9.0 E-01
Radium-226	^{226}Ra	7.0 E-01
Radium-226 + C-3	^{226}Ra + C	6.0 E-01
Actinium-227	^{227}Ac	5.0 E-01
Actinium-227 + C	^{227}Ac + C	5.0 E-01

Table B.2 Screening Values (pCi/g) of Common Radionuclides for Soil Surface Contamination Levels (continued)

Radionuclide	Symbol	Surface Soil Screening Values[a]
Thorium-228	^{228}Th	4.7 E+00
Thorium-228 + C[b]	^{228}Th + C	4.7 E+00
Thorium-230	^{230}Th	1.8 E+00
Thorium-230 + C	^{230}Th + C	6.0 E-01
Thorium-232	^{232}Th	1.1 E+00
Thorium-232 + C	^{232}Th + C	1.1 E+00
Protactinium-231	^{231}Pa	3.0 E-01
Protactinium-231 + C	^{231}Pa + C	3.0 E-01
Uranium-234	^{234}U	1.3 E+01
Uranium-235	^{235}U	8.0 E+00
Uranium-235 + C	^{235}U + C	2.9 E-01
Uranium-238	^{238}U	1.4 E+01
Uranium-238 + C	^{238}U + C	5.0 E-01
Plutonium-238	^{238}Pu	2.5 E+00
Plutonium-239	^{239}Pu	2.3 E+00
Plutonium-241	^{241}Pu	7.2 E+01
Americium-241	^{241}Am	2.1 E+00
Curium-242	^{242}Cm	1.6 E+02
Curium-243	^{243}Cm	3.2 E+00

These values represent surficial surface soil concentrations of individual radionuclides that would be deemed in compliance with the 25 mrem/y (0.25 mSv/y) unrestricted release dose limit in 10 CFR 20.1402. For radionuclides in a mixture, the "sum of fractions" rule applies; see Part 20, Appendix B, Note 4.

Notes:

a   Screening values are in units of (pCi/g) equivalent to 25 mrem/y (0.25 mSv/y). To convert from pCi/g to units of becquerel per kilogram (Bq/kg), divide each value by 0.027. These values were derived using DandD screening methodology (NUREG/CR–5512, Volume 3). They were derived based on selection of the 90th percentile of the output dose distribution *for each specific radionuclide* (or radionuclide with the specific decay chain). Behavioral parameters were set at the mean of the distribution of the assumed critical group. The metabolic parameters were set at "Standard Man" or at the mean of the distribution for an average man.

b   "Plus Chain (+C) " indicates a value for a radionuclide with its decay progeny present in equilibrium. The values are concentrations of the parent radionuclide, but account for contributions from the complete chain of progeny in equilibrium with the parent radionuclide (NUREG/CR–5512 Volumes 1, 2, and 3).

B.1   DERIVED CONCENTRATION GUIDELINE LEVELS (DCGLs)

The $DCGL_W$ is the concentration of a radionuclide which, if distributed uniformly across a survey unit, would result in an estimated dose equal to the applicable dose limit. The $DCGL_{EMC}$ is the concentration of a radionuclide which, if distributed uniformly across a smaller limited area within a survey unit, would result in an estimated dose equal to the applicable dose limit.

Two approaches are possible for developing DCGLs: screening and site-specific analysis. Site-specific DCGLs are discussed in Volume 2 of this guidance.

B.2   SCREENING DCGLs

NRC has published radionuclide-specific screening DCGLs in the *Federal Register* for residual building-surface radioactivity and residual surface-soil radioactivity. The DCGLs in the *Federal Register* are $DCGL_W$ values, in that they are intended to be concentrations which, if distributed uniformly across a building or soil surface, would individually result in a dose equal to the dose criterion. The licensee may adopt these screening DCGLs without additional dose modeling, if the site is suitable for screening analysis (see Chapter 2 of this document). Alternatively, the licensee may use the DandD computer code to develop screening DCGLs. The licensee would use the code to determine the dose attributable to a unit concentration of a radionuclide and scale the result to determine the DCGL for the radionuclide. Either of these methods for identifying screening DCGLs requires only that the licensee both (a) identify the radionuclides of concern for the site and (b) demonstrate that the source term and model screening assumptions are satisfied. Thus, this approach requires essentially no source-term abstraction. The screening process and the source-term screening assumptions are discussed in detail in Chapter 2 of this document.

Before designing a final status survey, the licensee will likely need to identify a $DCGL_{EMC}$ for each radionuclide over a range of smaller limited areas. Since the conservative screening models of DandD are not appropriate for modeling small limited areas of contamination, use of the DandD screening code would likely result in $DCGL_{EMC}$ values that are overly conservative. Therefore, licensees will likely use other codes or approaches to develop $DCGL_{EMC}$ values. These would be considered "site-specific" analyses in that they would not be using the DandD code with the default screening values.

B.3   SCREENING ANALYSES

In the case of screening, the decisions involved in identifying the appropriate scenario and critical group, with their corresponding exposure pathways, have already been made. Scenario descriptions acceptable to NRC for use in generic screening are developed and contained in NUREG/CR–5512, Volume 1. It and NUREG–1549 provide the rationale for applicability of the generic scenarios, critical groups, and pathways at a site; the rationale and assumptions for scenarios and pathways included (and excluded); and the associated parameter values or ranges. A summary of the scenarios is in Figure B.1. The latest version of the DandD computer code should contain the latest default data values for the critical group's habits and characteristics.

BUILDING OCCUPANCY SCENARIO

This scenario accounts for exposure to fixed and removable residual radioactivity on the walls, floor, and ceiling of a decommissioned facility. It assumes that the building will be used for commercial or light industrial activities (e.g., an office building or warehouse).

Pathways include:

- external exposure from building surfaces;

- inhalation of (re)suspended removable residual radioactivity; and inadvertent ingestion of removable residual radioactivity.

RESIDENT FARMER SCENARIO

This scenario accounts for exposure involving residual radioactivity that is initially in the surficial soil. A farmer moves onto the site and grows some of his or her diet and uses water tapped from the aquifer under the site.

Pathways include:

- external exposure from soil;

- inhalation to (re)suspended soil;

- ingestion of soil;

- ingestion of drinking water from aquifer;

- ingestion of plant products grown in contaminated soil and using aquifer to supply irrigation needs;

- ingestion of animal products grown onsite (using feed and water derived from potentially contaminated sources); and

- ingestion of fish from a pond filled with water from the aquifer.

Figure B.1 Pathways for Generic Scenarios.

B.4   SCREENING

An acceptable dose assessment analysis need not incorporate all the physical, chemical, and biological processes at the site. The scope of the analysis, and accordingly the level of sophistication needed in the conceptual model, should be based on the overall objective of the analysis. A performance assessment conceptual model can be simple if it still provides satisfactory confidence in site performance. For an initial screening analysis, little may be known about the site from which to develop a conceptual model. Computer codes used for screening analyses are generally intended to provide a generic and conservative representation of processes and conditions expected for a wide array of sites. Accordingly, the generic conceptual model in such codes may not provide a close representation of conditions and processes at a specific site. Such a generic representation is still acceptable as long as it provides a conservative assessment of the performance of the site.

The DandD code has two default land-use scenarios: a building occupancy and a resident farmer scenario. The building occupancy scenario is intended to account for exposure to both fixed and removable residual radioactive contamination within a building. Exposure pathways included in the building occupancy scenario include: external exposure to penetrating radiation, inhalation of resuspended surface contamination, and inadvertent ingestion of surface contamination. The resident farmer scenario is intended to account for exposure to residual radioactive contamination in soil. Exposure pathways included in the resident farmer scenario include: external exposure to penetrating radiation; inhalation exposure to resuspended soil; ingestion of soil; and ingestion of contaminated drinking water, plant products, animal products, and fish. The predefined conceptual models within DandD are geared toward assessing releases of radioactivity, transport to, and exposure along, these pathways. Technical details of the conceptual model for applying the screening criteria are contained in Volume 2 of this guidance.

In general, the conceptual models within DandD are expected to provide a conservative representation of site features and conditions. Therefore, for screening analyses, NRC will consider such generic conceptual models to be acceptable provided it is acceptable to assume that the initial radioactivity is contained in the top layer (building surface or soil) and the remainder of the unsaturated zone and ground water are initially free of contamination. In using DandD for site-specific analyses, it is important to ensure that a more realistic representation of the site that is consistent with what is known about the site would not lead to higher doses. Some site features and conditions that may be incompatible with the generic conceptual models within DandD are listed in Figure B.2.

SITE FEATURES

Sites with highly heterogeneous radioactivity;

- Sites with wastes other than soils (e.g., slags and equipment); Sites that have multiple source areas;

- Sites that have radionuclides that may generate gases (e.g., H-3 and C-14);

- Sites that have contaminated zones thicker than 15 cm (6 in); Sites with chemicals or a chemical environment that could facilitate radionuclide releases (e.g., colloids);

- Sites with soils that have preferential flow conditions that could lead to enhanced infiltration;

- Sites with a perched water table, surface ponding, or no unsaturated zone; Sites where the ground water discharges to springs or surface seeps; Sites with existing ground water contamination;

- Sites where the potential ground water use is not expected to be located immediately below the contaminated zone;

- Sites with significant transient flow conditions;

- Sites with significant heterogeneity in subsurface properties; Sites with fractured or karst formations;

- Sites where the ground water dilution would be less than 2000 m^3 (70,000 ft^3);

- Sites where overland transport of contaminants is of potential concern; and,

- Sites with stacks or other features that could transport radionuclides off the site at a higher concentration than onsite.

Figure B.2   Site Features and Conditions that May Be Incompatible with Those Assumed in DandD.

For any site where it is known that one or more of these conditions or features are present, the licensee should provide an appropriate rationale on why the use of the DandD will not result in an underestimation of potential doses at the specific site.

As an example, it may be possible to demonstrate the acceptable use of DandD for analyzing sites that contain H-3 and C-14, although both radionuclides may occur as a gas. The following approach can be used to demonstrate the acceptable use of DandD for analyzing sites that contain either H-3 or C-14 (Haaker, 1999): (1) determine the area of the contaminated zone; (2) run DandD for the site with only H-3 or C-14; (3) read the associated activity ratio factor for the given area from Figure I.4 from Volume 2 of this NUREG series; and (4) estimate the potential missed dose by multiplying the inhalation dose calculated from DandD by the activity ratio factor.

B.5 SCREENING ANALYSIS VERSUS SITE-SPECIFIC ANALYSES

A licensee may perform a screening analysis to demonstrate compliance with the radiological criteria for license termination specified in Part 20, Subpart E. The screening analysis described in Chapter 2 of this document requires that the licensee either: (1) refer to radionuclide-specific screening values listed in the *Federal Register* (63 FR 64132 and 64 FR 68395); or (2) use the DandD computer code. A licensee pursuing the screening option may find that implementation of the DandD code is necessary if radionuclides not included in the *Federal Register* listings must be considered.

The staff should ensure that a licensee performing a screening analysis using the DandD code limit parameter modification to identifying radionuclides of interest and specifying the radionuclide concentrations. The staff should verify that the licensee has not modified any other input parameter values. The output file generated by DandD identifies all parameter values that have been modified. Modifying any input parameter value from a default value will constitute a site-specific analysis. A full discussion of the use of screening criteria to evaluate site conditions can be found in Volume 2 of this NUREG.

B.6 DEFAULT VALUES VERSUS SITE-SPECIFIC VALUES

DandD and many other computer codes used for dose assessment provide the user with default values for the input parameters. Often, the user only needs to select radionuclides to execute the code. This allows the user to quickly obtain results with very little time expended in developing input data sets. This is basically how DandD, Version 2, was envisioned to be used for screening analyses.

Codes with default parameters, while developed to be run with little user input or thought, require several considerations that should be made and justified to NRC staff. In actuality, they may be inappropriate for site conditions, scenario, time period, and so forth. Basically, in using an off-the-shelf computer code and its default parameters, the user agrees with (a) the conceptual model used by the computer code, (b) the exposure scenario, and (c) the process used to select the default parameters so that they are appropriate for the site being modeled.

Users of computer codes should have an understanding of the conceptual and numerical modeling approaches of the code through the process of developing or justifying data input sets. If default parameter values are unavailable or inappropriate, the user should address each and every input parameter by (a) determining what characteristics of the modeled system the parameter represents and how the parameter is used in the code and (b) developing a value for the input parameter that is appropriate for both the system being modeled and for the conceptual and numerical models implemented by the code. In fact, many default data values in the computer code may be simply "placeholders" for site data.

NRC realizes that the theoretical approach is quite intensive and probably inappropriate, based on the risk from some sites. Experience has shown that the availability of default values for input parameters can result in the user performing a "site-specific" analysis to modify values for parameters for which site data are readily available and accept the default values as appropriate for the remaining parameters, without an adequate understanding of the parameters and the implications of accepting the default values. Therefore, for site-specific analyses, NRC requests that the licensee provide justification for using both the model and the default parameters, along with any justification for site-specific modifications. The level of justification appropriate for the parameter value is not, necessarily, constant for all parameters. This is why Volume 2 discusses uncertainty and sensitivity analyses to provide a means to focus both licensees and NRC staff resources on the important parameters.

NRC staff have reviewed, and considered appropriate for dose assessments using these codes, default parameter ranges for both DandD, Version 2 and RESRAD, Version 6. This supports decommissioning by (a) promoting consistency among analyses (where appropriate); (b) focusing licensee and NRC staff resources on parameters considered significant with respect to the dose assessment results; and (c) facilitating review of the licensee's dose assessment by NRC staff. Therefore, most licensees could use the code and its default parameter ranges with little justification. If parameters have been modified, the licensee may need to provide some more justification for default parameters associated with the site-specific parameters. While these are default data for the associated computer code, that does not mean that they can be transferred to another computer code for use in it without justification.

To benefit from the advantages while minimizing the disadvantages, the staff should ensure that the licensee employs default parameter values or ranges in a manner consistent with the guidance provided in this section.

Appendix C
Notification Checklist

C.1 CHECKLIST OF ACTIONS TO BE COMPLETED BY NRC STAFF UPON RECEIPT OF LICENSED FACILITIES NOTIFICATION OF INTENT TO CEASE LICENSED OPERATIONS

Facility Information

Facility Name: _____

Address: _____

License No.: _____

Docket No.: _____

Reviewer : _____

Date of Notification: _____

☐ Refer to Chapters 8 through 14 of the Consolidated Decommissioning Guidance, Volume 1, "Decommissioning Process for Materials Licensees."

— Licensee has complied with NRC's notification requirements.

☐ Refer to Chapter 5 of the Consolidated Decommissioning Guidance, Volume 1, "Decommissioning Process for Materials Licensees" and 10 CFR 30.36(d), 40.42(d), 70.38(d), or 72.54(d).

— Technical Assistance Control (TAC) number for the decommissioning action assigned, if warranted.

— Notification is placed in the licensee's docket file and in ADAMS.

— Written acknowledgment of the notification sent to licensee. Decommissioning of the facility, including the subjects outlined below, discussed with the licensee and documentation placed in docket.

☐ For Groups 1 and 2, the acceptable methods for demonstrating the suitability of the site for unrestricted use described in Chapters 8 and 9 of the Consolidated Decommissioning Guidance, Volume 1, "Decommissioning Process for Materials Licensees."

☐ For Groups 3 through 7, the information to be included in decommissioning plans (DPs) provided as described in Chapters 10 through 14 of the Consolidated Decommissioning Guidance, Volume 1, "Decommissioning Process for Materials Licensees."

☐ Any additional information NRC will require to support the licensee's request to terminate the license.

☐   NRC requirements for providing the public with the opportunity to observe meetings between the staff and licensees, as well as any potential hearing or public meeting requirements applicable to the decommissioning of the facility.

☐   Decommissioning schedule—refer to Chapter 5 of the Consolidated Decommissioning Guidance, Volume 1, "Decommissioning Process for Materials Licensees" and NRC's regulations in 10 CFR 30.36(d–h), 40.42(d–h), 70.38(d–h), or 72.54(d–j).

    — Contact made with other State or Federal regulatory authorities or other groups that have an interest in the decommissioning of the facility.

    — External distribution list for documents pertaining to the decommissioning developed.

    — Need to notice the licensee's proposed action in the FR determined and a notice prepared in accordance with 10 CFR Parts 2.102–2.108, as appropriate.

C.2   CRITERIA FOR DETERMINING DECOMMISSIONING GROUPS

The types of licensees for each of the seven Groups are shown below:

Group 1 Licensees

☐   Licensees that possessed and used only sealed sources and whose most recent leak tests are current and demonstrate that the sealed sources did not leak while in the licensee's possession.

☐   Licensees that possessed and used relatively short-lived radioactive material (i.e., $T_{1/2}$ less than or equal to 120 days) in an unsealed form, the maximum activity authorized under the license has decayed to less than the quantity specified in 10 CFR Part 20, Appendix C, and the licensee's survey performed in accordance with 10 CFR Part 30.36 does not identify any residual levels of radiological contamination greater than decommissioning screening criteria.

☐   Licensees decommissioning under Group 1 would not be required to develop a DP.

Group 2 Licensees

☐   Licensees that can demonstrate compliance with 10 CFR Part 20.1402 (Radiological criteria for unrestricted use) using the screening methodology.

☐   Licensees that possess and use only sealed sources that cannot demonstrate current leak tight integrity.

☐   Licensees decommissioning under Group 2 would not be required to develop a DP.

Group 3 Licensees

☐ Same provisions as for Group 2, except licensee must submit a simplified DP.

Group 4 Licensees

☐ Facilities decommissioned under Group 4 used licensed material in a manner that resulted in its release into the environment, activated adjacent materials, or resulted in persistent contamination of work areas, but did not result in contamination of ground water.

☐ These licensees cannot meet, or choose not to use, screening criteria so they must demonstrate that any residual radioactive material remaining at their site is within the levels specified in NRC's criteria for unrestricted use by applying a comprehensive dose analysis.

☐ DP is required for Group 4.

Group 5 Licensees

☐ Facilities that decommission under Group 5 have used licensed material in a manner that resulted in its release into the environment, activated adjacent materials or resulted in persistent contamination of work areas, and resulted in contamination of ground water.

☐ Group 5 decommissioning includes licensees that intend to decommission their facilities in accordance with NRC criteria for unrestricted use as described in 10 CFR 20.1402.

☐ DP is required.

Group 6 Licensees

☐ Facilities that decommission under Group 6 have used licensed material in a manner that resulted in releases to the environment, activated adjacent materials, or resulted in persistent contamination of work areas or ground water.

☐ Group 6 decommissioning includes licensees that intend to decommission its facility in accordance with NRC criteria for restricted use as described in 10 CFR 20.1403.

☐ DP is required.

Group 7 Licensees

☐ Facilities that have residual radiological contamination present in building surfaces, soils, and possibly ground water.

☐   These licensees intend to decommission their facilities such that residual radioactive material remaining at their site is in excess of the levels specified in NRC's criteria for unrestricted use.

☐   The licensees will apply site-specific criteria in a comprehensive dose analysis in accordance with alternate criteria for license termination (10 CFR 20.1404).

☐   A site DP that identifies the land use, exposure pathways, institutional controls, and critical group for the dose analysis is required.

☐   These sites require extensive NRC review and are handled on a case-by-case basis with license termination specifically approved by a vote of the NRC Commissioners.

Appendix D
DP Evaluation Checklist

D.1 ROADMAP GUIDELINE OF DECOMMISSIONING PLAN CHECKLIST

Introduction

The following table maps the application of the Decommissioning Plan Checklist to the various decommissioning groups. In general, larger group numbers require more information in the decommissioning plan (DP). The applicable boxes are color/shade-coded and contain a number which indicates the relative amount of information normally expected in each DP section.

Due to the diverse conditions found at decommissioning sites, even when categorized by group, it is not useful to attempt to indicate the expected length of each section of the DP. Additionally, any such estimate would necessarily make assumptions about the brevity and style of the DP authors. Therefore, a qualitative approach is taken, as described below. For complex sites, the actual DP roadmap should be developed through coordination with the NRC, using this Roadmap Guideline and DP checklist .

Qualitative Approach

The first qualifier is the site group determination. The licensee's proposed group selection is confirmed by NRC during document reviews and during discussions with the licensee. The group determination provides broad expectation of the content and detail needed in a DP. The table below provides the broad expectation.

In order to establish site-specific DP content expectations, the licensee should initiate a historical site assessment. This preliminary assessment should be of sufficient depth and quality to identify:

- potential, likely, and known sources of radioactive material and contamination, within the existing or historical site boundaries;

- any current or historical site conditions, operations, facilities, or improvements that could result in accumulation or migration of contaminants; and

- any potential threat to human health or the environment.

The level of detail must be sufficient to allow NRC staff to review the data and independently confirm the licensee's conclusions. The amount of information and data required to meet this burden will vary significantly from site to site, based on the complexity of the site history, site contamination, and the associated risks to human health and the environment.

The level of detail required for the remaining required portions of a DP (program organization, decommissioning procedures, inspections, surveys, dose calculations, et. al.) must be sufficient to address the concerns raised in the historical site assessment. The burden is to ensure that the information about contaminants, their potential locations, and public health and safety concerns contained in the site assessment are addressed, and the licensee's programs, methods and conclusions are able to be independently verified by NRC staff. Additionally, in cases where the historical site assessment does not provide a clear understanding of site conditions, the other elements of the DP demonstrate how the licensee will fill in the information gaps. The integrated result should provide a robust, confirmable understanding of what is at the site, how it will be remediated, and how the site cleanup will be verified safely. For complex sites, establishing the qualitative standard should involve prior coordination between the licensee and NRC staff. The first step for a complex decommissioning site is to meet with NRC staff and establish the scope and contents anticipated in the DP.

Table D.1 reflects that site conditions for simpler sites do not require as detailed information to support NRC analysis. The table is to be used with the DP Checklist in Appendix D.2, as a guide to assist licensee and NRC staff in developing the expected DP contents and scope at the beginning of the decommissioning process.

- For the blocks labeled with 1's, only a minimal amount of information is normally expected; this information is usually in existing documentation.

- For blocks marked with 2's, additional information would normally be needed to allow NRC staff to complete their independent assessment—some specific data and short analysis may be required.

- For blocks marked with 3's, a complete discussion is needed to explain the topic—significant data and analysis may be required. Such information is obtained through detailed site characterization and planning for remediation.

For Decommissioning Groups 1 and 2, the basic qualitative approach for required information is the same, but a formal DP is not required. A list of the information required for Groups 1 and 2 is provided in Chapters 8 and 9, respectively.

Table D.1 Application of Checklist to Decommissioning Groups

Checklist Sect. Group	3	4	5	6	7
EXECUTIVE SUMMARY	1	1	2	3	3
FACILITY OPERATING HISTORY					
License Number/Status/Authorized Activities	1	1	2	3	3
License History	1	1	2	3	3
Previous Decommissioning Activities	1	1	2	3	3
Spills	1	1	2	3	3
Prior Onsite Burials	1	1	2	3	3

Table D.1   Application of Checklist to Decommissioning Groups (continued)

Checklist Sect. Group	3	4	5	6	7
FACILITY DESCRIPTION					
Site Location and Description	1	2	2	3	3
Population Distribution	1	1	2	3	3
Current/Future Land Use	1	1	2	3	3
Meteorology and Climatology	1	1	2	3	3
Geology and Seismology	1	1	3	3	3
Surface Water Hydrology	1	1	3	3	3
Ground Water Hydrology	1	1	3	3	3
Natural Resources	1	2	3	3	3
RADIOLOGICAL STATUS OF FACILITY					
Contaminated Structures	2	2	2	3	3
Contaminated Systems and Equipment	2	2	2	3	3
Surface Soil Contamination	1	1	3	3	3
Subsurface Soil Contamination	N/A	N/A	3	3	3
Surface Water	1	1	3	3	3
Ground water	1	1	3	3	3

Table D.1 Application of Checklist to Decommissioning Groups (continued)

Checklist Sect. Group	3	4	5	6	7
DOSE MODELING					
Unrestricted Release Using Screening Criteria	2	3	N/A	N/A	N/A
Unrestricted release using screening criteria for building surface residual radioactivity:	2	3	N/A	N/A	N/A
Unrestricted release using screening criteria for surface soil residual radioactivity:	2	3	N/A	N/A	N/A
Unrestricted Release Using Site-Specific Information	N/A	N/A	3	N/A	N/A
Restricted Release Using Site-Specific Information	N/A	N/A	N/A	3	3
ALARA Analysis	1	1	2	3	3
PLANNED DECOMMISSIONING ACTIVITIES					
Contaminated Structures	1	2	2	3	3
Contaminated Systems and Equipment	1	2	2	3	3
Soil	1	2	3	3	3
Surface and Ground Water	N/A	N/A	2	3	3
Schedules	1	2	2	3	3
PROJECT MANAGEMENT AND ORGANIZATION					
Decommissioning Management Organization	1	1	2	3	3
Decommissioning Task Management	1	1	2	3	3
Decommissioning Management Positions and Qualifications	1	1	2	3	3
Radiation Safety Officer	1	1	2	3	3
Training	1	1	2	3	3
Contractor Support	1	1	2	3	3

Table D.1   Application of Checklist to Decommissioning Groups (continued)

Checklist Sect. Group	3	4	5	6	7
HEALTH AND SAFETY PROGRAM DURING DECOMMISSIONING					
Radiation Safety Controls and Monitoring for Workers	2	2	2	3	3
Air Sampling Program	2	2	2	3	3
Respiratory Protection Program	2	2	2	3	3
Internal Exposure Determination	2	2	2	3	3
External Exposure Determination	2	2	2	3	3
Summation of Internal and External Exposures	2	2	2	3	3
Contamination Control Program	2	2	2	3	3
Instrumentation Program	2	3	3	3	3
Nuclear Criticality Safety (if applicable)	2/3	2/3	2/3	2/3	2/3
Health Physics Audits, Inspections, and Recordkeeping Program	2	2	2	3	3
ENVIRONMENTAL MONITORING AND CONTROL PROGRAM					
Environmental ALARA Evaluation Program	1	1	2	3	3
Effluent Monitoring Program	1	1	2	3	3
Effluent Control Program	1	1	2	3	3
RADIOACTIVE WASTE MANAGEMENT PROGRAM					
Solid Radwaste	1	2	2	3	3
Liquid Radwaste	1	2	2	3	3
Mixed Waste	1	2	2	3	3

Table D.1 Application of Checklist to Decommissioning Groups (continued)

Checklist Sect. Group	3	4	5	6	7
QUALITY ASSURANCE PROGRAM					
Organization	1	2	2	3	3
Quality Assurance Program	1	2	2	3	3
Document Control	1	2	2	3	3
Control of Measuring and Test Equipment	1	2	2	3	3
Corrective Action	1	2	2	3	3
Quality Assurance Records	1	2	2	3	3
Audits and Surveillances	1	2	2	3	3
FACILITY RADIATION SURVEYS					
Release Criteria	1	2	2	3	3
Characterization Surveys	1	2	2	3	3
In-Process Surveys	1	2	2	3	3
Final Status Survey Design	1	2	2	3	3
Final Status Survey Report	1	2	2	3	3
FINANCIAL ASSURANCE					
Cost Estimate	1	2	2	3	3
Certification Statement	1	2	2	3	3
Financial Mechanism	1	2	2	3	3

Table D.1 Application of Checklist to Decommissioning Groups (continued)

Checklist Sect. Group	3	4	5	6	7
RESTRICTED USE/ALTERNATE CRITERIA					
Restricted Use	N/A	N/A	N/A	3	3
Eligibility Demonstration	N/A	N/A	N/A	3	3
Institutional Controls	N/A	N/A	N/A	3	3
Site Maintenance and Financial Assurance	N/A	N/A	N/A	3	3
Obtaining Public Advice	N/A	N/A	N/A	3	3
Dose Modeling and ALARA Demonstration	N/A	N/A	N/A	3	3
Alternate Criteria	N/A	N/A	N/A	N/A	3

D.2   DECOMMISSIONING PLAN CHECKLIST

Licensee Name: _____

License Number: _____ Docket Number: _____

Facility: _____

Decommissioning Plan Dated/Version: _____

For the acceptance review, NRC staff will use this checklist to conduct a limited technical review the decommissioning plan (DP). The detailed technical review assesses the technical adequacy and completeness of the information.

Staff should use the checklist first during the initial meeting with the licensee to discuss the scope and content of the DP. In most cases, licensees will not be required to submit all of the information in this checklist. The staff, in conjunction with the licensee, should determine what information should be submitted for the site, based on the uses of radioactive material at the site, the extent and types of radioactive material contamination, the manner in which the licensee intends to decommissioning the facility, and other factors affecting the potential for increased risk to the public or workers from the decommissioning operations. This information should be documented by modifying the acceptance review checklist. Copies of the modified checklist should be provided to the licensee and maintained by the NRC reviewer. When the DP is submitted, the reviewer should use the modified checklist to perform the acceptance review.

During the acceptance review, the staff will review the DP table of contents and the individual DP chapters or sections to ensure that the licensee has included this information in the DP. In addition, the staff may use Chapters 16 and 17 of this guidance to determine if the level of detail of the information appears to be adequate for the staff to perform a detailed technical review. Staff should recognize that failure to supply an item included in the checklist does not necessarily constitute grounds for rejecting the DP. Rather, the staff should determine if the licensee can supply the information in a timely manner and, if so, communicate the additional information needs to the licensee in a deficiency letter. Only in those cases where a detailed technical review cannot begin without the required information should the DP be rejected. For example, if the licensee is requesting restricted release and has not obtained the appropriate input from community interests who could be affected by the decommissioning, the DP should be rejected during the acceptance review. Questions regarding whether to reject a DP based on the results of the acceptance review should be forwarded to the Decommissioning Directorate, Division of Waste Management and Environmental Protection.

For the detailed technical review, staff should assess the technical accuracy and completeness of the information using the modified checklist.

I.   EXECUTIVE SUMMARY

☐   The name and address of the licensee or owner of the site

☐   The location and address of the site

☐   A brief description of the site and immediate environs

☐   A summary of the licensed activities that occurred at the site

☐   The nature and extent of contamination at the site

☐   The decommissioning objective proposed by the licensee (i.e., restricted or unrestricted use)

☐   The DCGLs for the site, the corresponding doses from these DCGLs, and the method that was use to determine the DCGLs

☐   A summary of the ALARA evaluations performed to support the decommissioning

☐   If the licensee requests license termination under restricted conditions, the restrictions the licensee intends to use to limit doses as required in 10 CFR Part 20.1403 or 20.1404, and a summary of institutional controls and financial assurance

☐   If the licensee requests license termination under restricted conditions or using alternate criteria, a summary of the public participation activities undertaken by the licensee to comply with 10 CFR Part 20.1403(d) or 20.1404(a)(4)

☐   The proposed initiation and completion dates of decommissioning

☐   Any post-remediation activities (such as ground water monitoring) that the licensee proposes to undertake prior to requesting license termination

☐   A statement that the licensee is requesting that its license be amended to incorporate the DP

II.   FACILITY OPERATING HISTORY

II.a.   LICENSE NUMBER/STATUS/AUTHORIZED ACTIVITIES

☐   The radionuclides and maximum activities of radionuclides authorized and used under the current license

☐   The chemical forms of the radionuclides authorized and used under the current license

☐   A detailed description of how the radionuclides are currently being used at the site

☐   The location(s) of use and storage of the various radionuclides authorized under current licenses

☐   A scale drawing or map of the building or site and environs showing the current locations of radionuclide use at the site

☐   A list of amendments to the license since the last license renewal

II.b.   LICENSE HISTORY

☐   The radionuclides and maximum activities of radionuclides authorized and used under all previous licenses

☐   The chemical forms of the radionuclides authorized and used under all previous licenses

☐   A detailed description of how the radionuclides were used at the site

☐   The location(s) of use and storage of the various radionuclides authorized under all previous licenses

☐   A scale drawing or map of the site, facilities, and environs showing previous locations of radionuclide use at the site

II.c.   PREVIOUS DECOMMISSIONING ACTIVITIES

☐   A list or summary of areas at the site that were remediated in the past

☐   A summary of the types, forms, activities, and concentrations of radionuclides that were present in previously remediated areas

☐   The activities that caused the areas to become contaminated

☐   The procedures used to remediate the areas, and the disposition of radioactive material generated during the remediation

☐   A summary of the results of the final radiological evaluation of the previously remediated area

☐   A scale drawing or map of the site, facilities, and environs showing the locations of previous remedial activity

II.d.   SPILLS

☐   A summary of areas at the site where spills (or uncontrolled releases) of radioactive material occurred in the past

☐   The types, forms, activities, and concentrations of radionuclides involved in the spill or uncontrolled release

☐   A scale drawing or map of the site, facilities, and environs showing the locations of spills

II.e.   PRIOR ONSITE BURIALS

☐   A summary of areas at the site where radioactive material has been buried in the past

☐ The types, forms, activities and concentrations of waste and radionuclides in the former burial

☐ A scale drawing or map of the site, facilities, and environs showing the locations of former burials

III. FACILITY DESCRIPTION

III.a. SITE LOCATION AND DESCRIPTION

☐ The size of the site in acres or square meters

☐ The State and county in which the site is located

☐ The names and distances to nearby communities, towns, and cities

☐ A description of the contours and features of the site

☐ The elevation of the site

☐ A description of property surrounding the site, including the location of all off-site wells used by nearby communities or individuals

☐ The location of the site relative to prominent features such as rivers and lakes

☐ A map that shows the detailed topography of the site using a contour interval

☐ The location of the nearest residences and all significant facilities or activities near the site

☐ A description of the facilities (e.g., buildings, parking lots, and fixed equipment) at the site

III.b. POPULATION DISTRIBUTION

☐ A summary of the current population in and around the site, by compass vectors

☐ A summary of the projected population in and around the site by compass vectors

III.c. CURRENT/FUTURE LAND USE

☐ A description of the current land uses in and around the site

☐ A summary of anticipated land uses

III.d. METEOROLOGY AND CLIMATOLOGY

☐ A description of the general climate of the region

☐ Seasonal and annual frequencies of severe weather phenomena

☐ Weather-related radionuclide transmission parameters

- ☐ Routine weather-related site deterioration parameters
- ☐ Extreme weather-related site deterioration parameters
- ☐ A description of the local (site) meteorology
- ☐ The National Ambient Air Quality Standards Category of the area in which the facility is located and, if the facility is not in a Category 1 zone, the closest and first downwind Category 1 Zone

III.e.   GEOLOGY AND SEISMOLOGY

- ☐ A detailed description of the geologic characteristics of the site and the region around the site
- ☐ A discussion of the tectonic history of the region, regional geomorphology, physiography, stratigraphy, and geochronology
- ☐ A regional tectonic map showing the site location and its proximity to tectonic structures
- ☐ A description of the structural geology of the region and its relationship to the site geologic structure
- ☐ A description of any crustal tilting, subsidence, karst terrain, landsliding, and erosion
- ☐ A description of the surface and subsurface geologic characteristics of the site and its vicinity
- ☐ A description of the geomorphology of the site
- ☐ A description of the location, attitude, and geometry of all known or inferred faults in the site and vicinity
- ☐ A discussion of the nature and rates of deformation
- ☐ A description of any man-made geologic features such as mines or quarries
- ☐ A description of the seismicity of the site and region
- ☐ A complete list of all historical earthquakes that have a magnitude of 3 or more, or a modified Mercalli intensity of IV or more within 200 miles of the site

III.f.   SURFACE WATER HYDROLOGY

- ☐ A description of site drainage and surrounding watershed fluvial features
- ☐ Water resource data including maps, hydrographs, and stream records from other agencies (e.g., U.S. Geological Survey and U.S. Army Corps of Engineers)
- ☐ Topographic maps of the site that show natural drainages and man-made features
- ☐ A description of the surface water bodies at the site and surrounding areas

☐ A description of existing and proposed water control structures and diversions (both upstream and downstream) that may influence the site

☐ Flow-duration data that indicate minimum, maximum, and average historical observations for surface water bodies in the site areas

☐ Aerial photography and maps of the site and adjacent drainage areas identifying features such as drainage areas, surface gradients, and areas of flooding

☐ An inventory of all existing and planned surface water users, whose intakes could be adversely affected by migration of radionuclides from the site

☐ Topographic and/or aerial photographs that delineate the 100-year floodplain at the site

☐ A description of any man-made changes to the surface water hydrologic system that may influence the potential for flooding at the site

III.g. GROUND WATER HYDROLOGY

☐ A description of the saturated zone

☐ Descriptions of monitoring wells

☐ Physical parameters

☐ A description of ground water flow directions and velocities

☐ A description of the unsaturated zone

☐ Information on all monitor stations including location and depth

☐ A description of physical parameters

☐ A description of the numerical analyses techniques used to characterize the unsaturated and saturated zones

☐ The distribution coefficients of the radionuclides of interest at the site

III.h. NATURAL RESOURCES

☐ A description of the natural resources occurring at or near the site

☐ A description of potable, agricultural, or industrial ground or surface waters

☐ A description of economic, marginally economic, or subeconomic known or identified natural resources as defined in U.S. Geological Survey Circular 831

☐ Mineral, fuel, and hydrocarbon resources near and surrounding the site which, if exploited, would effect the licensee's dose estimates

IV. RADIOLOGICAL STATUS OF FACILITY

IV.a. CONTAMINATED STRUCTURES

☐ A list or description of all structures at the facility where licensed activities occurred that contain residual radioactive material in excess of site background levels

☐ A summary of the structures and locations at the facility that the licensee has concluded have not been impacted by licensed operations and the rationale for the conclusion

☐ A list or description of each room or work area within each of these structures

☐ A summary of the background levels used during scoping or characterization surveys

☐ A summary of the locations of contamination in each room or work area

☐ A summary of the radionuclides present at each location, the maximum and average radionulide activities in dpm/100cm^2, and, if multiple radionuclides are present, the radionuclide ratios

☐ The mode of contamination for each surface (i.e., whether the radioactive material is present only on the surface of the material or if it has penetrated the material)

☐ The maximum and average radiation levels in mrem/hr in each room or work area

☐ A scale drawing or map of the rooms or work areas showing the locations of radionuclide material contamination

IV.b. CONTAMINATED SYSTEMS AND EQUIPMENT

☐ A list or description and the location of all systems or equipment at the facility that contain residual radioactive material in excess of site background levels

☐ A summary of the radionuclides present in each system or on the equipment at each location, the maximum and average radionulide activities in dpm/100cm^2, and, if multiple radionuclides are present, the radionuclide ratios

☐ The maximum and average radiation levels in mrem/hr at the surface of each piece of equipment

☐ A summary of the background levels used during scoping or characterization surveys

☐ A scale drawing or map of the rooms or work areas showing the locations of the contaminated systems or equipment

IV.c. SURFACE SOIL CONTAMINATION

☐ A list or description of all locations at the facility where surface soil contains residual radioactive material in excess of site background levels

☐ A summary of the background levels used during scoping or characterization surveys

☐ A summary of the radionuclides present at each location, the maximum and average radionuclide activities in pCi/gm, and, if multiple radionuclides are present, the radionuclide ratios

☐ The maximum and average radiation levels in mrem/hr at each location

☐ A scale drawing or map of the site showing the locations of radionuclide material contamination in surface soil

IV.d. SUBSURFACE SOIL CONTAMINATION

☐ A list or description of all locations at the facility where subsurface soil contains residual radioactive material in excess of site background levels

☐ A summary of the background levels used during scoping or characterization surveys

☐ A summary of the radionuclides present at each location, the maximum and average radionulide activities in pCi/gm, and, if multiple radionuclides are present, the radionuclide ratios

☐ The depth of the subsurface soil contamination at each location

☐ A scale drawing or map of the site showing the locations of subsurface soil contamination

IV.e. SURFACE WATER

☐ A list or description of all surface water bodies at the facility that contain residual radioactive material in excess of site background levels

☐ A summary of the background levels used during scoping or characterization surveys

☐ A summary of the radionuclides present in each surface water body and the maximum and average radionuclide activities in becquerel per liter (Bq/L) (picocuries per liter (pCi/L))

IV.f. GROUND WATER

☐ A summary of the aquifer(s) at the facility that contain residual radioactive material in excess of site background levels

☐ A summary of the background levels used during scoping or characterization surveys

☐ A summary of the radionuclides present in each aquifer and the maximum and average radionulide activities in becquerel per liter (Bq/L) (picocuries per liter (pCi/L))

V. DOSE MODELING

V.a. UNRESTRICTED RELEASE USING SCREENING CRITERIA

V.a.1. Unrestricted Release Using Screening Criteria for Building Surface Residual Radioactivity

☐ The general conceptual model (for both the source term and the building environment) of the site

☐ A summary of the screening method (i.e., running DandD or using the look-up tables) used in the DP

V.a.2. Unrestricted Release Using Screening Criteria for Surface Soil Residual Radioactivity

☐ Justification on the appropriateness of using the screening approach (for both the source term and the environment) at the site

☐ A summary of the screening method (i.e., running DandD or using the look-up tables) used in the DP

V.b. UNRESTRICTED RELEASE USING SITE-SPECIFIC INFORMATION

☐ Source term information including nuclides of interest, configuration of the source, and areal variability of the source

☐ Description of the exposure scenario including a description of the critical group

☐ Description of the conceptual model of the site including the source term, physical features important to modeling the transport pathways, and the critical group

☐ Identification/description of the mathematical model used (e.g., hand calculations, DandD Screen v1.0, and RESRAD v5.81)

☐ Description of the parameters used in the analysis

☐ Discussion about the effect of uncertainty on the results

☐ Input and output files or printouts, if a computer program was used

V.c. RESTRICTED RELEASE USING SITE-SPECIFIC INFORMATION

☐ Source term information including nuclides of interest, configuration of the source, areal variability of the source, and chemical forms

☐ A description of the exposure scenarios, including a description of the critical group for each scenario

☐ A description of the conceptual model(s) of the site that includes the source term, physical features important to modeling the transport pathways, and the critical group for each scenario

☐ Identification/description of the mathematical model(s) used (e.g., hand calculations and RESRAD v5.81)

☐ A summary of parameters used in the analysis

☐ A discussion about the effect of uncertainty on the results

☐ Input and output files or printouts, if a computer program was used

V.d. RELEASE INVOLVING ALTERNATE CRITERIA

☐ Source term information including nuclides of interest, configuration of the source, areal variability of the source, and chemical forms

☐ A description of the exposure scenarios, including a description of the critical group for each scenario

☐ A description of the conceptual model(s) of the site that includes the source term, physical features important to modeling the transport pathways, and the critical group for each scenario

☐ Identification/description of the mathematical model(s) used (e.g., hand calculations and RESRAD v5.81)

☐ A summary of parameters used in the analysis

☐ A discussion about the effect of uncertainty on the results

☐ Input and output files or printouts, if a computer program was used

VI. ENVIRONMENTAL INFORMATION

☐ Environmental information described in NUREG–1748

☐ For an EIS, the environmental information is reviewed by the EPAD EIS project manager

VII. ALARA ANALYSIS

☐ A description of how the licensee will achieve a decommissioning goal below the dose limit

☐ A quantitative cost benefit analysis

☐ A description of how costs were estimated

☐ A demonstration that the doses to the average member of the critical group are ALARA

VIII. PLANNED DECOMMISSIONING ACTIVITIES

VIII.a. CONTAMINATED STRUCTURES

☐ A summary of the remediation tasks planned for each room or area in the contaminated structure, in the order in which they will occur

☐ A description of the remediation techniques that will be employed in each room or area of the contaminated structure

☐ A summary of the radiation protection methods and control procedures that will be employed in each room or area

☐ A summary of the procedures already authorized under the existing license and those for which approval is being requested in the DP

☐ A commitment to conduct decommissioning activities in accordance with written, approved procedures

☐ A summary of any unique safety or remediation issues associated with remediating the room or area

☐ For Part 70 licensees, a summary of how the licensee will ensure that the risks addressed in the facility's Integrated Safety Analysis will be addressed during decommissioning

VIII.b. CONTAMINATED SYSTEMS AND EQUIPMENT

☐ A summary of the remediation tasks planned for each system in the order in which they will occur, including which activities will be conducted by licensee staff and which will be performed by a contractor

☐ A description of the techniques that will be employed to remediate each system in the facility or site

☐ A description of the radiation protection methods and control procedures that will be employed while remediating each system

☐ A summary of the equipment that will be removed or decontaminated and how the decontamination will be accomplished

☐ A summary of the procedures already authorized under the existing license and those for which approval is being requested in the DP

☐ A commitment to conduct decommissioning activities in accordance with written, approved procedures

☐ A summary of any unique safety or remediation issues associated with remediating any system or piece of equipment

☐ For Part 70 licensees, a summary of how the licensee will ensure that the risks addressed in the facility's Integrated Safety Analysis will be addressed during decommissioning

VIII.c. SOIL

☐ A summary of the removal/remediation tasks planned for surface and subsurface soil at the site in the order in which they will occur, including which activities will be conducted by licensee staff and which will be performed by a contractor

☐ A description the techniques that will be employed to remove or remediate surface and subsurface soil at the site

☐ A description of the radiation protection methods and control procedures that will be employed during soil removal/remediation

☐ A summary of the procedures already authorized under the existing license and those for which approval is being requested in the DP

☐ A commitment to conduct decommissioning activities in accordance with written, approved procedures

☐ A summary of any unique safety or removal/remediation issues associated with remediating the soil

☐ For Part 70 licensees, a summary of how the licensee will ensure that the risks addressed in the facility's Integrated Safety Analysis will be addressed during decommissioning

VIII.d. SURFACE AND GROUND WATER

☐ A summary of the remediation tasks planned for ground and surface water in the order in which they will occur, including which activities will be conducted by licensee staff and which will be performed by a contractor

☐ A description of the remediation techniques that will be employed to remediate the ground or surface water

☐ A description of the radiation protection methods and control procedures that will be employed during ground or surface water remediation

☐ A summary of the procedures already authorized under the existing license and those for which approval is being requested in the DP

☐ A commitment to conduct decommissioning activities in accordance with written, approved procedures

☐ A summary of any unique safety or remediation issues associated with remediating the ground or surface water

VIII.e. SCHEDULES

☐ A Gantt or PERT chart detailing the proposed remediation tasks in the order in which they will occur

☐ A statement acknowledging that the dates in the schedule are contingent upon NRC approval of the DP

☐ A statement acknowledging that circumstances can change during decommissioning, and, if the licensee determines that the decommissioning cannot be completed as outlined in the schedule, the licensee will provide an updated schedule to NRC

☐ If the decommissioning is not expected to be completed within the timeframes outlined in NRC regulations, a request for alternative schedule for completing the decommissioning

IX.   PROJECT MANAGEMENT AND ORGANIZATION

IX.a.   DECOMMISSIONING MANAGEMENT ORGANIZATION

☐ A description of the decommissioning organization

☐ A description of the responsibilities of each of these decommissioning project units

☐ A description of the reporting hierarchy within the decommissioning project management organization

☐ A description of the responsibility and authority of each unit to ensure that decommissioning activities are conducted in a safe manner and in accordance with approved written procedures

IX.b.   DECOMMISSIONING TASK MANAGEMENT

☐ A description of the manner in which the decommissioning tasks are managed

☐ A description of how individual decommissioning tasks are evaluated and how the Radiation Work Permits (RWPs) are developed for each task

☐ A description of how the RWPs are reviewed and approved by the decommissioning project management organization

☐ A description of how RWPs are managed throughout the decommissioning project

☐ A description of how individuals performing the decommissioning tasks are informed of the procedures in the RWP

IX.c. DECOMMISSIONING MANAGEMENT POSITIONS AND QUALIFICATIONS

☐ A description of the duties and responsibilities of each management position in the decommissioning organization and the reporting responsibility of the position

☐ A description of the duties and responsibilities of each chemical, radiological, physical, and occupational safety-related position in the decommissioning organization and the reporting responsibility of each position

☐ A description of the duties and responsibilities of each engineering, quality assurance, and waste management position in the decommissioning organization and the reporting responsibility of each position

☐ The minimum qualifications for each of the positions describe above, and the qualifications of the individuals currently occupying the positions

☐ A description of all decommissioning and safety committees

IX.d. RADIATION SAFETY OFFICER

☐ A description of the health physics and radiation safety education and experience required for individuals acting as the licensee's RSO

☐ A description of the responsibilities and duties of the RSO

☐ A description of the specific authority of the RSO to implement and manage the licensee's radiation protection program

IX.e. TRAINING

☐ A description of the radiation safety training that the licensee will provide to each employee

☐ A description of any daily worker "jobside" or "tailgate" training that will be provided at the beginning of each workday or job task to familiarize workers with job-specific procedures or safety requirements

☐ A description of the documentation that will be maintained to demonstrate that training commitments are being met

IX.f. CONTRACTOR SUPPORT

☐ A summary of decommissioning tasks that will be performed by contractors

☐ A description of the management interfaces that will be in place between the's management and onsite supervisors, and contractor management and onsite supervisors

☐ A description of the oversight responsibilities and authority that the licensee will exercise over contractor personnel

☐ A description of the training that will be provided to contractor personnel by the licensee and the training that will be provided by the contractor

☐ A commitment that the contractor will comply with all radiation safety and license requirements at the facility

X. HEALTH AND SAFETY PROGRAM DURING DECOMMISSIONING: RADIATION SAFETY CONTROLS AND MONITORING FOR WORKERS

X.a. AIR SAMPLING PROGRAM

☐ A description which demonstrates that the air sampling program is representative of the workers breathing zones

☐ A description of the criteria which demonstrates that air samplers with appropriate sensitivities will be used, and that samples will be collected at appropriate frequencies

☐ A description of the conditions under which air monitors will be used

☐ A description of the criteria used to determine the frequency of calibration of the flow meters on the air samplers

☐ A description of the action levels for air sampling results

☐ A description of how minimum detectable activities (MDA) for each specific radionuclide that may be collected in air samples are determined

X.b. RESPIRATORY PROTECTION PROGRAM

☐ A description of the process controls, engineering controls, or procedures to control concentrations of radioactive materials in air

☐ A description of the evaluation which will be performed when it is not practical to apply engineering controls or procedures

☐ A description of the considerations used which demonstrates respiratory protection equipment is appropriate for a specific task based on the guidance on assigned protection factors

☐ A description of the medical screening and fit testing required before workers will use any respirator that is assigned a protection factor

☐ A description of the written procedures maintained to address all the elements of the respiratory protection program

☐ A description of the use, maintenance, and storage of respiratory protection devices

☐ A description of the respiratory equipment users training program

☐ A description of the considerations made when selecting respiratory protection equipment

X.c. INTERNAL EXPOSURE DETERMINATION

☐ A description of the monitoring to be performed to determine worker exposure

☐ A description of how worker intakes are determined using measurements of quantities of radionuclides excreted from, or retained in the human body

☐ A description of how worker intakes are determined by measurements of the concentrations of airborne radioactive materials in the workplace

☐ A description of how worker intakes for an adult, a minor, and a declared pregnant woman (DPW) are determined using any combination of the measurements above, as may be necessary

☐ A description of how worker intakes are converted into committed effective dose equivalent

X.d. EXTERNAL EXPOSURE DETERMINATION

☐ A description of the individual-monitoring devices which will be provided to workers

☐ A description of the type, range, sensitivity, and accuracy of each individual-monitoring device

☐ A description of the use of extremity and whole body monitors when the external radiation field is non-uniform

☐ A description of when audible-alarm dosimeters and pocket dosimeters will be provided

☐ A description of how external dose from airborne radioactive material is determined

☐ A description of the procedure to insure that surveys necessary to supplement personnel monitoring are performed

☐ A description of the action levels for worker's external exposure, and the technical bases and actions to be taken when they are exceeded

X.e. SUMMATION OF INTERNAL AND EXTERNAL EXPOSURES

☐ A description of how the internal and external monitoring results are used to calculate TODE and TEDE doses to occupational workers

☐ A description of how internal doses to the embryo/fetus, which is based on the intake of an occupationally-exposed DPW will be determined

☐ A description of the monitoring of the intake of a DPW, if determined to be necessary

☐ A description of the program for the preparation, retention, and reporting of records for occupational radiation exposures

X.f. CONTAMINATION CONTROL PROGRAM

☐ A description of the written procedures to control access to, and stay time in, contaminated areas by workers, if they are needed

☐ A description of surveys to supplement personnel monitoring for workers during routine operations, maintenance, clean-up activities, and special operations

☐ A description of the surveys which will be performed to determine the baseline of background radiation levels and radioactivity from natural sources for areas where decommissioning activities will take place

☐ A description in matrix or tabular form which describes contamination action limits (that is, actions taken to either decontaminate a person, place, or area, restrict access, or modify the type or frequency of radiological monitoring)

☐ A description (included in the matrix or table mentioned above) of proposed radiological contamination guidelines for specifying and modifying the frequency for each type of survey used to assess the reduction of total contamination

☐ A description of the procedures used to test sealed sources, and to insure that sealed sources are leaked tested at appropriate intervals

X.g. INSTRUMENTATION PROGRAM

☐ A description of the instruments to be used to support the health and safety program

☐ A description of instrumentation storage, calibration, and maintenance facilities for instruments used in field surveys

☐ A description of the method used to estimate the MDC or MDA (at the 95 percent confidence level) for each type of radiation to be detected

☐ A description of the instrument calibration and quality assurance procedures

☐ A description of the methods used to estimate uncertainty bounds for each type of instrumental measurement

☐ A description of air sampling calibration procedures or a statement that the instruments will be calibrated by an accredited laboratory

X.h.   NUCLEAR CRITICALITY SAFETY

☐ A description of how the NCS functions, including management responsibilities and technical qualifications of safety personnel, will be maintained when needed throughout the decommissioning process

☐ A description of how an awareness of procedures and other items relied on for safety will be maintained throughout decommissioning among all personnel, with access to systems that may contain fissionable material in sufficient amounts for criticality

☐ A summary of the review of NCSA's or the ISA indicating either that the process needs no new safety procedures or requirements, or that new requirements or analysis have been performed

☐ A summary of any generic NCS requirements to be applied to general decommissioning, decontamination, or dismantlement operations, including those dealing with systems that may unexpectedly contain fissionable material

X.i.   HEALTH PHYSICS AUDITS, INSPECTIONS, AND RECORDKEEPING PROGRAM

☐ A general description of the annual program review conducted by executive management

☐ A description of the records to be maintained of the annual program review and executive audits

☐ A description of the types and frequencies of surveys and audits to be performed by the RSO and RSO staff

☐ A description of the process used in evaluating and dealing with violations of NRC requirements or license commitments identified during audits

☐ A description of the records maintained of RSO audits

XI.   ENVIRONMENTAL MONITORING AND CONTROL PROGRAM

XI.a.   ENVIRONMENTAL ALARA EVALUATION PROGRAM

☐ A description of ALARA goals for effluent control

☐ A description of the procedures, engineering controls, and process controls to maintain doses ALARA

☐ A description of the ALARA reviews and reports to management

XI.b. EFFLUENT MONITORING PROGRAM

☐ A demonstration that background and baseline concentrations of radionuclides in environmental media have been established through appropriate sampling and analysis

☐ A description of the known or expected concentrations of radionuclides in effluents

☐ A description of the physical and chemical characteristics of radionuclides in effluents

☐ A summary or diagram of all effluent discharge locations

☐ A demonstration that samples will be representative of actual releases

☐ A summary of the sample collection and analysis procedures

☐ A summary of the sample collection frequencies

☐ A description of the environmental monitoring recording and reporting procedures

☐ A description of the quality assurance program to be established and implemented for the effluent monitoring program

XI.c. EFFLUENT CONTROL PROGRAM

☐ A description of the controls that will be used to minimize releases of radioactive material to the environment

☐ A summary of the action levels and a description of the actions to be taken should a limit be exceeded

☐ A description of the leak detection systems for ponds, lagoons, and tanks

☐ A description of the procedures to ensure that releases to sewer systems are controlled and maintained to meet the requirements of 10 CFR 20.2003

☐ A summary of the estimates of doses to the public from effluents and a description of the method used to estimate public dose

XII. RADIOACTIVE WASTE MANAGEMENT PROGRAM

XII.a. SOLID RADWASTE

☐ A summary of the types of solid radwaste that are expected to be generated during decommissioning operations

☐ A summary of the estimated volume, in cubic feet, of each solid radwaste type summarized in Line 1 above

☐ A summary of the radionuclides (including the estimated activity of each radionuclide) in each estimated solid radwaste type summarized in Line 1 above

☐ A summary of the volumes of Class A, B, C, and Greater-than-Class-C solid radwaste that will be generated by decommissioning operations

☐ A description of how and where each of the solid radwaste summarized in Line 1 above will be stored onsite prior to shipment for disposal

☐ A description of how the each of the solid radwastes summarized in Line 1 above will be treated and packaged to meet disposal site acceptance criteria prior to shipment for disposal

☐ If appropriate, how the licensee intends to manage volumetrically contaminated material

☐ A description of how the licensee will prevent contaminated soil, or other loose solid radwaste, from being re-disbursed after exhumation and collection

☐ The name and location of the disposal facility that the licensee intends to use for each solid radwaste type summarized in Line 1 above

XII.b. LIQUID RADWASTE

☐ A summary of the types of liquid radwaste that are expected to be generated during decommissioning operations

☐ A summary of the estimated volume, in liters, of each liquid radwaste type summarized in Line 1 above

☐ A summary of the radionuclides (including the estimated activity of each radionuclide) in each liquid radwaste type summarized in Line 1 above

☐ A summary of the estimated volumes of Class A, B, C, and Greater-than-Class-C liquid radwaste that will be generated by decommissioning operations

☐ A description of how and where each of the liquid radwastes summarized in Line 1 above will be stored onsite prior to shipment for disposal

☐ A description of how the each of the liquid radwastes summarized in Line 1 above will be treated and packaged to meet disposal site acceptance criteria prior to shipment for disposal

☐ The name and location of the disposal facility that the licensee intends to use for each liquid radwaste type summarized in Line 1 above

XII.c. MIXED WASTE

☐ A summary of the types of solid and liquid mixed waste that are expected to be generated during decommissioning operations

☐ A summary of the estimated volumes in cubic feet of each solid mixed waste type summarized in Line 1 above, and in liters for each liquid mixed waste

☐ A summary of the radionuclides (including the estimated activity of each radionuclide) in each type of mixed waste type summarized in Line 1 above

☐ A summary of the estimated volumes of Class A, B, C, and Greater-than-Class-C mixed waste that will be generated by decommissioning operations

☐ A description of how and where each of the mixed wastes summarized in Line 1 above will be stored onsite prior to shipment for disposal

☐ A description of how the each of the mixed wastes summarized in Line 1 above will be treated and packaged to meet disposal site acceptance criteria prior to shipment for disposal

☐ The name and location of the disposal facility that the licensee intends to use for each mixed waste type summarized in Line 1 above

☐ A discussion of the requirements of all other regulatory agencies having jurisdiction over the mixed waste

☐ A demonstration the that the licensee possesses the appropriate EPA or State permits to generate, store, and/or treat the mixed wastes

XIII. QUALITY ASSURANCE PROGRAM

XIII.a. ORGANIZATION

☐ A description of the QA program management organization

☐ A description of the duties and responsibilities of each unit within the organization and how delegation of responsibilities is managed within the decommissioning program

☐ A description of how work performance is evaluated

☐ A description of the authority of each unit within the QA program

☐ An organization chart of the QA program organization

XIII.b. QUALITY ASSURANCE PROGRAM

☐ A commitment that activities affecting the quality of site decommissioning will be subject to the applicable controls of the QA program and activities covered by the QA program are identified on program defining documents

☐ A brief summary of the company's corporate QA policies

☐ A description of provisions to ensure that technical and quality assurance procedures required to implement the QA program are consistent with regulatory, licensing, and QA program requirements and are properly documented and controlled

☐ A description of the management reviews, including the documentation of concurrence in these quality-affecting procedures

☐ A description of the quality-affecting procedural controls of the principal contractors

☐ A description of how NRC will be notified of changes (a) for review and acceptance in the accepted description of the QA program as presented or referenced in the DP before implementation and (b) in organizational elements within 30 days after the announcement of the changes

☐ A description is provided of how management regularly assesses the scope, status, adequacy, and compliance of the QA program

☐ A description of the instruction provided to personnel responsible for performing activities affecting quality

☐ A description of the training and qualifications of personnel verifying activities

☐ For formal training and qualification programs, documentation includes the objectives and content of the program, attendees, and date of attendance

☐ A description of the self-assessment program to confirm that activities affecting quality comply with the QA program

☐ A commitment that persons performing self-assessment activities are not to have direct responsibilities in the area they are assessing

☐ A description of the organizational responsibilities for ensuring that activities affecting quality are (a) prescribed by documented instructions, procedures, and drawings and (b) accomplished through implementation of these documents

☐ A description of the procedures to ensure that instructions, procedures, and drawings include quantitative acceptance criteria and qualitative acceptance criteria for determining that important activities have been satisfactorily performed

XIII.c. DOCUMENT CONTROL

☐ A summary of the types of QA documents that are included in the program

☐ A description of how the licensee develops, issues, revises, and retires QA documents

XIII.d. CONTROL OF MEASURING AND TEST EQUIPMENT

☐ A summary of the test and measurement equipment used in the program

☐ A description of how and at what frequency the equipment will be calibrated

☐ A description of the daily calibration checks that will be performed on each piece of test or measurement equipment

☐ A description of the documentation that will be maintained to demonstrate that only properly calibrated and maintained equipment was used during the decommissioning

XIII.e. CORRECTIVE ACTION

☐   A description of the corrective action procedures for the facility, including a description of how the corrective action is determined to be adequate

☐   A description of the documentation maintained for each corrective action and any follow-up activities by the QA organization after the corrective action is implemented

XIII.f. QUALITY ASSURANCE RECORDS

☐   A description of the manner in which the QA records will be managed

☐   A description of the responsibilities of the QA organization

☐   A description of the QA records storage facility

XIII.g. AUDITS AND SURVEILLANCES

☐   A description of the audit program

☐   A description of the records and documentation generated during the audits and the manner in which the documents are managed

☐   A description of all follow-up activities associated with audits or surveillances

☐   A description of the trending/tracking that will be performed on the results of audits and surveillances

XIV. FACILITY RADIATION SURVEYS

XIV.a. RELEASE CRITERIA

☐   A summary table or list of the $DCGL_W$ for each radionuclide and impacted media of concern

☐   If Class 1 survey units are present, a summary table or list of area factors that will be used for determining a $DCGL_{EMC}$ for each radionuclide and media of concern

☐   If Class 1 survey units are present, the $DCGL_{EMC}$ values for each radionuclide and medium of concern

☐   If multiple radionuclides are present, the appropriate $DCGL_W$ for the survey method to be used

XIV.b. CHARACTERIZATION SURVEYS

☐ A description and justification of the survey measurements for impacted media

☐ A description of the field instruments and methods that were used for measuring concentrations and the sensitivities of those instruments and methods

☐ A description of the laboratory instruments and methods that were used for measuring concentrations and the sensitivities of those instruments and methods

☐ The survey results, including tables or charts of the concentrations of residual radioactivity measured

☐ Maps or drawings of the site, area, or building, showing areas classified as non-impacted or impacted

☐ Justification for considering areas to be non-impacted

☐ A discussion of why the licensee considers the characterization survey to be adequate to demonstrate that it is unlikely that significant quantities of residual radioactivity have gone undetected

☐ For areas and surfaces that are inaccessible or not readily accessible, a discussion of how they were surveyed or why they did not need to be surveyed

☐ For sites, areas, or buildings with multiple radionuclides, a discussion justifying the ratios of radionuclides that will be assumed in the final status survey or an indication that no fixed ratio exists and each radionuclide will be measured separately

XIV.c. IN-PROCESS SURVEYS

☐ A description of field screening methods and instrumentation

☐ A demonstration that field screening should be capable of detecting residual radioactivity at the DCGL

XIV.d. FINAL STATUS SURVEY DESIGN

☐ A brief overview describing the final status survey design

☐ A description and map or drawing of impacted areas of the site, area, or building classified by residual radioactivity levels (Class 1, 2, or 3) and divided into survey units with an explanation of the basis for division into survey units

☐ A description of the background reference areas and materials, if they will be used, and a justification for their selection

☐ A summary of the statistical tests that will be used to evaluate the survey results

☐ A description of scanning instruments, methods, calibration, operational checks, coverage, and sensitivity for each media and radionuclide

☐ For in-situ sample measurements made by field instruments, a description of the instruments, calibration, operational checks, sensitivity, and sampling methods, with a demonstration that the instruments and methods have adequate sensitivity

☐ A description of the analytical instruments for measuring samples in the laboratory, as well as calibration, sensitivity, and methods with a demonstration that the instruments and methods have adequate sensitivity

☐ A description of how the samples to be analyzed in the laboratory will be collected, controlled, and handled

☐ A description of the final status survey investigation levels and how they were determined

☐ A summary of any significant additional residual radioactivity that was not accounted for during site characterization

☐ A summary of direct measurement results and/or soil concentration levels in units that are comparable to the DCGL, and if data is used to estimate or update the survey unit

☐ A summary of the direct measurements or sample data used to both evaluate the success of remediation and to estimate the survey unit variance

XIV.e. FINAL STATUS SURVEY REPORT

☐ An overview of the results of the final status survey

☐ A discussion of any changes that were made in the final status survey from what was proposed in the DP or other prior submittals

☐ A description of the method by which the number of samples was determined for each survey unit

☐ A summary of the values used to determine the number of samples and a justification for these values

☐ The survey results for each survey unit include:

— The number of samples taken for the survey unit;

— A description of the survey unit, including (a) a map or drawing of the survey unit showing the reference system and random start systematic sample locations for Class 1 and 2 survey units and random locations shown for Class 3 survey units and reference areas, and (b) a discussion of remedial actions and unique features;

— The measured sample concentrations in units that are comparable to the DCGL;

— The statistical evaluation of the measured concentrations;

— Judgmental and miscellaneous sample data sets reported separately from those samples collected for performing the statistical evaluation;

— A discussion of anomalous data, including any areas of elevated direct radiation detected during scanning that exceeded the investigation level or measurement locations in excess of $DCGL_W$; and

— A statement that a given survey unit satisfied the $DCGL_W$ and the elevated measurement comparison if any sample points exceeded the $DCGL_W$.

☐ A description of any changes in initial survey unit assumptions relative to the extent of residual radioactivity (e.g., material not accounted for during site characterization)

☐ A description of how ALARA practices were employed to achieve final activity levels

☐ If a survey unit fails, a description of the investigation conducted to ascertain the reason for the failure and a discussion of the impact that the failure has on the conclusion that the facility is ready for final radiological surveys and that it satisfies the release criteria

☐ If a survey unit fails, a discussion of the impact that the reason for the failure has on other survey unit information

XV. FINANCIAL ASSURANCE

XV.a. COST ESTIMATE

☐ A cost estimate that appears to be based on documented and reasonable assumptions

XV.b. CERTIFICATION STATEMENT

☐ The certification statement is based on the licensed possession limits and the applicable quantities specified in 10 CFR 30.35, 40.36, or 70.25

☐ The licensee is eligible to use a certification of financial assurance and, if eligible, that the certification amount is appropriate

XV.c. FINANCIAL MECHANISM

☐ The financial assurance mechanism supplied by the licensee consists of one or more of the following instruments:

— Trust fund;

— Escrow account;

— Government fund;

— Certificate of deposit;

— Deposit of government securities;

— Surety bond;

— Letter of credit;

— Line of credit;

— Insurance policy;

— Parent company guarantee;

— Self guarantee;

— External sinking fund;

— Statement of intent; or

— By special arrangements with a government entity assuming custody or ownership of the site.

☐ The financial assurance mechanism is an originally signed duplicate

☐ The wording of the financial assurance mechanism is identical to the recommended wording provided in Appendix F of this document

☐ For a licensee regulated under 10 CFR Part 72, a means is identified in the DP for adjusting the financial assurance funding level over any storage and surveillance period

☐ The amount of financial assurance coverage provided by the licensee for site control and maintenance is at least as great as that calculated using the formula provided in this NUREG

XVI. RESTRICTED USE/ALTERNATE CRITERIA

XVI.a. RESTRICTED USE

XVI.a.1.   Eligibility Demonstration

☐ A demonstration that the benefits of dose reduction are less than the cost of doses, injuries, and fatalities

☐ A demonstration that the proposed residual radioactivity levels at the site are ALARA

XVI.a.2.   Institutional Controls

☐ A description of the legally enforceable institutional control(s) and an explanation of how the institutional control is a legally enforceable mechanism

☐ A description of any detriments associated with the maintenance of the institutional control(s)

☐ A description of the restrictions on present and future landowners

☐ A description of the entities enforcing, and their authority to enforce, the institutional control(s)

☐ A description of the design features of the site that support institutional controls

☐ A discussion of the durability of the institutional control(s), including the performance of any engineered barriers used

☐ A description of the activities that the entity with the authority to enforce the institutional controls may undertake to enforce the institutional control(s)

☐ A description of the manner in which the entity with the authority to enforce the institutional control(s) will be replaced if that entity is no longer willing or able to enforce the institutional control(s) (this may not be needed for Federal or State entities)

☐ A description of the duration of the institutional control(s), the basis for the duration, the conditions that will end the institutional control(s), and the activities that will be undertaken to end the institutional control(s)

☐ A description of the plans for corrective actions that may be undertaken in the event the institutional control(s) fail

☐ A description of the records pertaining to the institutional controls, how and where will they will be maintained, and how the public will have access to the records

XVI.a.3.　Site Maintenance and Financial Assurance

☐ A demonstration that an appropriately qualified entity has been provided to control and maintain the site

☐ A description of the site maintenance and control program and the basis for concluding that the program is adequate to control and maintain the site

☐ A description of the arrangement or contract with the entity charged with carrying out the actions necessary to maintain control at the site

☐ A demonstration that the contract or arrangement will remain in effect for as long as feasible, and include provisions for renewing or replacing the contract

☐ A description of the manner in which independent oversight of the entity charged with maintaining the site will be conducted and what entity will conduct the oversight

☐ A demonstration that the entity providing the oversight has the authority to replace the entity charged with maintaining the site

☐ A description of the authority granted to the third party to perform, or have performed, any necessary maintenance activities

☐ Unless the entity is a government entity, a demonstration that the third party is not the entity holding the financial assurance mechanism

☐ A demonstration that sufficient records evidencing to official actions and financial payments made by the third party are open to public inspection

☐ A description of the periodic site inspections that will be performed by the third party, including the frequency of the inspections

☐ A copy of the financial assurance mechanism provided by the licensee

☐ A demonstration that the amount of financial assurance provided is sufficient to allow an independent third party to carry out any necessary control and maintenance activities

XVI.a.4.   Obtaining Public Advice

☐ A description of how individuals and institutions that may be affected by the decommissioning were identified and informed of the opportunity to provide advice to the licensee

☐ A description of the manner in which the licensee obtained advice from these individuals or institutions

☐ A description of how the licensee provided for participation by a broad cross-section of community interests in obtaining the advice

☐ A description of how the licensee provided for a comprehensive, collective discussion on the issues by the participants represented

☐ A copy of the publicly available summary of the results of discussions, including individual viewpoints of the participants on the issues, and the extent of agreement and disagreement among the participants

☐ A description of how this summary has been made available to the public

☐ A description of how the licensee evaluated the advice, and the rationale for incorporating or not incorporating the advice from affected members of the community into the DP

XVI.a.5.   Dose Modeling and ALARA Demonstration

☐ A summary of the dose to the average member of the critical group when radionuclide levels are at the DCGL with institutional controls in place, as well as the estimated doses if they are no longer in place

☐ A summary of the evaluation performed pursuant to Chapter 6 of Volume 2 of this NUREG series, demonstrating that these doses are ALARA

☐ If the estimated dose to the average member of the critical group could exceed 100 mrem/y (but would be less than 500 mrem/y) when the radionuclide levels are at the DCGL, a demonstration that the criteria in 10 CFR 20.1403(e) have been met

XVI.b. ALTERNATE CRITERIA

☐ A summary of the dose in TEDE(s) to the average member of the critical group when the radionuclide levels are at the DCGL (considering all man-made sources other than medical)

☐ A summary of the evaluation performed pursuant to Chapter 6 of Volume 2 of this NUREG series demonstrating that these doses are ALARA

☐ An analysis of all possible sources of exposure to radiation at the site and a discussion of why it is unlikely that the doses from all man-made sources, other than medical, will be more than 1 mSv/y (100 mrem/y)

☐ A description of the legally enforceable institutional control(s) and an explanation of how the institutional control is a legally enforceable mechanism

☐ A description of any detriments associated with the maintenance of the institutional control(s)

☐ A description of the restrictions on present and future landowners

☐ A description of the entities enforcing and their authority to enforce the institutional control(s)

☐ A discussion of the durability of the institutional control(s)

☐ A description of the activities that the party with the authority to enforce the institutional controls will undertake to enforce the institutional control(s)

☐ A description of the manner in which the entity with the authority to enforce the institutional control(s) will be replaced if that entity is no longer willing or able to enforce the institutional control(s)

☐ A description of the duration of the institutional control(s), the basis for the duration, the conditions that will end the institutional control(s), and the activities that will be undertaken to end the institutional control(s)

☐ A description of the corrective actions that will be undertaken in the event the institutional control(s) fail

☐ A description of the records pertaining to the institutional controls, how and where they will be maintained, and how the public will have access to the records

☐ A description of how individuals and institutions that may be affected by the decommissioning were identified and informed of the opportunity to provide advice to the licensee

☐ A description of the manner in which the licensee obtained advice from affected individuals or institutions

☐ A description of how the licensee provided for participation by a broad cross-section of community interests in obtaining the advice

☐   A description of how the licensee provided for a comprehensive, collective discussion on the issues by the participants represented

☐   A copy of the publicly available summary of the results of discussions, including individual viewpoints of the participants on the issues and the extent of agreement and disagreement among the participants

☐   A description of how this summary has been made available to the public

☐   A description of how the licensee evaluated advice from individuals and institutions that could be affected by the decommissioning and the manner in which the advice was addressed

Appendix E
Checklist for Use of Generic Environmental Impact Statement for License Termination and Sample Environmental Assessment for Sites that Use Screening Criteria

E.1 LICENSE TERMINATION RULE—GENERIC ENVIRONMENTAL IMPACT STATEMENT (GEIS) REFERENCE FACILITIES[1, 2] CHECKLIST

The GEIS reference facilities were developed to broadly and generically represent categories of licensee facilities. Specific facilities will not exactly match the descriptions of the reference facilities. The primary purpose of comparing a specific facility to the reference facility with regard to dose assessment is to determine whether the specific facility has important contaminants, potential scenarios, or pathways that were not analyzed for the reference facilities or which may be sufficiently different from those in the GEIS to change conclusions regarding environmental impacts. In general, if a specific facility has contaminants, concentrations, and spacial distributions less than or generally equivalent to those used for the reference facilities, the GEIS should be applicable. Potential limitations of the GEIS dose assessments, as well as a summary of the characteristics of the reference facilities, are shown below.

GEIS Dose Assessment Scenarios: Potential Limitations

1.0   Building Occupancy (structures)

    1.1   Structures are assumed to have a 70-year life span following license termination. A shorter expected life span is acceptable. Expected life spans significantly longer than 70 years may require additional analysis if long-lived radionuclides are involved.

    1.2   Contamination significantly more extensive than that analyzed in the GEIS should be evaluated on a site-specific basis. Areas and concentrations analyzed in the GEIS are shown in the tables in the following sections.

    1.3   Radionuclides present on the site that contribute significantly to dose but which were not analyzed in the GEIS for the subject facility type will need to be evaluated separately.

Checklist for Structures

Yes No

☐   ☐   Additional analysis required due to expected >70 year building lifespan following decommissioning <u>and</u> long-lived contaminants

---

[1]   Overview from NUREG–1496, Volume 1, Section 3.

[2]   Note that the GEIS does not apply to uranium mills or tailings, low level waste, or high level waste.

☐ ☐ Contamination significantly more extensive than that shown in Tables 1 through 6 [Table E.1] in the following sections

☐ ☐ Radionuclides present that contribute significantly to dose, were not analyzed in the GEIS, and could change the conclusions in the GEIS regarding environmental impacts

2.0 Residual (soil)

2.1 Assumes people live and work onsite over a 1,000 year period.

2.2 If the site is subject to weather or other events (tornadoes, flash floods, etc) that could result in extensive redistribution or mass movement of contaminates, additional analysis may be required.

2.3 Pre-existing contamination of ground water must be evaluated on a site-specific basis.

2.4 10 CFR 20.302/20.2002 or other burials or disposal areas may need additional site-specific evaluation.

Checklist for Soil

Yes No

☐ ☐ Site subject to weather or other events that could redistribute contaminants in ways not analyzed in the GEIS

☐ ☐ Contaminated ground water present

☐ ☐ Onsite burials or disposal areas

2. Example fuel cycle facilities: power, test, and research reactors; uranium fuel fabrication; uranium hexafluoride conversion facilities; and independent spent fuel storage installations (ISFSI).

The power, test, and research reactors, and the ISFSI have been consolidated into a single analysis in the GEIS based on common radionuclide contaminants (Co-60 and Cs-137), and are represented by the analysis for the power reactor.

The uranium fabrication facility is used as the reference for both the fabrication and hexafluoride facilities.

Table E.1   Facility Characteristics Applicable to Dose Modeling[‡]

1. Soil Surface Activities for the Radionuclides of Interest[a]	
Radionuclide	Surface Concentration (pCi/g)
Co-60	60
Cs-137	20
Uranium	1000

2. Total and Contaminated Surface Areas for Structures and Soils at Reference Site[b]

Reference Facility	Structures Radionuclide Activity[c], dpm/100 cm^2	Structures Surface Areas				Soil Surface Area, ft^2	
		ft^2		% Contaminated			
		Floor	Wall	Floor	Wall	Total Site	Contaminated
PWR	7.5 x 10^6 Co60 2.4 x 10^6 Cs137	250000	300000	10	2	50 x 10^6	3000
Uranium Fuel Fab	18000	240000	240000	50	5	4.7 x 10^6	100000

3. Contamination Distribution Used in the GEIS[d]

Reference Facility	Soil Area	Soil Depth	Soil Volume	Below-Building Soil Depth	Below-Building Soil Volume
	ft^2	cm	m^3	cm	m^3
Nuclear Power Plant	3000	4 – 100	12 – 250	3 – 21	15 – 100
Uranium Fuel Fabrication	100000	44 – 300	4,000 – 28,000	18 – 29	82 – 129

‡   Example Non-Fuel-Cycle facilities: universities; medical institutions; sealed source manufactures; industrial users of radioisotopes; research and development laboratories; and rare metal refineries.

Notes:

a   See NUREG–1748 for detailed guidance

b   The estimated surface areas listed above (reproduced from NUREG–1496, Appendix C) are based on limited information and in many cases represent an engineering judgment based on the size of the building structural facilities and types of operation These estimates are considered to be conservatively large, i e , they probably overestimate the actual areas involved

c   Radionuclide activity shown is for building surfaces Radionuclide activity for soil surfaces is given below

d   Data obtained from NUREG–1496, Table C 1 10 and C 2 6

Table E.1 Facility Characteristics Applicable to Dose Modeling (continued)[‡]

4. Total and Contaminated Surface Areas for Structures and Soils at Reference Sites[e]

| Reference Facility | Structures Radionuclide Activity[f], dpm/100 cm^2 | Structures Surface Areas | | | | Soil Surface Area, ft^2 | |
| | | ft^2 | | % Contaminated | | | |
		Floor	Wall	Floor	Wall	Total Site	Contaminated
Sealed Source Manufacturer	102,000 Co60 33,300 Cs137	6000	4600	10	5	40000	5000
Rare Metal Extraction	18,000 Thorium	150000	180000	40	10	740000	100000

5. Soil Surface Activities for the Radionuclides of Interest[g]

Radionuclide	Surface Concentration (pCi/g)
Co-60	60
Cs-137	20
Thorium	200

6. Contamination Distribution Used in the GEIS[h]

| Reference Facility | Soil Area | Soil Depth | Soil Volume | Below-Building Soil Depth | Below-Building Soil Volume |
	ft^2	cm	m^3	cm	m^3
Sealed Source	5000	4 – 90	20 – 425	3 – 21	0 – 2
Rare Metals Extraction	100000	10 – 60	1,000 – 5,700	0 – 2	0 – 6
Slag Pile Volume: 7,000 m^3					

---

‡   Example Non-Fuel-Cycle facilities: universities; medical institutions; sealed source manufactures; industrial users of radioisotopes; research and development laboratories; and rare metal refineries.

Notes (continued):

e   The estimated surface areas listed above (reproduced from NUREG–1496, Appendix C) are based on limited information and in many cases represent an engineering judgment based on the size of the building structural facilities and types of operation These estimates are considered to be conservatively large, i e , they probably overestimate the actual areas involved

f   Radionuclide activity shown is for building surfaces Radionuclide activity for soil surfaces is shown below

g   Data obtained from NUREG–1496, Table C 7 1 2

h   Data obtained from NUREG–1496, Table C 3 6 and C 4 6

The sealed source manufacturers and R&D laboratories are consolidated into a single analysis. The analysis of the rare metals processing facility is used to represent all other non-fuel-cycle facilities with low to medium to significant contamination.

Materials licensees who use only sealed sources or short-lived radioactive materials are not expected to require decontamination of buildings or soil, and therefore the impacts and costs of decommissioning are expected to be minimal. The GEIS does not include a detailed analysis of these licensees. If a licensee in this category does require more extensive analysis, the applicability of the GEIS should be evaluated by comparison to the other non-fuel-cycle reference facilities based on the radioisotopes and contamination levels involved.

E.2 SAMPLE ENVIRONMENTAL ASSESSMENT FOR RELYING ON THE LICENSE TERMINATION RULE GEIS TO SATISFY NEPA OBLIGATIONS FOR SITES THAT USE SCREENING CRITERIA

NOTE: This is an example of an EA. Depending upon the nature of the action involved, a specific EA might require some adjustments to this model. Therefore, NUREG–1748, Chapter 3 should be reviewed in detail prior to preparation of the EA in order to ensure that all necessary issues are addressed. While the conclusion to this example EA is a Finding of No Significant Impact (FONSI), the actual FONSI is published separately in the *Federal Register*. Note that this has not been formatted for inclusion in a *Federal Register* notice.

U.S. NUCLEAR REGULATORY COMMISSION

DOCKET NO. **XXXX**

[Date]

Environmental Assessment

Related to Issuance of a License Amendment

of U.S. Nuclear Regulatory Commission **XXXX** Materials License No. **XXXXXX,**

[name of licensee]

The U.S. Nuclear Regulatory Commission (NRC) has performed an environmental review of the ABC Corporation's (ABC's or the licensee's) decommissioning plan for its Anytown site. The XYZ facility is operated by ABC in Anytown, State. ABC was authorized by NRC from 1973 to 1998 to use radioactive materials for nuclear medicine purposes at the site. In 1998, ABC ceased operations at the XYZ facility and requested that NRC terminate its license. ABC has conducted characterization surveys of the facilities and identified carbon-14 (C-14) and tritium (H-3) contamination in the XYZ nuclear medicine facilities. The NRC staff has evaluated ABC's request and has developed an environmental assessment (EA) to support the review of ABC's proposed decommissioning plan and license amendment request, in accordance with the requirements of 10 CFR Part 51. Based on the staff evaluation, the conclusion of the EA is a Finding of No Significant Impact (FONSI) on human health and the environment for the proposed licensing action.

Introduction

[Describe the proposal. Briefly characterize the location and contamination and reference the decommissioning plan or license termination request.]

The XYZ facility incorporates 10 buildings on 40 acres located at 123 East Main Street in Anytown. ABC conducted a characterization survey of the affected areas and developed a decommissioning plan. The survey confirmed the presence of H-3 contamination in portions of the facility and was used as the basis for development of the decommissioning plan. The affected area of the XYZ facility consists of the former nuclear medicine laboratory and associated rooms in the basement of one building, identified as Building One. ABC proposed to use the screening values developed by NRC as the derived concentration guideline levels (DCGLs) for decommissioning and as the basis for demonstrating that the site meets NRC's radiological cleanup criteria.

The Proposed Action

[Describe the proposal. Briefly summarize the remediation activities and reference the decommissioning plan or license termination request for a more thorough description.]

The proposed action is to amend NRC Radioactive Materials License Number 31–XXXX to incorporate appropriate and acceptable DCGLs into the license. The licensee's objective for the decommissioning project, as stated in the decommissioning plan, is to decontaminate and remediate the affected areas of Building One sufficiently to enable unrestricted use, while ensuring exposures to occupational workers and the public during the decommissioning are maintained as low as reasonably achievable (ALARA). ABC's decommissioning plan for the XYZ facility proposes to use DCGLs that are screening values developed by NRC (65 FR 37186, June 13, 2000) to demonstrate compliance with the radiological criteria for license termination in 10 CFR 20.1402. The DCGLs will define the maximum amount of residual radioactivity on building surfaces, equipment and materials and in soils, that will satisfy the NRC requirements in Subpart E, 10 CFR Part 20, "Radiological Criteria for License Termination." The DCGLs proposed to be incorporated into the license are as follows:

Radionuclide	Release of equipment & materials (surfaces)	Building surfaces	Soil
C-14			
H-3			

Need for the Proposed Action

The purpose of the proposed action is to reduce residual radioactivity at the XYZ facility to a level that permits release of the property for unrestricted use and termination of the license. NRC is fulfilling its responsibilities under the Atomic Energy Act to make a decision on a proposed license amendment for decommissioning that ensures protection of the public health and safety and environment.

Environmental Impacts of the Proposed Action

[Briefly summarize special environmental or cultural issues that may be associated with a decommissioning action and may require a particular analysis. Include radiological and nonradiological direct and indirect impacts — including: ecological; aesthetic; historical; cultural; socioeconomic; and health. Also, include a paragraph on adverse impacts, cumulative impacts and the evaluation of the significance of the impacts (see NUREG–1748 Chapter 3).]

The NRC staff has reviewed the decommissioning plan for the XYZ facility and examined the impacts of decommissioning. Based on its review, the staff has determined that the affected environment and the environmental impacts associated with the decommissioning of the XYZ facility are bounded by the impacts evaluated by the "Generic Environmental Impact Statement in Support of Rulemaking on Radiological Criteria for License Termination of NRC–Licensed Nuclear Facilities" (NUREG–1496). The staff also finds that the proposed decommissioning of the XYZ facility is in compliance with 10 CFR 20.1402, the radiological criteria for unrestricted use.

Since ceasing operations, the XYZ site has been stabilized to prevent contamination from spreading beyond its current locations. Access to the contaminated areas is controlled to assure the health and safety of workers and the public. No ongoing licensed activities are occurring in the facilities.

Contamination controls will be implemented during decommissioning to prevent airborne and surface contamination from escaping the remediation work areas, and therefore no release of airborne contamination is anticipated. However, the potential will exist for generating airborne radioactive material during decontamination, removal and handling of contaminated materials. If produced, any effluent from the proposed decommissioning activities will be limited in accordance with NRC requirements in 10 CFR Part 20 or contained onsite or treated to reduce contamination to acceptable levels before release, and shall be maintained ALARA. Release of contaminated liquid effluents are not expected to occur during the work.

ABC and subcontractors will perform the remediation under the XYZ license, with ABC overseeing the activities and maintaining primary responsibility. The XYZ facility has adequate radiation protection procedures and capabilities, and will implement an acceptable program to keep exposure to radioactive materials ALARA. As noted above, ABC has prepared a decommissioning plan describing the work to be performed, and work activities are not anticipated to result in a dose to workers or the public in excess of the 10 CFR Part 20 limits. Past experiences with decommissioning activities at sites similar to the XYZ facility indicate that public and worker exposure will be far below the limits found in 10 CFR Part 20.

Environmental Impacts of the Alternatives to the Proposed Action

[Describe reasonable alternatives. A no-action alternative should always be considered.]

The only alternative to the proposed action of allowing decommissioning of the site is no action. The no-action alternative is not acceptable because it will result in violation of NRC's Timeliness Rule (10 CFR 30.36), which requires licensees to decommission their facilities when licensed activities cease, and to request termination of their radioactive materials license.

Agencies and Persons Consulted

This EA was prepared by NRC staff and coordinated with the following agencies: State Department of Environmental Quality, State Office of Historical Preservation, State Fish and Wildlife Service, and the U.S. Fish and Wildlife Service.

NRC staff provided a draft of its Environmental Assessment to [**State agency**] for review.

On [**provide date**], the [**State agency**] responded by [**telephone, letter, etc**] and stated that it had no comments. [**Or explain comments. Additionally, identify any sources used, if applicable**.]

Conclusion

The NRC staff has concluded that the proposed action complies with 10 CFR Part 20. Decommissioning of the site to the DCGLs proposed for this action will result in reduced residual contamination levels in the facility, enabling release of the facility for unrestricted use and termination of the radioactive materials license. No radiologically contaminated effluents are expected during the decommissioning. Occupational doses to decommissioning workers are expected to be low and well within the limits of 10 CFR Part 20. No radiation exposure to any member of the public is expected, and public exposure will therefore also be less than the applicable public exposure limits of 10 CFR Part 20.

NRC has prepared this EA in support of the proposed license amendment to incorporate appropriate and acceptable DCGLs and to use the proposed DCGLs for the planned decommissioning by the licensee at the XYZ facility. On the basis of the EA, NRC has concluded that the environmental impacts from the proposed action are expected to be insignificant and has determined not to prepare an environmental impact statement for the proposed action.

List of Preparers

Sources Used

Appendix F
Master Inspection Plan

NRC will develop a Master Inspection Plan utilizing the inspection procedures listed below. NRC's Inspection Manual Chapters (MCs), Inspection Procedures (IPs), and Temporary Instructions (TIs) listed below are especially applicable and are recommended to be used for inspections at sites undergoing decommissioning. These documents should be used as guidelines for inspectors in determining the inspection requirements for decommissioning and radiological safety aspects of various types of licensee activities. Recommended core chapters and procedures for the decommissioning inspection program are starred (*). These documents are available through NRC's Web site.

Document No.	Title – Subject Area Applicable to Decommissioning
MC 0610	"Inspection Reports" – Documentation of inspections.
MC 2600*	"Fuel Cycle Facility Operational Safety and Safeguards Inspection Program" – Program requirements applicable to decommissioning: s 2600–01 through 2600–07; Appendix A, Parts I and IV.
MC 2602*	"Decommissioning Inspection Program for Fuel Cycle Facilities and Material Licenses"
MC 2605*	"Decommissioning Procedures for Fuel Cycle and Materials Licenses"
MC 2681*	"Physical Protection and Transport of SNM and Irradiated Fuel Inspection of Fuel Facilities" – Safeguards and physical security of the site including: Sections 2681–01 through 2681–03; the physical protection inspection programs in Exhibits 1 through 6; and the material control and accounting inspection program in Exhibit 8.
MC 2800*	"Materials Inspection Program" – Program requirements applicable to decommissioning: all sections, for licensee activities and NRC inspections that carry over from licensee operations.
IP 36100	"10 CFR Part 21 Inspection at Nuclear Power Reactors" – Inspection of equipment used during decommissioning.
IP 83822*	"Radiation Protection" – Radiation protection.
IP 83890*	"Closeout Inspection and Survey" – Confirmatory surveys.
IP 83895	"Radiation Protection – Follow up on Expired Licenses"–Radiation protection.
IP 84850*	"Radioactive Waste Management – Inspection of Waste Generator Requirements of 10 CFR 20 and 10 CFR 61" – Waste management.
IP 84900	"Low-Level Radioactive Waste Storage" – Waste storage.
IP 86740*	"Inspection of Transportation Activities" – Transportation of waste.

Document No.	Title – Subject Area Applicable to Decommissioning
IP 87103	"Inspection of Materials Licensees Involved in an Incident or Bankruptcy Filing" – Response to incidents or bankruptcy.
IP 87104*	"Decommissioning Inspection Procedure for Materials Licensees"
IP 88005*	"Management Organization and Controls" – Quality assurance program; records control; internal review and audit; procedure control; safety committee.
IP 88015*	"Headquarters Nuclear Criticality Safety Program" – Criticality for fuel cycle facilities.
IP 88020 & IP 88025	"Regional Criticality Safety Inspection Program" and "Maintenance and Surveillance Testing" – Surveillance testing and safety limits.
IP 88035*	Radioactive Waste Management" – Waste management.
IP 88045*	"Environmental Protection" – Releases to the environment.
IP 88050* & IP 88055*	"Emergency Preparedness" and "Fire Protection" – Emergency planning.
IP 88104*	"Decommissioning Inspection Procedure for Fuel Cycle Facilities"
IP 93001	"OSHA Interface Activities" – Interface with Other Agencies.
TI 2800/026	"Follow up Inspection of Formerly Licensed Sites Identified as Potentially Contaminated"

Appendix G
Safety Evaluation Report Outline and Template

G.1 OUTLINE FOR A SAFETY EVALUATION REPORT

The following outline for an SER is shown as a checklist in Appendix D of this report. The checklist shows the finding NRC must reach before a DP is approved. Note that some sections may not apply to all facilities and DP's. For example, the discussion of Institutional Controls does not apply to sites planning release for unrestricted use.

I. EXECUTIVE SUMMARY

II. FACILITY OPERATING HISTORY

1. License Number/Status/Authorized Activities
2. License History
3. Previous Decommissioning Activities
4. Spills
5. Prior Onsite Burials

III. FACILITY DESCRIPTION

1. Site Location and Description
2. Population Distribution
3. Current/Future Land Use
4. Meteorology and Climatology
5. Geology and Seismology
   6. Surface Water Hydrology
   7. Ground Water Hydrology
   8. Natural Resources

IV. RADIOLOGICAL STATUS OF FACILITY

1. Contaminated Structures
2. Contaminated Systems and Equipment
3. Surface Soil Contamination
4. Subsurface Soil Contamination
5. Surface Water
6. Ground Water

V. DOSE MODELING

1. Unrestricted Release Using Screening Criteria
2. Unrestricted release using screening criteria for building surface residual radioactivity

XI. ENVIRONMENTAL MONITORING AND CONTROL PROGRAM

1.   Environmental ALARA Evaluation Program
2.   Effluent Monitoring Program
3.   Effluent Control Program

XII. RADIOACTIVE WASTE MANAGEMENT PROGRAM

1.   Solid Radwaste
2.   Liquid Radwaste
3.   Mixed Waste

XIII. QUALITY ASSURANCE PROGRAM

1.   Organization
2.   Quality Assurance Program
3.   Document Control
4.   Control of Measuring and Test Equipment
5.   Corrective Action
6.   Quality Assurance Records
7.   Audits and Surveillances

XIV. FACILITY RADIATION SURVEYS

1.   Release Criteria
2.   Characterization Surveys
3.   In-Process Surveys
4.   Final Status Survey Design
5.   Final Status Survey Report

XV. FINANCIAL ASSURANCE

1.   Cost Estimate
2.   Certification Statement
3.   Financial Mechanism

XVI. RESTRICTED USE/ALTERNATE CRITERIA

1.   Restricted Use
2.   Eligibility Demonstration
3.   Institutional Controls
4.   Site Maintenance and Financial Assurance

G.2 TEMPLATE FOR A SAFETY EVALUATION REPORT

The template and data file below demonstrate the correct format and language for SERs. This template and a sample data file contain the areas of review and the findings required before approval of the DP can be issued. They are available to NRC staff electronically as SER1.dat and SER-1.frm (in WordPerfect 8 format) on the shared network drive. These electronic files are combined using the WP merge function to generate the outline of a site-specific SER.

1.0   Executive Summary

2.0   Facility Operating History

      2.1   License Number/Status/Authorized Activities

NRC staff has reviewed the information in the "Facility Operating History" section of the Decommissioning Plan for the [*facility name*], license number 040–0XXXX located at [*facility location*] according to the Consolidated Decommissioning Guidance, Volume 1, Section 16.2 (Facility Operating History). Based on this review, NRC staff has determined that the licensee [*licensee name*] has provided sufficient information to aid NRC staff in evaluating the licensee's determination of the radiological status of the facility and the licensee's planned decommissioning activities, to ensure that the decommissioning can be conducted in accordance with NRC requirements. *(Note to reviewers — this finding incorporates the results of the staff's assessment under Sections 2.2, 2.3, 2.4, and 2.5, below)*

      2.2   License History

      2.3   Previous Decommissioning Activities

      2.4   Spills

      2.5   Prior Onsite Burials

      2.6   Prior Partial Site Releases

3.0   Facility Description

      3.1   Site Location and Description

      3.2   Population Distribution

      3.3   Current/Future Land Use

      3.4   Meteorology and Climatology

3.5   Geology and Seismology

3.6   Surface Water Hydrology

3.7   Ground water Hydrology

3.8   Natural Resources

4.0 Radiological Status of Facility

4.1   Contaminated Structures

[The staff may combine the evaluation finding for the licensee's description of contaminated structures with the findings for the remaining areas in this section of Volume 1 as follows.]

NRC staff has reviewed the information in the "Facility Radiological Status" section of the Decommissioning Plan for the [*facility name*], license number 040–0XXXX located at [*facility location*] according to the Consolidated Decommissioning Guidance, Volume 1, Section 16.4 (Radiological Status of Facility). Based on this review, NRC staff has determined that the licensee has described the types and activity of radioactive material contamination at its facility sufficiently to allow the NRC staff to evaluate the potential safety issues associated with remediating the facility, whether the remediation activities and radiation control measures proposed by the licensee are appropriate for the type of radioactive material present at the facility, whether the licensee's waste management practices are appropriate, and whether the licensee's cost estimates are plausible, given the amount of contaminated material that will need to be removed or remediated.

4.2   Contaminated Systems and Equipment

4.3   Surface Soil Contamination

4.4   Subsurface Soil Contamination

4.5   Surface Water

4.6   Ground Water

5.0 Dose Analysis

5.1   Unrestricted Release using Screening Criteria

5.1.1   Building Surfaces

The staff has reviewed the dose modeling analyses for *[identifier/name of decommissioning option]* as part of the review of the [*licensee name*] decommissioning plan, using the Consolidated Decommissioning Guidance, Volume 2, Section 5.1.1 (Building Surface Evaluation Criteria).

The staff concludes that the dose estimate calculated using the default screening analysis is appropriate for the decommissioning option and exposure scenario assumed. In addition, this dose estimate provides reasonable assurance that the dose criterion in 10 CFR 20.1402 will be met. This conclusion is based on the modeling effort performed by the staff in initially developing the default screening analysis.

In determining the dose to the average member of the critical group, the licensee has used the assumptions inherent in the screening analysis and the parameter uncertainties have been previously evaluated on a generic basis by the staff as part of establishing the default screening analysis.

5.1.2   Surface Soil

The staff has reviewed the dose modeling analyses for *[identifier/name of decommissioning option]* as part of the review of the [*licensee name*]'s decommissioning plan, using the Consolidated Decommissioning Guidance, Volume 2, Section 5.1.2 (Surface Soil Evaluation Criteria).

The staff concludes that the dose estimate calculated using the default screening analysis is appropriate for the decommissioning option and exposure scenario assumed. In addition, this dose estimate provides reasonable assurance that the dose criterion in 10 CFR 20.1402 will be met. This conclusion is based on the modeling effort performed by the staff in initially developing the default screening analysis.

In determining the dose to the average member of the critical group, the licensee has used the assumptions inherent in the screening analysis and the parameter uncertainties have been previously evaluated on a generic basis by the staff as part of establishing the default screening analysis.

5.2   Unrestricted Release using Site-Specific Information

The staff has reviewed the dose modeling analyses for *[identifier/name of decommissioning option]* as part of the review of the [*licensee name*]'s decommissioning plan, using the Consolidated Decommissioning Guidance, Volume 2, Section 5.2 (Unrestricted Release Using Site-Specific Information).

The staff concludes that the dose modeling completed for [option description] is reasonable and is appropriate for the exposure scenario under consideration. In addition, the dose estimate provides reasonable assurance that the dose to the average member of the critical group is not likely to exceed the 0.25 mSv (25 mrem) annual dose criterion in 10 CFR 20.1402. This conclusion is based on the modeling effort performed by the licensee and the independent analysis performed by the staff.

In determining the dose, the licensee has a combination of the conceptual model, exposure scenario, mathematical model and input parameters to calculate a reasonable estimate of dose. The licensee has adequately considered the uncertainties inherent in the modeling analysis.

5.3   Restricted Release using Site-Specific Information

The staff has reviewed the dose modeling analyses for *[identifier/name of decommissioning option]* as part of the review of the *[licensee name]*'s decommissioning plan, using the Consolidated Decommissioning Guidance, Volume 2, Section 5.3 (Restricted Release).

The staff concludes that the dose modeling completed for *[option description]* is reasonable and is appropriate for the exposure scenarios under consideration. The dose estimates provide reasonable assurance that if the restrictions work as proposed, the dose to the average member of the critical group is not likely to exceed the 0.25-mSv (25-mrem) annual dose limit in 10 CFR 20.1403(b), and if they fail, the dose to the average member of the critical group is not likely to exceed the annual dose limit in 10 CFR 20.1403(e). This conclusion is based on the modeling effort performed by the licensee and the independent analyses and review performed by the staff.

In determining the dose, the licensee has used a combination of the conceptual model(s), exposure scenarios, mathematical model(s), and input parameters to calculate a reasonable estimate of dose. The licensee has adequately considered the uncertainties inherent in the modeling analysis.

[The staff's technical evaluation report should include: (1) a brief summary of the exposure scenarios used to evaluate compliance with 10 CFR 20.1403; (2) a brief summary of any independent analyses conducted by the staff; (3) reference to the mathematical method(s) used; and (4) a comparison of the dose value(s) computed by the staff with those of the licensee.]

5.4   Release Involving Alternate Criteria

The staff has reviewed the dose modeling analyses for *[identifier/name of decommissioning option]* as part of the review of the *[licensee name]*'s decommissioning plan, using the Consolidated Decommissioning Guidance, Volume 2, Section 5.4 (Release Involving Alternate Criteria).

The staff concludes that the dose modeling completed for [*option description*] is reasonable and is appropriate for the exposure scenarios under consideration. This conclusion is based on the modeling effort performed by the licensee and the independent analyses and review performed by the staff.

In determining the dose, the licensee has used a combination of the conceptual model(s), exposure scenarios, mathematical model(s), and input parameters to calculate a reasonable estimate of dose. The licensee has adequately considered the uncertainties inherent in the modeling analysis.

[The staff's technical evaluation report should include: (1) a brief summary of the exposure scenarios used; (2) a brief summary of any independent analyses conducted by the staff; (3) reference to the mathematical method(s) used; and (4) a comparison of the dose value(s) computed by the staff with those of the licensee.]

6.0   Planned Decommissioning Activities

    6.1   Contaminated Structures

[The staff may combine the evaluation finding for the licensee's description of the planned decommissioning activities with the findings for the remaining areas in this section of Volume 1 as follows.]

The NRC staff has reviewed the decommissioning activities described in the Decommissioning Plan for the [*facility name*], license number 040–0XXXX located at [*facility location*] according to the Consolidated Decommissioning Guidance, Volume 1, Section 17.1 (Planned Decommissioning Activities). Based on this review the NRC staff has determined that the licensee, [*licensee name*], has provided sufficient information to allow the NRC staff to evaluate the licensee's planned decommissioning activities to ensure that the decommissioning can be conducted in accordance with NRC requirements.

    6.2   Contaminated Systems and Equipment

    6.3   Soil

    6.4   Surface and Ground Water

    6.5   Schedules

7.0   Project Management and Organization

      7.1   Decommissioning Management Organization

The NRC staff has reviewed the description of the decommissioning project management organization, position descriptions, management and safety position qualification requirements and the manner in which the licensee [*licensee name*], license number 040–0XXXX will use contractors during the decommissioning of its facility located at [*location of facility*] according to the Consolidated Decommissioning Guidance, Volume 1, Section 17.2 (Project Management and Organization). Based on this review, the NRC staff has determined that the licensee, [*licensee name*], has provided sufficient information to allow the NRC staff to evaluate the licensee's decommissioning project management organization and structure to determine if the decommissioning can be conducted safely and in accordance with NRC requirements. *(Note that this finding incorporates the results of the staff's assessment under Sections 7.2–7.5, below.)*

      7.2   Decommissioning Task Management

      7.3   Decommissioning Management Positions and Qualifications

            7.3.1   Radiation Safety Officer

      7.4   Training

      7.5   Contractor Support

8.0   Radiation Safety and Health Program

      8.1   Radiation Safety Controls and Monitoring for Workers

            8.1.1   WorkplaceAir Sampling Program

The NRC staff has reviewed the information in the Decommissioning Plan for the [*facility name*], license number 040–0XXXX located at [*facility location*] according to the Consolidated Decommissioning Guidance, Volume 1, Section 17.3.1.1 (Workplace Air Sampling Program). Based on this review, the NRC staff has determined that the licensee, [*licensee name*], has provided sufficient information on when air samples will be taken in work areas, the types of air sample equipment to be used and where they will be located in work areas, calibration of flow meters, minimum detectable activities (MDA) of equipment to be used for analyses of radionuclides collected during air sampling, action levels for airborne radioactivity (and corrective actions to be taken when these levels are exceeded) to allow the NRC staff to conclude that the licensee's air sampling program will comply with 10 CFR 20.1204, 20.1501(a)–(b), 20.1502(b), 20.1703(a)(3)(i)–(ii), and Regulatory Guide 8.25.

8.1.2   Respiratory Protection Program

The NRC staff has reviewed the information in the Decommissioning Plan for the [*facility name*], license number 040–0XXXX located at [*facility location*] according to the Consolidated Decommissioning Guidance, Volume 1, Section 17.3.1.2 (Respiratory Protection Program). Based on this review, the NRC staff has determined that the licensee, [*licensee name*], has provided sufficient information to implement an acceptable respiratory protection program so as to allow the NRC staff to conclude that the licensee's program will comply with 10 CFR 20.1101(b), and 10 CFR 20.1701 to 20.1704 and Appendix A of 10 CFR Part 20.

8.1.3   Internal Exposure Determination

The NRC staff has reviewed the information in the Decommissioning Plan for the [*facility name*], license number 040–0XXXX located at [*facility location*] according to the Consolidated Decommissioning Guidance, Volume 1, Section 17.3.1.3 (Internal Exposure Determination). Based on this review, the NRC staff has determined that the licensee, [*licensee name*], has provided sufficient information on methods to calculate internal dose of a worker based upon measurements from air samples or bioassay samples to allow the NRC staff to conclude that the licensee's program to determine internal exposure will comply with 10 CFR 20.1101(b), 20.1201(a)(1), (d) and (e), 20.1204 and 20.1502(b).

8.1.4   External Exposure Determination

The NRC staff has reviewed the information in the Decommissioning Plan for the [*facility name*], license number 040–0XXXX located at [*facility location*] according to the Consolidated Decommissioning Guidance, Volume 1, Section 17.3.1.4 (External Exposure Determination). Based upon this review, the NRC staff has determined that the licensee, [*licensee name*], has provided sufficient information on methods to measure or calculate the external dose of a worker to allow the NRC staff to conclude that the licensee's program to determine external exposure will comply with the requirements of 10 CFR 20.1101(b), 20.1201(c), 20.1203, 20.1501(a)(2)(i) and (c), 20.1502(a), and 20.1601.

8.1.5   Summation of Internal and External Exposures

The NRC staff has reviewed the information in the Decommissioning Plan for the [*facility name*], license number 040–0XXXX located at [*facility location*] according to the Consolidated Decommissioning Guidance, Volume 1, Section 17.3.1.5 (Summation of Internal and External Exposures). Based on this review, the NRC staff has determined that the licensee, [*licensee name*], has provided sufficient information to conclude that the licensee's program for summation of internal and external exposures will comply with 10 CFR 20.1202 and 20.1208(c)(1) and (2), and 20.2106.

8.1.6   Contamination Control Program

The NRC staff has reviewed the information in the Decommissioning Plan for the [*facility name*], license number 040–0XXXX located at [*facility location*] according to the Consolidated Decommissioning Guidance, Volume 1, Section 17.3.1.6 (Contamination Control Program). Based on this review, the NRC staff has determined that the licensee, [*licensee name*], has provided sufficient information to control contamination on skin, on protective and personal clothing, on fixed and removable contamination on work surfaces, on transport vehicles, on equipment (including ventilation hoods), and on packages to allow the NRC staff to conclude that the licensee's contamination control program will comply with 20.1501(a), 20.1702, 20.1906 (b), (d), and (f) of 10 CFR Part 20. The staff has verified that the information summarized under "Evaluation Criteria" above is included in the licensee's description of the methodology used to control contamination at the facility.

8.1.7   Instrumentation Program

The NRC staff has reviewed the information in the Decommissioning Plan for the [*facility name*], license number 040–0XXXX located at [*facility location*] according to the Consolidated Decommissioning Guidance, Volume 1, Section 17.3.1.7 (Instrumentation Program). Based on this review, the NRC staff has determined that the licensee, [*licensee name*], has provided sufficient information on the sensitivity and the calibration of instruments and equipment to be used to make quantitative measurements of ionizing radiation during surveys to allow the NRC staff to conclude that the licensee's instrumentation program will comply with 10 CFR 20.1501(b) and (c).

8.2   Nuclear Criticality Safety

The results of staff's review of the licensee's submittal should be stated in the form of findings of fact and acceptability for compliance with the regulations as guided by this NUREG series. In particular, the evaluation should make findings as to the acceptability and adequacy of the items addressed by this NUREG series to provide reasonable assurance of protection of public health and safety from the risk of nuclear criticalities during decommissioning.

8.3   Health Physics Audits and Recordkeeping Program

The NRC staff has reviewed the description of the licensee's, [*facility name*], license number 040–0XXXX audit and recordkeeping program which the licensee will utilize during the decommissioning of its facility located at [*location of facility*] according to the Consolidated Decommissioning Guidance, Volume 1, Section 17.3.3 (Health Physics Audits, Inspections, and Recordkeeping Program). Based on this review, the NRC staff has determined that the licensee, [*licensee name*], has provided sufficient information to allow the NRC staff to evaluate the licensee's executive management and RSO audit and recordkeeping program to determine if the decommissioning can be conducted safely and in accordance with NRC requirements.

9.0   Environmental Monitoring and Control Program

9.1   Environmental ALARA Evaluation Program

The NRC staff has reviewed the information in the Decommissioning Plan for the [*facility name*], license number 040–0XXXX located at [*facility location*] according to the Consolidated Decommissioning Guidance, Volume 1, Section 17.4 (Environmental Monitoring and Control Program) . Based on this review, the NRC staff has determined that the licensee, [*licensee name*], has provided sufficient information on the staff to conclude that the licensee's program will comply with 10 CFR Part 20.

Note that the results from the staff's evaluation of the Environmental ALARA, Environmental Monitoring, and Effluent Control programs should be combined in this finding.

9.2   Effluent Monitoring Program

9.3   Effluent Control Program

10.0   Radioactive Waste Management Program

10.1   Solid Radioactive Waste

[The staff may combine the evaluation finding for the licensee's description of solid radioactive waste management programs with the findings for the remaining areas in this section of Volume 1, as follows.]

The NRC staff has reviewed the licensee's descriptions of the radioactive waste management program for the [*facility name*], license number 040–0XXXX located at [*facility location*] according to the Consolidated Decommissioning Guidance, Volume 1, Section 17.5 (Radioactive Waste Management Program) . Based on this review, the NRC staff has determined that the licensee's, [*licensee name*], programs for the management of radioactive waste generated during decommissioning operations ensure that the waste will be managed in accordance with NRC requirements and in a manner that is protective of the public health and safety.

10.2   Liquid Radioactive Waste

10.3   Mixed Waste

11.0   Quality Assurance Program

11.1   Organization

The NRC staff has reviewed the Quality Assurance Program for the [*facility name*], license number 040–0XXXX located at [*facility location*] according to the Consolidated

Decommissioning Guidance, Volume 1, Section 17.6 (Quality Asssurance Program). Based on this review, the NRC staff has determined that the licensee's, [*licensee name*], QA program is sufficient to ensure that information submitted to support the decommissioning of its facility should be of sufficient quality to allow the staff to determine if the licensee's planned decommissioning activities can be conducted in accordance with NRC requirements. *(Note that this finding incorporates the results of the staff's assessment of the entire QA program as described in the following subsections of Section 17.6.)*

        11.2   Quality Assurance Program

        11.3   Document Control

        11.4   Control of Measuring and Test Equipment

        11.5   Corrective Action

        11.6   Quality Assurance Records

        11.7   Audits and Surveillances

12.0   Facility Radiation Surveys

        12.1   Release Criteria

The NRC staff has reviewed the information in the Decommissioning Plan *(or the Final Status Survey Report)* for the [*facility name*], license number 040–0XXXX according to the Consolidated Decommissioning Guidance, Volume 2, Section 4.1 (Release Criteria)). Based on this review, the NRC staff have determined that [*licensee name*] has summarized the DCGL(s) and area factors used for survey design and for demonstrating compliance with the radiological criteria for license termination.

        12.2   Characterization Surveys

The NRC staff has reviewed the information in the Decommissioning Plan (or Final Status Survey Report) for the [*facility name*], license number 040–0XXXX according to the Consolidated Decommissioning Guidance, Volume 2, Section 4.2 (Characterization Surveys). This review has determined that the radiological characterization of the site, area, or building is adequate to permit planning for a remediation that will be effective and will not endanger the remediation workers, to demonstrate that it is unlikely that significant quantities of residual radioactivity has not gone undetected, and to provide information that will be used to design the final status survey.

12.3   Remedial Action Support Surveys

The staff should combine the findings from Section 12.3 with those from Sections 12.1 and 12.2.

12.4   Final Status Survey Design

The NRC staff has reviewed the information in the Decommissioning Plan (or the Final Status Survey Report) for the [*facility name*], license number 040–0XXXX according the Consolidated Decommissioning Guidance, Volume 2, Section 4.4 (Final Status Survey Design). Based on this review, the NRC staff has determined that [*licensee name*] final status survey design is adequate to demonstrate compliance with radiological criteria for license termination.

12.5   Final Status Survey Report

The NRC staff has reviewed the final status survey results for the [*facility name*], license number 040–0XXXX according the Consolidated Decommissioning Guidance, Volume 2, Section 4.5 (Final Status Survey Report). Based on this review, the NRC staff has determined that [*licensee name*] has demonstrated that the licensee's site (or area or building) meets the radiological criteria for license termination.

13.0 Financial Assurance

13.1   Cost Estimate

13.1.1   Evaluation Criteria Applicable to all Cost Estimates For Restricted or Unrestricted Use

13.1.2   Additional Information Criteria Applicable to Cost Estimates for Restricted Use

The NRC staff has reviewed the cost estimate for the [*facility name*], license number 040–0XXXX located at [*facility location*] according to the Consolidated NMSS Decommissioning Guidance, Volume 3, Section 4.1 (Cost Estimate (as Contained in a Decommissioning Funding Plan or Decommissioning Plan)). Based on this review, the NRC staff has determined that the cost estimate submitted by the licensee adequately reflects the costs to carry out all required decommissioning activities prior to license termination and, if the license is being terminated under restricted conditions, to enable an independent third party to assume and carry out responsibilities for any necessary control and maintenance of the site.

13.2   Prescribed Amount

The NRC staff has reviewed the prescribed amount for the [*facility name*], license number 040–0XXXX located at [*facility location*] according to the Consolidated NMSS Decommissioning Guidance, Volume 3, Section 4.2 (Prescribed Amount). Based on this review, the NRC staff has determined that the certification statement submitted by the licensee specifies the appropriate information and level of financial assurance coverage.

13.3   Financial Assurance Mechanism

13.3.3   Evaluation Criteria for Specific Financial Assurance Mechanisms (Unrestricted and Restricted Use)

13.3.3.1   Trust Funds

13.3.3.2   Escrow Accounts

13.3.3.3   Government Funds

13.3.3.4   Certificates of Deposit

13.3.3.5   Deposits of Government Securities

13.3.3.6   Surety Bonds

13.3.3.7   Letters of Credit

13.3.3.8   Lines of Credit

13.3.3.9   Insurance Policies

13.3.3.10   Parent Company Guarantees

13.3.3.11   Self Guarantees

13.3.3.12   External Sinking Funds

13.3.3.13   Statements of Intent

13.3.3.14   Special Arrangements with a Government Entity

13.3.3.15   Standby Trust Funds

The NRC staff has reviewed the financial assurance mechanism(s) for the [*facility name*], license number 040–0XXXX located at [*facility location*] according to the Consolidated NMSS Decommissioning Guidance, Volume 3, Section 4.3 (Financial Assurance Mechanisms). Based on this review, the NRC staff has determined that the financial assurance mechanism(s) submitted by the licensee is *(are)* adequate to ensure that sufficient funds will be available to carry out all required decommissioning activities prior to license termination and, if the license is being terminated under restricted conditions, to enable an independent third party to assume and carry out responsibilities for any necessary control and maintenance of the site.

14.0 Restricted Use/Alternate Criteria

14.1   Restricted Use

14.1.1   Initial Eligibility Demonstration

The NRC staff has reviewed the licensee's justification for requesting license termination under restricted conditions in the Decommissioning Plan for the [*facility name*], license number 040–0XXXX located at [*facility location*] according to the Consolidated Decommissioning Guidance, Volume 1, Section 17.7.2 (Initial Eligibility Demonstration).

Based on this review, the NRC staff has determined that the licensee [*insert name and license number*] has adequately demonstrated that [*insert one: [the benefits of dose reduction are less than the cost of doses, injuries and fatalities] or [further reductions in radioactivity levels at the site are unnecessary because they are ALARA]*].

14.1.2   Institutional Controls and Engineered Barriers

The NRC staff has reviewed the description of the institutional controls and engineered barriers in the Decommissioning Plan for the [*facility name*], license number 040–0XXXX located at [*facility location*] according to the Consolidated Decommissioning Guidance, Volume 1, Section 17.7.3 (Institutional Controls and Engineered Barriers) and considered public comments made pursuant to 10 CFR 20.1405. The NRC staff has determined that the licensee, [*licensee name*], has adequately demonstrated that institutional controls are enforceable, durable and should ensure that doses to the public comply with the criteria in 10 CFR 20.1403. In addition, the licensee has made adequate provisions to replace the entity charged with enforcing the institutional control in the event that the entity is no longer willing or able to enforce the institutional control and has made provisions to address corrective actions at the site.

14.1.3   Site Maintenance and Long-Term Monitoring

The NRC staff has reviewed the information regarding site maintenance, long-term monitoring, and financial assurance in the Decommissioning Plan for the [*facility name*], license number 040–0XXXX located at [*facility location*] according to the Consolidated Decommissioning Guidance, Volume 1, Section 17.7.4 (Site Maintenance and Long-Term Monitoring). Based on this review, the NRC staff has determined that the licensee, [*licensee name*], has adequately demonstrated that the site maintenance arrangements and financial assurance mechanism are adequate to ensure that the site will be maintained in accordance with the institutional controls described in the decommissioning plan and that sufficient funds are available to allow an independent third party to assume and carry out responsibilities for any necessary control and maintenance of the site after the NRC has terminated the license.

14.1.4   Obtaining Public Advice

The NRC staff has reviewed the information regarding how advice from individuals and institutions that may be affected by the decommissioning was obtained and summarized in the Decommissioning Plan for the [*facility name*], license number 040–0XXXX located at [*facility location*] according to the Consolidated Decommissioning Guidance, Volume 1, Section 17.7.5 (Obtaining Public Advice). Based on this review, the NRC staff has determined that the licensee, [*licensee name*], has demonstrated that advice from individuals and institutions that may be affected by the decommissioning was sought, obtained, evaluated, and, as appropriate, incorporated into the licensee's plans for decommissioning its facility, in accordance with NRC requirements at 10 CFR 20.1403(d).

14.1.5   Dose Modeling and ALARA Demonstration

The NRC staff has reviewed the information regarding compliance with 10 CFR 20.1403(e) summarized in the Decommissioning Plan for the [*facility name*], license number 040–0XXXX located at [*facility location*] according to the Consolidated Decommissioning Guidance, Volume 1, Section 17.7.6 (Dose Modeling and ALARA Demonstration). Based on this review, the NRC staff has determined that the licensee, [*licensee name*], has demonstrated that doses to the public from residual radioactive material after the license is terminated should not exceed 0.25 mSv/yr (25 mrem/yr), with restriction in place or [*insert one: 1 mSv/yr (100 mrem/yr) if restrictions are removed, or 5 mSv/yr (500 mrem/yr), with conditions, if restrictions are removed*].

[If doses are estimated to be in excess of 1mSv/yr (100 mrem/yr), but less than 5 mSv/yr (500 mrem/yr) with institutional controls removed, insert the following.]

In addition the licensee, [*licensee name*], has demonstrated that further reductions in residual radioactivity necessary to comply with the 1 mSv/yr (100 mrem/yr requirement) [*select as appropriate*: are not technically achievable, are prohibitively expensive, or result in net public or environmental harm]. The licensee has also established durable institutional controls for the site. Finally, the licensee has provided sufficient financial assurance to allow an independent third party to carry out rechecks at the site at no less than every 5 years and the amount of financial assurance is sufficient to assume and carry out responsibilities for any necessary control and maintenance of the controls at the site.

14.2   Alternate Criteria

NRC staff has reviewed the information regarding the licensee's, [*licensee name*], request to decommission its facility pursuant to 10 CFR 20.1404, summarized in the Decommissioning Plan for the [*facility name*], license number 040–0XXXX located at [*facility location*] according to the Consolidated Decommissioning Guidance, Volume 1, Section 17.8 (Alternate Criteria) and considered public comments made pursuant to 10 CFR 20.1405. Based on this review, NRC staff has determined that the licensee, [*licensee name*], has demonstrated that doses to the public from residual radioactive material after the license is terminated should be less than the NRC limits of

1 mSv/yr (100 mrem/yr) and are ALARA. In addition, the licensee has adequately demonstrated that it has provided appropriate restrictions according to the provisions of 10 CFR 20.1403 and has adequately sought, managed and addressed advice from individuals and institutions that may be affected by the decommissioning.

Appendix H
EPA/NRC
Memorandum of Understanding

MEMORANDUM OF UNDERSTANDING BETWEEN THE ENVIRONMENTAL PROTECTION AGENCY AND THE NUCLEAR REGULATORY COMMISSION

CONSULTATION AND FINALITY ON DECOMMISSIONING AND DECONTAMINATION OF CONTAMINATED SITES

I.   Introduction

The Environmental Protection Agency (EPA) and the Nuclear Regulatory Commission (NRC), in recognition of their mutual commitment to protect the public health and safety and the environment, are entering into this Memorandum of Understanding (MOU) in order to establish a basic framework for the relationship of the agencies in the radiological decommissioning and decontamination of NRC–licensed sites. Each Agency is entering into this MOU in order to facilitate decision-making. It does not establish any new requirements or rights on parties not subject to this agreement.

II.   Purpose

The purpose of this MOU is to identify the interactions of the two agencies for the decommissioning and decontamination of NRC–licensed sites and to indicate the way in which those interactions will take place. Except for Section VI, addressing corrective action under the Resource Conservation and Recovery Act (RCRA), this MOU is limited to the coordination between EPA, when acting under its Comprehensive Environmental Response, Compensation and Liability Act (CERCLA) authority, and NRC, when a facility licensed by the NRC is undergoing decommissioning, or when a facility has completed decommissioning, and the NRC has terminated its license. It continues a basic policy of EPA deferral to NRC decision-making in the decommissioning of NRC–licensed sites except in certain circumstances, and establishes the procedures to govern the relationship between the agencies in connection with the decommissioning of sites at which those circumstances arise.

III.   Background

An August 3, 1999, report (106–286) from the House Committee on Appropriations to accompany the bill covering EPA's FY1999 Appropriations/FY 2000 budget request states:

> Once again the Committee notes that the Nuclear Regulatory Commission (NRC) has and will continue to remediate sites under its jurisdiction to a level that fully protects public health and safety, and believes that any reversal of the long-standing policy of the Agency to defer to the NRC for cleanup of NRC's licensed sites is not a good use of public or private funds. The interaction of the EPA with the NRC, NRC licensees, and others, with regard to sites being remediated under NRC regulatory requirements—when not specifically requested by the NRC—has created stakeholder concerns regarding the authority and finality of NRC licensing decisions, the duration and costs of site cleanup,

and the potential future liability of parties associated with affected sites. However, the Committee recognizes that there may be circumstances at specific NRC licensed sites where the Agency's expertise may be of critical use to the NRC. In the interest of ensuring that sites do not face dual regulation, the Committee strongly encourages both agencies to enter into an MOU which clarifies the circumstances for EPA's involvement at NRC sites when requested by the NRC. The EPA and NRC are directed to report to the Committee on Appropriations no later than May 1, 2000, on the status of the development of such an MOU.

Since September 8, 1983, EPA has generally deferred listing on the CERCLA National Priorities List (NPL) those sites that are subject to NRC's licensing authority, in recognition that NRC's actions are believed to be consistent with the CERCLA requirement to protect human health and the environment. However, as EPA indicated in the Federal Register notice announcing the policy of CERCLA deferral to NRC, if EPA "determines that sites which it has not listed as a matter of policy are not being properly responded to, the Agency will consider listing those sites on the NPL" (see 48 FR 40658).

EPA reaffirms its previous 1983 deferral policy. EPA expects that any need for EPA CERCLA involvement in the decommissioning of NRC licensed sites should continue to occur very infrequently because EPA expects that the vast majority of facilities decommissioned under NRC authority will be decommissioned in a manner that is fully protective of human health and the environment. By this MOU, EPA agrees to a deferral policy regarding NRC decision-making without the need for consultation except in certain limited circumstances as specified in paragraphs V.C.2 and V.C.3.

One set of circumstances in which continued consultation should occur, pursuant to the procedures defined herein, relates to sites at which the NRC determines during the license termination process that there is radioactive ground-water contamination above certain limits. Pursuant to its License Termination rule, NRC applies a dose criterion that encompasses all pathways, including ground water. In its cleanup of sites pursuant to CERCLA, by contrast, EPA customarily establishes a separate ground-water cleanup standard in which it applies certain Maximum Contaminant Levels (MCLs, found at 40 CFR 141) promulgated for radionuclides and other substances pursuant to the Safe Drinking Water Act. NRC has agreed in this MOU to consult with EPA on the appropriate approach in responding to the circumstances at particular sites with ground-water contamination at the time of license termination in excess of EPA's MCLs or those sites for which NRC contemplates either restricted release or the use of alternate criteria for license termination, or radioactive contamination at the time of license termination exceeds the corresponding levels in Table 1 [Table H.1] as provided in Section V.C.2.

IV.   Principles

In carrying out their respective responsibilities, the EPA and the NRC will strive to:

1.   Establish a stable and predictable regulatory environment with respect to EPA's CERCLA authority in and NRC's decommissioning of contaminated sites.

2.   Ensure, to the extent practicable, that the responsibilities of the NRC under the AEA and the responsibilities of EPA under CERCLA are implemented in a coordinated and consistent manner.

V.   Implementation

A.   Scope

This MOU is intended to address issues related to the EPA involvement under CERCLA in the cleanup of radiologically contaminated sites under the jurisdiction of the NRC. EPA will continue its CERCLA policy of September 8, 1983, which explains how EPA implements deferral decisions regarding listing on the NPL of any sites that are subject to NRC's licensing authority. The NRC's review of sites under NRC jurisdiction indicates that few of these sites have radioactive ground-water contamination in excess of the EPA's MCLs. At those sites at which NRC determines during the license termination process that there is radioactive ground-water contamination above the relevant EPA MCLs, NRC will consult with EPA and, if necessary, discuss with EPA the use of flexibility under EPA's phased approach to addressing ground-water contamination. NRC has agreed in this MOU to consult with EPA on the appropriate approach in responding to the circumstances at particular sites where ground-water contamination will exceed EPA's MCLs, NRC contemplates either restricted release or the use of alternate criteria for license termination, or radioactive contamination at the time of license termination exceeds the corresponding levels in Table 1 [Table H.1] as provided in Section V.C.2.

B.   General

Each agency will keep the other agency generally informed of its relevant plans and schedules, will respond to the other agency's requests for information to the extent reasonable and practicable, and will strive to recognize and ameliorate to the extent practicable any problems arising from implementation of this MOU.

C.   NRC Responsibilities

1.   NRC will continue to ensure remediation of sites under its jurisdiction to a level that fully protects public health and safety.

2.   For NRC–licensed sites at which NRC determines during the license termination process that there is radioactive ground-water contamination in excess of EPA's MCLs, or for

which NRC contemplates either restricted release (10 CFR 20.1403) or the use of alternate criteria for license termination (10 CFR 20.1404), NRC will seek EPA's expertise to assist in NRC's review of a decommissioning or license termination plan. In addition, NRC will consult with EPA if either the planned level of residual radioactive soil concentrations in the proposed action or the actual residual level of radioactive soil concentrations found in the final site survey exceed the radioactive soil concentration in Table 1 [Table H.1]. With respect to all such sites, the NRC will consult with EPA on the application of the NRC decommissioning requirements and will take such action as the NRC determines to be appropriate based on its consultation with EPA. For example, if NRC determines during the license termination process that there will be radioactive ground-water contamination in excess of EPA's MCLs at the time of license termination, then NRC will discuss with EPA the use of flexibility under EPA's phased approach for addressing ground-water contamination. If NRC does not adopt recommendations provided by the EPA, NRC will inform EPA of the basis for its decision not to do so.

3. NRC will defer to EPA regarding matters involving hazardous materials not under NRC's jurisdiction.

D. EPA Responsibilities

1. If the NRC requests EPA's consultation on a decommissioning plan or license termination plan, EPA will provide, within 90 days of NRC's notice to EPA, written notification of its views on the matter.

2. Consistent with this MOU, EPA agrees to a policy of deferral to NRC decision making on decommissioning without the need for consultation on sites other than those presenting the circumstances described in Sections V.C.2 and V.C.3. The agencies will consult with each other pursuant to the provisions of this MOU with respect to those sites presenting the circumstances described in Sections V.C.2 and V.C.3. EPA does not expect to undertake CERCLA actions related to radioactive contamination at a site that has been decommissioned in compliance with the NRC's standards, including a site addressed under Section V.C.2, despite the agencies decision to engage in consultation on such sites. EPA's deferral policy, and its expectation of not taking CERCLA action, continues to apply to sites that are covered under Section V.C.2.

3. For NRC–licensed sites presenting the circumstances described in Section V.C.2 and for which NRC has not adopted the EPA recommendation, EPA will consult with NRC on any CERCLA actions EPA expects to take if EPA does not agree with the NRC's decision.

4. EPA will resolve any CERCLA concerns involving hazardous substances outside of NRC's jurisdiction at NRC licensed sites, including concerns involving hazardous constituents that are not under the authority of NRC. As provided in Section V.D.2, EPA under CERCLA will defer or consult with NRC as appropriate regarding matters involving AEA materials under NRC's jurisdiction.

E.   Other Provisions

1.   Nothing in this MOU shall be deemed to establish any right nor provide a basis for any action, either legal or equitable by any person, or class of persons challenging a government action or failure to act.

2.   Each agency will appoint a designated contact for implementation of this MOU. The designated individuals will meet at least annually or at the request of either agency to review NRC–licensed sites that meet the criteria for consultation pursuant to Section V.C.2. The NRC designated contact is the Director, Office of Nuclear Materials Safety and Safeguards, and the EPA designated contact is the Director Office of Emergency and Remedial Response, or as each designee delegates.

3.   This MOU will remain in effect until terminated by the written notice of either party submitted six months in advance of termination.

4.   Within six months of the execution of this MOU, each party will revise its guidance to its Headquarters and Regional Offices to reflect the terms of this MOU.

5.   If differences arise that cannot be resolved by senior EPA and NRC management within 90 days, then either senior EPA or NRC management may raise the issue to their respective agency head.

Section VI.   Corrective Action Under RCRA

Some NRC sites undergoing decommissioning may be subject to cleanup under RCRA corrective action authority. This authority, administered either by EPA or authorized states, requires cleanup of releases of hazardous waste or constituents at hazardous waste treatment, storage or disposal facilities. NRC sites subject to RCRA corrective action will be expected to meet RCRA cleanup standards for chemical contamination within EPA's jurisdiction. EPA Office of Solid Waste's policy is to encourage regional and State program implementers to coordinate RCRA cleanups with decommissioning, as appropriate, at those NRC sites subject to EPA's corrective action authority (See letter from Elizabeth Cotsworth, Acting Director, Office of Solid Waste to James R. Roewer, USWAG, dated March 5, 1997).

EPA will continue to support coordination of cleanups under the RCRA corrective action program with decommissioning at NRC sites consistent with its March 5, 1997 policy. In addition, under RCRA the majority of States are authorized to implement the corrective action requirements. States are not signatories to this MOU; however, EPA will encourage States to act in accordance with this policy where they have responsibility for RCRA corrective action at NRC sites undergoing decommissioning.

APPENDIX H

Items 1 and 3 of the "Other Provisions" of Section V.E apply to this section.

/RA/	09/30/2002	/RA/	10/09/2002
Christine T. Whitman Administrator	Date	Richard A. Meserve Chairman	Date
US Environmental Protection Agency		US Nuclear Regulatory Commission	

Table H.1 Consultation Triggers for Residential and Commercial/Industrial Soil Contamination (MOU Table 1)

Except for radium-226, thorium-232, or total uranium, concentrations should be aggregated using a sum of the fraction approach to determine site-specific consultation trigger concentrations. This table is based on single contaminant concentrations for residential and commercial/industrial land use when using generally accepted exposure parameters. Table users should select the appropriate column based on the site's reasonably anticipated land use.

Radionuclide	Residential Soil Concentration	Industrial/Commercial Soil Concentration
H-3	228 pCi/g	423 pCi/g
C-14	46 pCi/g	123,000 pCi/g
Na-22	9 pCi/g	14 pCi/g
S-35	19,600 pCi/g	32,200,000 pCi/g
Cl-36	6 pCi/g	10,700 pCi/g
Ca-45	13,500 pCi/g	3,740,000 pCi/g
Sc-46	105 pCi/g	169 pCi/g
Mn-54	69 pCi/g	112 pCi/g
Fe-55	269,000 pCi/g	2,210,000 pCi/g
Co-57	873 pCi/g	1,420 pCi/g
Co-60	4 pCi/g	6 pCi/g
Ni-59	20,800 pCi/g	1,230,000 pCi/g
Ni-63	9,480 pCi/g	555,000 pCi/g
Sr-90+D	23 pCi/g	1,070 pCi/g
Nb-94	2 pCi/g	3 pCi/g
Tc-99	25 pCi/g	89,400 pCi/g
I-129	60 pCi/g	1,080 pCi/g
Cs-134	16 pCi/g	26 pCi/g
Cs-137+D	6 pCi/g	11 pCi/g
Eu-152	4 pCi/g	7 pCi/g
Eu-154	5 pCi/g	8 pCi/g
Ir-192	336 pCi/g	544 pCi/g
Pb-210+D	15 pCi/g	123 pCi/g
Ra-226	5 pCi/g	5 pCi/g
Ac-227+D	10 pCi/g	21 pCi/g
Th-228+D	15 pCi/g	25 pCi/g
Th-232	5 pCi/g	5 pCi/g
U-234	401 pCi/g	3,310 pCi/g
U-235+D	20 pCi/g	39 pCi/g
U-238+D	74 pCi/g	179 pCi/g
total uranium	47 mg/kg	1230 mg/kg
Pu-238	297 pCi/g	1,640 pCi/g
Pu-239	259 pCi/g	1,430 pCi/g
Pu-241	40,600 pCi/g	172,000 pCi/g
Am-241	187 pCi/g	568 pCi/g
Cm-242	32,200 pCi/g	344,000 pCi/g
Cm-243	35 pCi/g	67 pCi/g

Appendix I
Using the Internet to Obtain Copies of NRC Documents and Other Information

In an effort to make NRC documents and information readily available to licensees and the general public, NRC is placing documents and information on its Internet Web site.

Many of the reference sections of the NUREG refer to a World Wide Web address on the Internet (e.g., http://www.nrc.gov). Applicants and licensees who have Internet access may use the referenced address to find more information on a topic, the referenced document, or information on obtaining the referenced document.

To access the referenced site, type the address into the location box of the Internet browser software and press the enter key. Sometimes the given address does not go directly to the necessary page; however, the addressed page will have links to the information referenced in this NUREG. Generally, links appear either as blue text or as a picture in the document. To use a link, place the pointer on the blue text or picture. The pointer will change from an arrow to a hand with the index finger extended. By single-clicking the mouse on the blue text or picture, the Internet browser will go to the selected page. For example, to review the definitions in 10 CFR Part 20, type http://www.nrc.gov in the location box of your browser and press the enter key. After NRC's homepage comes up, place the pointer on the "Electronic Reading Room." A drop-down list will appear. Next, place the pointer on the text, "Regulations (10 CFR)" and click the mouse. Place the pointer on the blue text "20" and single-click. Finally, place the pointer on the blue text "20.1003 Definitions" and single-click. This specific example regarding NRC's Web site is current as of August 12, 2003.

Appendix J
Sample Licenses

NRC FORM 374

U.S. NUCLEAR REGULATORY COMMISSION

MATERIALS LICENSE

CORRECTED COPY

Pursuant to the Atomic Energy Act of 1954, as amended, the Energy Reorganization Act of 1974 (Public Law 93-438), and Title 10, Code of Federal Regulations, Chapter I, Parts 30, 31, 32, 33, 34, 35, 36, 39, 40, and 70, and in reliance on statements and representations heretofore made by the licensee, a license is hereby issued authorizing the licensee to receive, acquire, possess, and transfer byproduct, source, and special nuclear material designated below; to use such material for the purpose(s) and at the place(s) designated below; to deliver or transfer such material to persons authorized to receive it in accordance with the regulations of the applicable Part(s). This license shall be deemed to contain the conditions specified in Section 183 of the Atomic Energy Act of 1954, as amended, and is subject to all applicable rules, regulations, and orders of the Nuclear Regulatory Commission now or hereafter in effect and to any conditions specified below.

Licensee

1. Fuel Renovation, Inc.

2. 1205 Flag Road
   Paul, BL XXXXX-XXXX

3. License Number SNM-XXX, Amendment 27

4. Expiration Date July 31, 2009

5. Docket No. 70-XXX
   Reference No.

6. Byproduct Source, and/or Special Nuclear Material	7. Chemical and/or Physical Form	8. Maximum amount that Licensee May Possess at Any One Time Under This License
A. Uranium enriched up to 100 w/% in the U235 isotope which may contain up to 10^{-6} grams plutonium per gram of uranium, 0.25 millicuries of fission products per gram of uranium, and 1.5×10^{-5} grams transuranic materials (including plutonium), per gram of uranium, as contaminants.	A. As described in AppendixB to Chapter 1 of the FR license application, excluding pyrophoric forms	A. 7000 kgs U235
B. Uranium enriched up to 100 w/% in the U233 isotope	B.1 Any form, but only as residual contamination from previous operations	B.1 One kg U233
	B.2 Any form, as received for analysis and/or for input into development studies	B.2 250 grams U233
C. Plutonium	C.1 As counting and calibration standards	C.1 10 millicuries
	C.2 As residual contamination and	C.2 As described in the license application and

APPENDIX J

NRC FORM 374A U.S. NUCLEAR REGULATORY COMMISSION		2
MATERIALS LICENSE SUPPLEMENTARY SHEET	License Number SNM–XXX	
	Docket or Reference Number 70–XX	
	Amendment 27 CORRECTED COPY	

holdup from previous operations.

an FR report to the NRC transmitted by letter dated January 21, XXXX (FR Document No. 28G94-001), and FR report dated October 17, 1988 (FR Document No. 28G88-007)

C.3 As received for analysis or for input into development studies, any form except pyrophoric

C.3 200 grams

C.4 As waste resulting from decontamination and volume reduction of equipment received from other organizations, any form except pyrophoric

C.4 200 grams

D. Transuranic Isotopes

D. As waste resulting from processing enriched uranium

D. 20 grams

E. Fission Products

E. As waste resulting from processing enriched uranium

E. 50 Curies each isotope, total not to exceed 500 Curies, Cs-137 not to exceed 5 Curies, Co-60 not to exceed 5 Curies, H-3 not to exceed 15 Curies, I-129 not to exceed 100 millicuries.

9. Authorized place of use: The licensee's existing facilities in Uncommon County, Bliss, as described in the referenced application.

NRC FORM 374A	U S. NUCLEAR REGULATORY COMMISSION	3

MATERIALS LICENSE SUPPLEMENTARY SHEET	License Number SNM–XXX
	Docket or Reference Number 70–XX
	Amendment 27
	CORRECTED COPY

10. This license shall be deemed to contain two sections: Safety Conditions and Safeguards Conditions. These sections are part of the license, and the licensee is subject to compliance with all listed conditions in each section.

FOR THE NUCLEAR REGULATORY COMMISSION

Date: _____   By: _____ . Chief
Fuel Cycle Licensing Branch
Division of Fuel Cycle Safety
and Safeguards
Washington, DC 20555

NRC FORM 374A	U.S. NUCLEAR REGULATORY COMMISSION		4
MATERIALS LICENSE SUPPLEMENTARY SHEET		License Number SNM–XXX	
		Docket or Reference Number 70–XX	
		Amendment 27 CORRECTED COPY	

SAFETY CONDITIONS

S-1:   For use in accordance with the statements, representations, and conditions in Chapters 1 through 8 of the application submitted by letter dated July 24, XXXX, and supplements dated May 9 and November 14, XXXX; March 13, March 25, June 23, July 23, August 7, August 14, August 28, September 4, September 11, September 15, September 25, September 28, October 19, October 21, October 22, October 23, November 6, November 13, November 16, November 20, November 24, December 18, and December 21, XXXX; January 29, February 4, February 10, February 16, February 24, April 20, April 23, May 21, July 30 (FR No. 21G–99–0058), July 30 (FR No. 21G–99–0093), August 13, December 10, December 21, and December 29, XXXX; and January 25, March 31, July 6, August 18, August 23, September 1, November 3, December 5, December 8, December 14, December 20, December 27, XXXX; and January 11, January 12, March 30, and May 11, XXXX.

S-2:   FR shall not operate the fuel manufacturing processes described in Sections xx.1 and x.x of the license application until an Integrated Safety Analysis (ISA) has been performed, including the appropriate nuclear criticality safety evaluations. A summary of the ISA shall be submitted to the NRC, in addition to an application for amendment to the license, at least 90 days prior to the FR planned restart of operations.

S-3:   Deleted by Amendment 5, dated May XXXX.

S-4:   FR shall not operate the LEU recovery facility described in Section xx.4 of the license application until an ISA has been performed, including the appropriate nuclear criticality safety evaluations. A summary of the ISA shall be submitted to the NRC, in addition to an application for amendment to the license, at least 90 days prior to the FR planned restart of operations.

S-5:   FR shall not operate the 300 complex incinerator system described in Section xx.4 of the license application until an ISA has been performed, including the appropriate nuclear criticality safety evaluations. A summary of the ISA shall be submitted to the NRC, in addition to an application for amendment to the license, at least 90 days prior to the FR planned restart of operations.

S-6:   Deleted by Amendment 2, dated February XXXX.

S-7:   Deleted by Amendment 2, dated February XXXX.

S-8:   FR shall conduct quarterly NCS audits of selected plant activities involving SNM such that SNM processing or storage areas are audited biennially. The purpose of the audits is to determine that: (a) site operations are conducted in compliance with license conditions, operating procedures, and posted limits, (b) administrative controls and postings are consistent with NCSE, (c) equipment and operations comply with NCSE, and (d) corrective actions relative to findings of NCS inspections are adequate.

S-9:   Subcritical parameter values based on experiments, unless they are from the ANSI/ANS series 8 standards, shall be not less than that corresponding to k_{eff} of 0.98 or, alternatively, the factors in Section x.x.x.x of the license application may be applied for uranium-water systems.

NRC FORM 374A	U.S. NUCLEAR REGULATORY COMMISSION		5
MATERIALS LICENSE SUPPLEMENTARY SHEET		License Number SNM–XXX	
		Docket or Reference Number 70–XX	
		Amendment 27 CORRECTED COPY	

S-10: Notwithstanding the description of setting failure limits in Section x.x.x. of the application, when determining subcriticality based on computer code calculations the failure limit shall be no greater than the value corresponding to: k_{eff} = .95 for systems containing uranium enriched in ^{235}U above 20%, k_{eff} = .95 for systems above 10% but below 20% enrichment that are not highly moderated, k_{eff} = .97 for systems above 10% but below 20% enrichment that are highly moderated, and k_{eff} = .97 for systems containing uranium enriched in ^{235}U less than 10%. As one acceptable method, the margin may be based on a validation against applicable benchmark experiments using a one-sided 95% tolerance limit at a 95% confidence level less an additional 0.015 Δk_{eff}. The k_{eff} values of .95 and .97 above are exact limit values, and do not imply that compliance need only be shown to 2 significant figures. Compliance with them shall allow for purely calculational inaccuracies, such as Monte Carlo variance, by meeting the limit with a margin in the conservative direction of at least two standard deviations. Any rounding shall be in the conservative direction.

S-11: Notwithstanding Section x.x..x of the application, for situations in which it is credible, and not unlikely, that critical masses or concentrations may accumulate in a solution confined to a favorable geometry or poisoned vessel, and then be released to vessels of unfavorable geometry, transfer shall be controlled by one of the following three general provisions for double contingency:

   (1) multiple engineered hardware controls capable of preventing unsafe transfer; or

   (2) at least one engineered hardware control capable of preventing unsafe transfer plus a determination of safe conditions and actuation of transfer by an individual; or

   (3) a design requiring independent actions by two individuals before transfer is possible, each action supported by independent measurements of material to be transferred, and a determination of safe conditions. In this case, physical impediments should be included in the system design which will prohibit either individual from performing both of the actions intended to be performed independently.

S-12: Prior to August 15, XXXX, FR will implement fire protection procedures to minimize the threat of fire, explosions, or related perils to process control and safety systems which could lead to an unacceptable release of hazardous material related to SNM or radiation that would threaten workers, the public health and safety, or the environment, as committed to in Section x.x of the license application.

S-13: Deleted by Amendment No. 4, March XXXX.

NRC FORM 374A   U.S. NUCLEAR REGULATORY COMMISSION

MATERIALS LICENSE
SUPPLEMENTARY SHEET

License Number	SNM–XXX
Docket or Reference Number	70–XX
Amendment 27	
CORRECTED COPY	

6

S-14: The 200 and 300 Complex vaults will be protected by barriers with an equivalent two hour fire resistance rating.

S-15: Active and administrative controls for flammable liquids and gasses must be operable in the fire area where flammable liquids and gases are present during CARP processing.

S-16: Prior to August 15, XXXX, CARP Process fire walls will be upgraded to meet FHA recommendations, as described in FR Document No. 21G–98–0198, *FR Response to Request for Additional Fire Safety Information for the CARP Process*, dated December 8, XXXX.

S-17: Prior to December 31, XXXX, FR shall protect CARP process areas and special nuclear material vaults from lightning by installing a lightning protection system in accordance with the standard "Lightning Protection Code," NFPA 780.

S-18: Prior to August 15, XXXX, fixed combustible gas detectors in the 600 and 800 Areas shall be capable of alarming locally and at a constantly manned location.

S-19: Prior to December 31, XXXX, FR will upgrade all process area sprinkler systems to alarm at a constantly manned location.

S-20: Deleted by Amendment 24, April XXXX.

S-21: FR will maintain an industrial fire brigade in accordance with industry standards (NFPA 600). FR will have a proceduralized method for the rapid response of external firefighting resources when sufficient fire brigade staffing is unavailable.

S-22: FR shall perform the following steps as detailed in the FR Bulk Chemical Tank Analysis (FR Document 21G–99–0207).

   A. By July 31, XXXX for 330–TANKXX–002 (sulfuric acid tank), FR shall:

      1. Perform a 100 percent visual internal tank inspection.
      2. Provide details of internal nozzle penetrations and welds, add these details to drawing, then recalculate estimated service life.
      3. Conduct liquid penetrant examinations of floor-to-shell welds.
      4. Perform a magnetic flux leakage inspection of 100 percent of the tank bottom to detect underside corrosion and pitting.

   B. By September 1, XXXX, FR shall provide a written plan that details the continued inspection and testing of bulk chemical storage tanks that will provide a documented safety basis for bulk storage tanks.

NRC FORM 374A	U.S. NUCLEAR REGULATORY COMMISSION	7
MATERIALS LICENSE SUPPLEMENTARY SHEET	License Number SNM–XXX	
	Docket or Reference Number 70–XX	
	Amendment 27 CORRECTED COPY	

C.   Prior to December 31, XXXX, FR shall conduct a second set of ultrasonic thickness tests for 312–TANKXX–013 (nitric acid), T–306–7 (ammonium hydroxide), T306–6 (ammonium hydroxide). These readings will provide data that will allow the corrosion rate and tank wall thickness to be determined. The nitric acid tank, 312–TANKXX–013, shall also have an internal inspection and a liquid penetrant examination of the floor-to-shell welds.

D.   As required by code, each tank shall have a permanent nameplate attached specifying tank operating conditions. The American Society of Mechanical Engineers, "Boiler and Pressure Vessel Code," Section VII, "Markings," lists necessary information for nameplates.

S-23:   FR shall inform the NRC within 30 days of receipt of a violation notice from the State of Bliss Division of Air Pollution or Water Pollution Control, or receipt of modified requirements of the state-issued National Pollutant Discharge Elimination System (NPDES) permit.

S-24:   The licensee shall maintain and execute the response measures in the Emergency Plan, Revision 4, dated September 27, XXXX, or as further revised by the licensee consistent with 10 CFR 70.32(i).

S-25   FR may make changes (modifications, additions, or removals) to the site, structures, processes, systems, equipment, components, computer programs, and activities of personnel without license amendment, provided that the proposed change does not involve:

(1) the creation of new types of accident sequences that, unless mitigated or prevented, would exceed the performance requirements of 10 CFR 70.61 and have not previously been described in the ISA summary;
(2) the usage of new processes, technologies, or controls for which FR has no prior experience;
(3) the removal, without at least an equivalent replacement of the safety function, of an item relied on for safety that is listed in the ISA summary;
(4) the alteration of any item relied on for safety, listed in the ISA summary, that is the sole item preventing or mitigating an accident sequence that exceeds the performance requirements of 10 CFR 70.61; and
(5) a change to the conditions of this license or Part I of the license application.

Proposed changes not meeting all of the above criteria shall be deemed to require NRC approval by amendment. As part of the application for amendment, FR shall perform an ISA for the change and submit either an ISA summary or applicable changes to a prior existing ISA summary. FR shall also provide any necessary revisions to its environmental report.

Proposed changes requiring revision of applicable safety or environmental bases, but not requiring an amendment to the license in accordance with the above criteria, shall be reviewed and approved by the FR safety review committee. The internally authorized change documentation shall provide the basis for determining that the change will be consistent with the criteria (1) through (5) above.

<table>
<tr><td>NRC FORM 374A</td><td>U.S. NUCLEAR REGULATORY COMMISSION</td><td>8</td></tr>
</table>

MATERIALS LICENSE SUPPLEMENTARY SHEET	License Number SNM–XXX
	Docket or Reference Number 70–XX
	Amendment 27 CORRECTED COPY

For any internally authorized change implemented by FR without NRC approval pursuant to this license condition, FR shall submit annually to the NRC applicable changes to the ISA summary of a prior existing ISA. In addition, FR will submit annually a brief summary of all internally authorized changes not requiring prior NRC approval. FR will submit by January 30th of each calendar year the revisions to the ISA summary and the summary of all internally authorized changes not requiring NRC approval.

S-26: Prior to engaging in the decommissioning activities specified in Section c.c.c of the license application dated November 16, XXXX, FR must determine the status of the procedures and activities planned with respect to 10 CFR 70.38(g)(1). If required, FR must submit a decommissioning plan to the NRC for review and approval prior to initiating such actions.

S-27: At not more than 1-year intervals from the issuance date of this license, the licensee shall update the demonstration sections of the license application to reflect the licensee's current operations and evaluations. The updates shall, as a minimum, include information for the health and safety section of the application as required by 10 CFR 70.22(a) through 70.22(f) and 70.22(i) and operational data or environmental releases as required by 70.21.

S-28: By May 1, XXXX, FR shall submit an evaluation of available seismology data for the facility site and specify the maximum earthquake magnitude, the peak ground acceleration, and the return period for an earthquake occurrence with a likelihood of one in 1,000 years.

S-29: By February 1, XXXX, FR shall provide design information (e.g., applicable building codes; other construction standards) pertinent to understanding the resistance of the CARP process facility, structures, and equipment to failures caused by external events.

S-30: By November 1, XXXX, FR shall improve the process descriptions in the ISA Summary Document to focus on the safety aspects of the CARP process and to facilitate an understanding of the results of the ISA and the selection of items relied on for safety. The process descriptions should identify and describe, at each point in the process, the significant hazards that are present, the design features of the process equipment that are relevant to protecting against these hazards, and the safety systems that have been implemented to prevent accidents or mitigate their consequences.

S-31: By August 1, XXXX, FR shall fully and explicitly identify, in the ISA Summary Document, the information it considers to be "process safety information" for the CARP process and shall commit to maintaining such information current and accurate utilizing the configuration management system.

S-32: By August 1, XXXX, FR shall state in its Safety Program Description that its ISA team for the CARP process shall have expertise in fire safety, and that the team shall address in the ISA potential accident sequences resulting from fires.

S-33: By August 1, XXXX, FR shall describe, in the ISA Summary Document for the CARP process, its approach for hazard identification and for evaluating the adequacy of items relied on for safety.

NRC FORM 374A	U.S. NUCLEAR REGULATORY COMMISSION		9
	MATERIALS LICENSE SUPPLEMENTARY SHEET	License Number SNM–XXX	
		Docket or Reference Number 70–XX	
		Amendment 27 CORRECTED COPY	

S-34: By August 1, XXXX, FR shall improve the ISA Summary Document for the CARP process to clearly identify and describe the potential accident sequences, including the initiating and subsequent events that result in the accident, the specific controls (i.e., items relied on for safety) that are used to prevent or mitigate such accidents, and the specific process materials that may be released during the accident.

S-35: By November 1, XXXX, FR shall identify specific values ((e.g., OSHA Permissible Exposure Limits (PELs), Emergency Response Planning Guidelines (ERPGs), Acute Exposure Guideline Levels (ERPGs), Threshold Limiting Values (TLVs), or the Immediately Dangerous to Life and Health values (IDLH)), used, in the ISA Summary Document for the CARP process, to define both intermediate and high consequence chemical accidents. If alternate values are used, FR shall provide justification for their choice. Also, FR shall include the environmental criterion, "a 24-hour averaged release of radioactive material outside the restricted area in concentrations exceeding 5000 times the values in Table 2 of Appendix B to 10 CFR Part 20," as a threshold for an intermediate consequence accident.

S-36: By August 1, XXXX, FR shall improve the ISA Summary Document for the CARP process to demonstrate that the potential effect on radiological safety resulting from accidental exposure of workers to hazardous chemicals is taken into account and that appropriate measures are taken to prevent or mitigate the consequences of such exposure.

S-37: By August 1, XXXX, FR shall, for each postulated accident sequence having (uncontrolled) intermediate or high consequences, identify in the ISA Summary Document for the CARP process the method(s) used to determine the consequences of the accident.

S-38: By November 1, XXXX, FR shall define in its ISA Summary Document for the CARP process, as part of FR safety program requirements: (1) qualitative or quantitative criteria for determining acceptable likelihoods for high and intermediate consequence accidents, and (2) methods used to determine compliance with these criteria for each potential accident. These criteria shall be consistent with an expectation that no high consequence accident would occur at the facility in 100 years. By November 1, 2003, FR shall apply these methods to each high and intermediate consequence accident sequence defined in the ISA, and shall determine that each meets the likelihood acceptance criteria.

S-39: For individual fire areas in the XXX Building area which contain more than 350g ^{235}U, FR shall complete a nuclear criticality safety analysis demonstrating that a criticality accident resulting from a credible fire, analyzed in the Fire Hazards Analysis, or from the consequences of fire-suppression activities, is highly unlikely. This may be done by: (i) demonstrating that a criticality resulting from an accident sequence initiated by a major fire would be highly unlikely, or (ii) demonstrating that a major fire is highly unlikely. FR shall also review all NCSAs potentially affected by the installation of automatic fire suppression systems and associated facility modifications to determine their effect on the safety basis. For the analyses specified by this safety condition, a major fire is defined as one which would affect two or more process Areas in Building XXX.

10

MATERIALS LICENSE
SUPPLEMENTARY SHEET

License Number
SNM–XXX

Docket or Reference Number
70–XX

Amendment 27
CORRECTED COPY

S-40:   By December 31, XXXX, for CARP process structures and equipment, FR shall classify all items relied on for nuclear criticality safety as either safety-related or configuration-controlled equipment. Safety-related equipment (SRE) is defined as active or passive engineered-controls that are relied on to prevent nuclear criticality in accordance with the double contingency principle, and whose operation can change with time such that the equipment might not perform its function. Configuration-controlled equipment (CCE) is defined as structures, systems, or components for which either:

(i) some characteristic is relied on for double contingency, which characteristic will not change with time as a result of accidents identified in the ISA, or

(ii) the control is supplemented by one or more controls as one leg of the double contingency principle.

For SRE items, maintenance, calibration, testing, and/or inspection shall be performed in accordance with written, approved procedures to assure continued reliability and functional performance. SRE that has undergone maintenance will be functionally tested, calibrated, or inspected (as applicable) prior to restart.

CCE will be functionally tested, maintained, calibrated, and/or inspected periodically in accordance with written, approved procedures, with the following exceptions:

CCE that has no credible mechanism to fail beyond the conditions assumed in the bounding normal case does not require functional testing, calibration, or preventive maintenance.

CCE that is tested by every use and that is used with sufficient frequency to ensure adequate reliability does not require functional testing or preventive maintenance, unless it contains parts that degrade over time.

CCE items will be inspected after initial installation, replacement, and by periodic NCS audits.

S-41:   FR shall provide an automatic fire suppression system to suppress and contain a fire involving extraction solvent (i.e., combustible liquids) of the uranium recovery process in Building XXX no later than June 30, XXXX. Until such time that an automatic fire suppression system has been provided, the compensatory measures described below shall be required. In addition, the duration of compensatory measures required for operating uranium recovery process Area E (column dissolvers), Area F and Area H (process involving extraction solvent), or Area G (uranyl nitrate solution evaporators) shall not exceed June 30, XXXX. Prior to June 30, XXXX, operations involving using extraction solvent shall be terminated and all extraction solvent safely removed from Building XXX unless by June 30, XXXX, the automatic fire suppression system is operational.

1.   During CARP processing, a continuously manned fire watch of at least 2 trained personnel will be located in the XXX Building. These may be operators who are suitably trained to extinguish Class B fires. Once HEU is entered into the recovery process (Areas D thru J), a continuously manned fire watch of at least 4 trained personnel must be located in the 300 Complex, 2 of the 4 must be located in the XXX Building. Fire watch personnel need to be suitably trained in the use of self-contained

NRC FORM 374A	U.S. NUCLEAR REGULATORY COMMISSION	11
	License Number SNM–XXX	
MATERIALS LICENSE SUPPLEMENTARY SHEET	Docket or Reference Number 70–XX	
	Amendment 27 CORRECTED COPY	

breathing apparatus (SCBA), and extinguishing Class B fires utilizing portable handheld extinguishers and Aqueous-Film Forming Foam (AFFF) extinguishers units. Operators may be utilized as fire watch personnel, if suitability trained. Non-moderating agents shall be used as a first recourse to extinguish a fire. Fire hoses should be used as a last resort, when all alternatives are not successful and the overall risk to personnel is minimized.

2. Within the xxx Building, portable fire extinguisher size and placement shall meet Class B Extra (High) Hazard Classification as specified in NFPA 10. Two extra AFFF extinguisher units, with a minimum UL Classification of 160B, shall be provided for immediate use at two separate locations outside the XXX Building doorways. SCBAs shall be co-located with the AFFF extinguisher units.

3. When the fire brigade is unavailable and the XXX Building smoke detection system annunciates, the fire department shall be immediately requested. If the smoke detectors are inoperable; solvent extraction process, furnace, and calciner operations shall be suspended.

4. Firefighters who may have to use fire hoses shall be trained in nuclear criticality safety to a level equivalent to that received by a general fissile material worker. This training shall be sufficient to acquaint these personnel with the criticality hazards in the facility and the credible effects of water in areas containing SNM. Personnel shall be trained in practices which minimize the potential for criticality to the extent practicable.

5. FR shall provide the following prior to operating uranium recovery process involving Area E (column dissolvers), Area F and Area H (process involving extraction solvent), or Area G (uranyl nitrate solution evaporators) in Building XXX:

   A. Two firefighters (professional firefighters or plant fire brigade members with enhance firefighting training) shall be stationed in or immediately outside of Building XXX. These individuals must be trained in interior structural firefighting to successfully perform fire fighting operations with a high assurance in mitigating a combustible liquid fire during the early stages of fire development in Building XXX. They shall be capable of responding with required personal protective equipment and self-contained breathing apparatus to begin firefighting operations in Building XXX within 2 minutes after detection of a fire. During the course of a work-shift, only one of the two firefighters may be temporarily relieved at any given time by another firefighter or a trained fire watch fuel manufacturing operator for authorized activities such as lunch, rest, or other breaks. In those occasions where the individual providing relief is fire watch trained but not a trained firefighter, the firefighter on authorized leave from his or her duty station shall be capable of responding within 2 minutes after detection of a fire to begin firefighting operations in Building XXX. The licensee shall minimize the use of fire watch trained individuals to relief firefighters. In addition, FR shall ensure that plant fire brigade staffing is adequate during operations described above to ensure that the two dedicated firefighters would not be called upon for emergency response to plant emergencies outside of Building XXX.

   U.S. NUCLEAR REGULATORY COMMISSION   12

MATERIALS LICENSE
SUPPLEMENTARY SHEET

License Number
SNM–XXX

Docket or Reference Number
70–XX

Amendment 27
CORRECTED COPY

B. A dedicated fire watch shall be stationed in Building XXX. The individual must be trained as a fire watch and the only duty perform is that of a fire watch during the operations of the processes described above. A firefighter may serve as the fire watch.

C. A nuclear criticality safety engineer shall be available in the 300 Complex with capability of responding to technically assist the on-scene incident commander or the emergency control director within 1 minute upon notification of a fire.

D. Combustible containers of fissile material (greater than contamination levels) in Building XXX may only be stored in ventilated process containment or in metal sleeved storage racks, birdcages, and carts that have been demonstrated to meet the nuclear criticality double contingency principles in the event of a fire.

6. The following conditions shall be met when uranium recovery process Area E (column dissolvers), Area F and Area H (process involving extraction solvent), or Area G (uranyl nitrate solution evaporators) is not in operation:

   a. No operations shall be conducted in Areas F and H involving the use or transfer of extraction solvent.
   b. Area G (uranyl nitrate solution evaporators) and Area E (column dissolvers) shall not be heated.
   c. Valves that isolate columns and tanks containing extraction solvent shall be closed.

S-42: Deleted by Amendment 5, dated April XXXX.

S-43: Deleted by Amendment 22, dated March XXXX.

S-44: Deleted by Amendment 22, dated March XXXX.

S-45: Prior to placing water in the Building XXX pre-action sprinkler system (except under fire emergency conditions):

   1. FR shall submit the detailed design of any safety features installed to prevent nuclear criticality in the event of a fire or activation of the fire suppression system, including the drainage rates from enclosures and equipment in which an unsafe depth of fissile material could accumulate, sprinkler spray patterns, and any other pertinent design information related to the sprinkler system which affects criticality safety, for NRC review and approval.

   2. FR shall install and functionally test rigid and passive engineered barriers to prevent moderator intrusion across the boundary of moderation control areas. These barriers shall be composed of fire resistant materials.

NRC FORM 374A   U S. NUCLEAR REGULATORY COMMISSION

13

MATERIALS LICENSE
SUPPLEMENTARY SHEET

License Number
SNM–XXX

Docket or Reference Number
70–XX

Amendment 27
CORRECTED COPY

3. FR shall ensure that all enclosures in areas not restricted by these moderation barriers have at least two drain holes of sufficient size and separation to ensure that a safe depth will not be exceeded.

4. FR shall ensure that there are no unfavorable geometry collection points where liquid water may accumulate. In the sump pit of Area 600, and in any other such areas where fissile material is handled, engineered measures, such as raschig rings, shall be used to ensure the enclosure volume remains subcritical or that liquid water from firefighting activities cannot intrude.

5. Firewater pipes and other pipes carrying moderating materials shall be prohibited from being routed over moderation control areas, unless they are double sleeved with a means provided to detect failure of the inner containment.

6. Enclosures in moderation control areas shall be analyzed to be safe under conditions of mist intrusion, unless demonstrated airtight under fire conditions.

7. Extraneous combustible materials (those not part of the materials of construction or explicitly considered in the S–39 NCS analysis) shall be prohibited from the operating floor. A fire watch shall be established if extraneous materials are introduced.

S-46: By August 1, XXXX, FR shall submit a Criticality Safety Upgrade Program (CSUP) Plan to NRC for review and approval. This CSUP shall address the following elements, at a minimum:

1. All Nuclear Criticality Safety Analyses (NCSAs) performed or revised after May 1, XXXX shall be upgraded as follows:

   A. the criticality safety basis shall be consolidated in a single integrated and self-consistent document;

   B. all engineered structures, systems, and components and operator actions relied on to meet the double contingency principle shall be clearly identified for each accident sequence leading to criticality;

   C. the basis for double contingency shall be clearly documented, including technical documentation of the independence and unlikelihood of control failure;

   D. normal and credible abnormal operating conditions shall be clearly identified; and

   E. all assumptions credited for criticality safety shall be supported by documentation consisting of a technical demonstration of the adequacy of the assumptions rather than reliance on engineering judgement or historical practices.

NRC FORM 374A	U.S. NUCLEAR REGULATORY COMMISSION		14
		License Number SNM–XXX	
MATERIALS LICENSE SUPPLEMENTARY SHEET		Docket or Reference Number 70–XX	
		Amendment 27 CORRECTED COPY	

2. By August 1, XXXX, management procedures defining the criticality safety program shall be upgraded to the following standards:

   A. the NCSAs consist of self-contained safety basis documents, sufficiently detailed to permit independent reconstruction of results by a knowledgeable criticality safety specialist without reliance on additional site-specific or historical knowledge;
   B. the standard technical practices used in designing calculational models are specified in sufficient detail to ensure that the resulting NCSAs are uniform with respect to modeling reflection, determining the optimal range of moderation, treating interactions, accounting for dimensional tolerances, and any bounding approximations in models;
   C. evaluation of accident sequences take potential interaction between fire and chemical safety and criticality safety into account;
   D. the scope, conduct, and documentation of independent reviews of NCSAs are specified;
   E. the applicability of code validation(s) to the specific cases being modeled is evaluated, including a determination of the adequacy of the subcritical margin;
   F. engineered as opposed to administrative controls are used as the preferred method of ensuring criticality safety, wherever practicable.
   G. the basis for using administrative instead of engineered controls is documented as part of the NCSA; and
   H. a problem reporting and corrective action program is established to ensure the effectiveness of the criticality safety program and criticality controls, and to ensure that effective corrective actions and lessons learned are flowed down into appropriate implementing documents. This program shall include the re-evaluation of the unlikelihood of control failure, as part of the double contingency safety basis, as control failure data is generated.

S-47: By June 29, XXXX, FR shall submit to NRC for approval the following information related to the North Site Decommissioning Plan:

   (a) area factors for volumetrically-contaminated soils and the technical basis for those area factors,
   (b) actual Minimum Detectable Concentrations (MDCs) for the NaI detector and the technical basis for those MDCs,
   (c) appropriate investigation levels (ILs) for static and scan survey measurements that will be performed in impacted areas.

SAFEGUARDS CONDITIONS

Section–x.0 – ABRUPT LOSS DETECTION (For SSNM Only):

SG-1.1. Notwithstanding the requirement of 10 CFR 74.53(b)(1) to have a process detection capability for each unit process, the process units listed in Section X.XXX of the Plan identified in Condition SG–5.1 shall be exempt from such detection capability, and the licensee's process monitoring

NRC FORM 374A	U.S. NUCLEAR REGULATORY COMMISSION	15

MATERIALS LICENSE SUPPLEMENTARY SHEET	License Number SNM–XXX
	Docket or Reference Number 70–XX
	Amendment 27
	CORRECTED COPY

system shall be comprised of the control units described in Section X.X (and all sub-sections therein) of the above mentioned Plan.

Section-x.0 – ITEM MONITORING (For SSNM Only):

SG-2.1. Notwithstanding the requirement of 10 CFR 74.55(b) for item monitoring tests for all item categories except those identified by 10 CFR 74.55(c), and notwithstanding statement #8 of Section x.x.3 of the Plan identified in Condition SG–5.1, the licensee is exempt from applying item monitoring tests on NDA calibration and control standards which are two liters or more in size and contain less than 0.10 formula kilogram. Such standards are not, however, exempted from physical inventory requirements.

Section–x.0 – ALARM RESOLUTION

SG-3.1. The licensee is authorized to continue material processing operations in Control Units 1, 3, 4, 5, and 15 under process monitoring alarm conditions. During the continuation of processing operations, the measures contained in Section x.1.1 of the Plan identified in Condition SG–5.1 shall be implemented.

Section–x.0 – QUALITY ASSURANCE (SSNM & LEU):

SG-4.1. Notwithstanding the requirements of 10 CFR 74.31(c)(2) for LEU and 10 CFR 74.59(d)(1) for SSNM to maintain a system of measurements to substantiate both the element and fissile isotope content of all SNM received, inventoried, shipped or discarded, SNM measured by the licensee for U-233, U-235, or Pu-239 by non-destructive assay techniques need not be measured for total element if the calculated element content is based on the measured isotope content which, in turn, is traceable to an isotopic abundance measurement at the area of generation.

SG-4.2. Notwithstanding the requirement of 10 CFR 74.59(e)(8) to establish and maintain control limits at the 0.05 and 0.001 levels of significance for all HEU related measurements, the licensee may use one and two scale divisions as being equivalent to the 0.05 and 0.001 control levels, respectively, for mass measurements.

SG-4.3 Notwithstanding Section x.x.x of the Plan identified in Condition SG–5.1, which states that a physical inventory of SSNM is conducted at an interval of at least every six calendar months with no more than 185 days elapsing between any two consecutive inventories, the licensee is granted an extension of time from April 3, XXXX, to June 2, XXXX, for conducting its SSNM physical inventory. This condition automatically expires on June 5, XXXX.

NRC FORM 374A	U.S. NUCLEAR REGULATORY COMMISSION	16

MATERIALS LICENSE
SUPPLEMENTARY SHEET

License Number
SNM–XXX

Docket or Reference Number
70–XX

Amendment 27
CORRECTED COPY

SG-4.4. Notwithstanding the requirement of 10 CFR 74.59(f)(2)(viii) to remeasure, at the time of physical inventory, any in-process SSNM for which the validity of a prior measurement has not been assured by tamper-safing, the licensee may book for HEU physical inventory purposes:

(1.) XXX XXX and Building XXX/XXX process holdup quantities determined by NDA measurements performed prior to the start of an inventory, in accordance with the controls described in Sections x.x.x.x and x.x.x.x.x of the Plan identified in Condition SG–5.1;

(2.) pre-listed feed material to the Building XXX/XXX process that is introduced into process prior to the start of an inventory, in accordance with the controls described in Section x.x.x.x.x of the Plan identified in Condition SG–5.1; and

(3.) Building XXX holdup quantities determined by the most recent NDA measurements, in accordance with the controls described in Section x.x.x.x.x of the Plan identified in Condition SG–5.1.

SG-4.5. Notwithstanding the requirements of 10 CFR 74.59(f)(1) and 74.59(f)(2)(viii) to measure and inventory all SSNM, the licensee may determine process exhaust ventilation system inventory quantities in accordance with Section x.x.x.x of the Plan identified in Condition SG–5.1.

SG-4.6. The restriction of 10 CFR 74.51(d)(2) is hereby removed, and based on process monitoring performance in MBA–6 acceptable to the NRC, the licensee is authorized to conduct HEU physical inventories in accordance with the requirements of 10 CFR 74.59(f)(1), provided HEU scrap recovery operations in MBA–5 are restricted to the last 60 calendar days of each physical inventory period.

SG-4.7. Notwithstanding the requirement of 10 CFR 74.59(d)(1) to substantiate the uranium and U-235 content of SSNM transferred between areas of custodial responsibility, the licensee may transfer scrap materials from MBA–6 to MBA–5 on estimated values provided (1) such estimates are based on historical factors (with a unique factor for each scrap category) which are updated at least once every six months, and (2) that the estimated transfer values are corrected upon obtaining "first dissolution plus residue" measurements.

SG-4.8. The SNM content of liquid waste discarded from collection tanks shall be analyzed and recorded at measured values. The measurement methods must have a greater sensitivity than the concentration of the sample aliquot analyzed, except when the quantity discarded does not exceed 50 grams U-235 per month from Plant I (HEU) and does not exceed 10 grams U-235 per month from MBA–4 (LEU) through those discard batches where the sample aliquot concentration is less than the sensitivity of the method.

SG-4.9. Notwithstanding the statement in Section .x. of the Plan identified in Condition SG–5.2, pertaining to bias corrections to inventory difference (ID) values, the licensee shall comply with Section x.x.x of such Plan with respect to determining any bias corrections to IDs.

NRC FORM 374A	U.S. NUCLEAR REGULATORY COMMISSION	17

MATERIALS LICENSE
SUPPLEMENTARY SHEET

License Number
SNM–XXX

Docket or Reference Number
70–XX

Amendment 27
CORRECTED COPY

SG-4.10. Notwithstanding the requirements of 10 CFR 74.59(e)(8) relative to actions to be taken when replicate measurement data exceed a 0.001 control limit, the licensee shall comply with Section x.x.x.x.x.4 of the Plan identified in Condition SG–5.1.

SG-4.11. Notwithstanding the requirement of 10 CFR 74.59(e)(4) that allows the pooling of data which has been shown to be not significantly different on the basis of appropriate statistical tests, the licensee may pool data from equivalent scales without testing.

SG-4.12. Notwithstanding the requirement of 10 CFR 74.59(e)(5) to evaluate all program data to establish random error variances, limits for systematic error, etc., the licensee may randomly select a partial quantity of bulk measurement program data, as described in Section 1.1.1(3) of the Plan identified in Condition SG–5.1, provided the partial data set is not statistically different from the total data population whenever the impact on SEID is greater than 1.0 percent.

SG-4.13. Notwithstanding the requirement of 10 CFR 74.59(f)(1)(i) to calculate the SEID associated with each HEU inventory difference (ID) value, the licensee need not determine such SEID for MBA–7 whenever its ID is less than 300 grams U-235.

SG-4.14. Notwithstanding the requirement of 10 CFR 74.31(c)(3) and of 74.59(e)(3)(i) to measure control standards for all measurement systems for the purpose of determining bias, and notwithstanding the requirement of 10 CFR 74.31(c)(4) and of 74.59(e)(8) to maintain a statistical control system to monitor such control standard measurements, the licensee need not measure nor monitor such control standards for point calibrated, bias-free, systems. To be regarded as bias-free, a measurement system must be calibrated by one or more measurements of a representative standard(s) each time process unknowns are measured, and the measurement value assigned to a given unknown is based on the associated calibration.

SG-4.15. All SNM not in transit shall be physically located within an MBA or ICA, except as specified in Condition SG–4.15.1.

SG-4.15.1. The requirement of Condition SG–4.15 shall not apply to HEU or LEU contained in, or precipitated from, measured liquid or gaseous waste discards.

SG-4.16. Solutions generated from the use of sinks, eye washers, safety showers, drinking fountains, etc., located within HEU MAAs shall be collected and measured prior to discarding.

SG-4.17. All HEU–bearing liquid effluents that are routed to the Waste Water Treatment Facility (WWTF) shall be measured for total uranium in the WWTF prior to commingling with LEU. Each WWTF HEU input batch measurement shall serve as an overcheck to the corresponding summation of accountability values. If for any material balance period, the WWTF total cumulative HEU over-check value does not agree within 500 grams HEU of the corresponding accountability

NRC FORM 374A

U.S. NUCLEAR REGULATORY COMMISSION

18

MATERIALS LICENSE
SUPPLEMENTARY SHEET

License Number
SNM–XXX

Docket or Reference Number
70–XX

Amendment 27
CORRECTED COPY

value, an investigation shall be conducted and documented as to the cause and corrective action taken, and the appropriate NRC safeguards licensing authority shall be notified within 30 days after the start of the associated physical inventory. The WWTF input overcheck measurement system shall be subject to all appropriate requirements of the Measurement Control Program as specified in Section 4.4 of the Plan identified in Condition SG–5.1.

SG-4.18. Notwithstanding the requirement of 10 CFR 74.15 to include limit of error data on DOE/NRC Form–741 for all SNM shipments, the licensee is exempt from including such data on 741 Forms associated with waste burial shipments.

SG-4.19. Whenever a SNM Material Superintendent or designated SNM Custodian is summoned to an MAA exit point to assist in resolving whether an item or container should be allowed to exit to the protected Area, in accordance with the currently approved "Physical Safeguards Plan," the Superintendent or Custodian shall document the basis for any decision allowing the item or container to leave the area.

SG-4.20. The licensee is exempted from calculating the standard error of inventory difference (SEID) and measurement system biases associated with LEU physical inventories provided that the calculated inventory difference does not exceed 1,000 grams U-235.

SG-4.21. Notwithstanding Section x.x of the Plan identified in Condition SG–5.2, which states that "confirmatory measurements of scrap receipts are performed after the scrap is dissolved," the term "scrap receipts" shall not apply to receipt materials whose SNM content can be determined on the as-received-material by weighing, sampling and analyses with a measurement uncertainty (at the 95% C.L.) of less than 2.00 percent (based on a single sample).

SG-4.22. Notwithstanding the heading "Typical MC&A Procedures" for Table c.c of the Plan identified in Condition SG–5.2, all procedures listed in Table 3.5 shall be officially designated as "Critical MC&A Procedures", and any revisions to these procedures shall be subject to the same review and approval requirements (as specified in Section.x of the Plan) that applied to the original procedures.

SG-4.23. Notwithstanding statements contained in Section c.c.c of the Plan identified in Condition SG–5.2, if the normal minimum number of control standard measurements per week, day, or shift of system use (depending on type of measurement system) does not generate at least 25 control standard measurements for a given LEU measurement system during any inventory period in which the active inventory is greater than 9,000 grams U-235, the licensee shall nevertheless generate at least 16 control standard measurements for each key measurement system utilized during the inventory period.

SG-4.24. Deleted by Amendment 3, March XXXX. This Condition expired May 15, XXXX.

NRC FORM 374A	U.S. NUCLEAR REGULATORY COMMISSION		19
	MATERIALS LICENSE SUPPLEMENTARY SHEET	License Number SNM–XXX	
		Docket or Reference Number 70–XX	
		Amendment 27 CORRECTED COPY	

SG-4.25.   Deleted by Amendment 16, January XXXX. This Condition expired July 8, XXXX.

SG-4.26.   Deleted by Amendment 21, March XXXX. This Condition expired February 11, XXXX.

SG-4.27   Notwithstanding the requirement of 10 CFR 74.17(c) and the commitments of Section XXX of the Fundamental Nuclear Material Control (FNMC) Plan identified in Condition SG–5.1, to submit a completed Special Nuclear Material Physical Inventory Summary Report on NRC Form 327, not later than 45 days from the start of the physical inventory, the licensee is exempted from the above stated requirements and shall have 18 additional days to complete the February 9, XXXX, physical inventory report. This exemption automatically expires on April 14, XXXX.

Section x.0 – PLANS AND SPECIAL ISSUES IN PLAN APPENDICES:

SG-5.1.   In order to achieve the performance objectives of 10 CFR 74.51(a) and maintain the system capabilities identified in 10 CFR 74.51(b), the licensee shall follow its "Fundamental Nuclear Material Control Plan" with respect to all activities involving strategic special nuclear material, except as noted in Condition SG–5.5. This Plan, as currently revised and approved, consists of:

      General Discussion ------------------------- Rev. 6 (dated February XXXX)
      Sec. 1 -- Process Monitoring ------------- Rev. 6 (dated February XXXX)
      Sec. 2 -- Item Monitoring ------------------ Rev. 3 (dated August XXXX)
      Sec. 3 -- Alarm Resolution ---------------- Rev. 3 (dated August XXXX)
      Sec. 4 -- QA & Accounting ---------------- Rev. 8 (dated February XXXX)
      Annex A ----------------------------------- Rev. 3 (dated August XXXX)
      Annex B ----------------------------------- Rev. 1 (dated August XXXX)
      Annex C ----------------------------------- Rev. 1 (dated August XXXX)
      Annex D ----------------------------------- Rev. 1 (dated February XXXX)
      Appendix G -- Pu Decommissioning --- Rev. 137 (dated April XXXX)

    Revisions to this Plan shall be made only in accordance with, and pursuant to, either 10 CFR 70.32(c) or 70.34.

SG-5.2.   In order to achieve the performance objectives of 10 CFR 74.31(a) and maintain the system capabilities identified in 10 CFR 74.31(c), the licensee shall follow its "Fundamental Nuclear Material Control Plan for SNM of Low Enriched Uranium" with respect to all activities involving SNM of low strategic significance. The Plan, as currently revised and approved, consists of:

    Sections x and x ----------------   Both labeled as Revision 3, and dated
                                    April XXXX

NRC FORM 374A	U.S. NUCLEAR REGULATORY COMMISSION	20

MATERIALS LICENSE SUPPLEMENTARY SHEET	License Number SNM–XXX
	Docket or Reference Number 70–XX
	Amendment 27 CORRECTED COPY

Sections x, and x through x -----   All labeled as Revision 2, and dated April XXXX

Sections x through x -----------   All labeled as Revision 1, and dated February XXXX

Annex --------------------------   Labeled as Revision 3, and dated April XXXX

Revisions to this Plan shall be made only in accordance with, and pursuant to, either 10 CFR 70.32(c) or 70.34.

SG-5.3.   Notwithstanding the requirement of 10 CFR 74.59(f)(1)(i) to estimate the standard error associated with SSNM inventory difference values, and notwithstanding the requirements of 10 CFR 74.59(e)(3) through (e)(8), the licensee may, in lieu of said requirements, follow Appendix G of the Plan identified in SG–Condition 5.1 with respect to plutonium measurements and measurement control associated with the plutonium decommissioning project.

SG-5.3.1.   With regard to the plutonium decommissioning project (described in Appendix G of the Plan identified in Condition SG–5.1), the licensee shall comply with the following:

(a)   For plutonium accountability measurements, the maximum measurement uncertainty (at the 95% confidence level) of measurement values equal to or greater than 100 grams Pu shall not exceed plus or minus 10.0%. For measurement values less than 100 grams Pu, but equal to or greater than 25 grams Pu, the maximum measurement uncertainty shall not exceed plus or minus 20.0% (at the 95% C.L.).

(b)   For net weight measurements utilized for establishing "nanocuries Pu per gram waste" values (which in turn are used for establishing the category of waste), the maximum measurement uncertainty (at the 95% C.L.) shall not exceed plus or minus 2.00%.

(c)   Sufficient control measurements shall be generated and documented so as to demonstrate compliance with 5.3.1(a) and (b) above.

(d)   For each inventory period during which plutonium decommissioning activities are conducted, the measurement uncertainty associated with the total quantity of plutonium in item form generated and measured during the period shall be derived from all relevant measurement control data generated during that inventory period.

(e)   For each inventory period during which plutonium decommissioning activities are conducted, plutonium "additions to" and "removals from material in process" (ATP and RFP) shall be calculated. Any measured Pu quantity, in item form, which is generated from existing residual holdup shall be regarded as an ATP at the time of its generation. Any measured Pu

NRC FORM 374A	U.S. NUCLEAR REGULATORY COMMISSION	21

MATERIALS LICENSE
SUPPLEMENTARY SHEET

License Number
SNM–XXX

Docket or Reference Number
70–XX

Amendment 27
CORRECTED COPY

quantity, in item form, which is tamper-safe sealed and which will not undergo any additional processing (such as washing, compaction, etc.) prior to shipment offsite shall be regarded as an RFP upon obtaining such status. The limit for total plutonium measurement uncertainty for each inventory period shall be the larger of (1) 250 grams plutonium or (2) 10.0 percent of the larger of ATP or RFP.

(f)   The licensee shall investigate any non-zero inventory difference, since a non-zero ID will be (for this operation) indicative of an item(s) discrepancy.

SG-5.3.2.   Storage of plutonium items generated during plutonium decommissioning activities shall be in accordance with the commitments contained in the licensee's Plan identified in Condition SG–6.1.

SG-5.4.   Operations involving special nuclear material which are not described in the appropriate Plan identified by either Condition SG-5.1 or SG-5.2 shall not be initiated until an appropriate safeguards plan (describing all new and/or modified security and MC&A measures to be implemented) has been approved by the appropriate NRC safeguards licensing authority.

SG-5.5.   Notwithstanding the requirements of 10 CFR 74.51(b) and (d), 74.53, and 74.59(d)(3), during periods of curtailed SSNM activities limited to (1) use of less than five (5.000) formula kilograms of SSNM contained in encapsulated or tamper-safe sealed standards; (2) use of less than five (5.000) formula kilograms of SSNM contained in materials associated with R&D activities and/or laboratory services; (3) vault storage of HEU oxides in item form except for samples utilized for independent receipt measurement; (4) storage of low level waste materials destined for ofFRite disposal; and (5) decontamination and decommissioning operations involving residual holdup and site remediation; the licensee is exempt from the above mentioned regulations and shall, in lieu of these regulations, follow sections 1.0 through 4.0 of its "Fundamental Nuclear Material Control Plan Applicable for Periods of Limited HEU Processing Activities." This Plan, as currently revised and approved, consists of:

General Discussion --- Revision 1 (dated October XXXX)
Section 1 ----------------- Revision 1 (dated October XXXX)
Section 2 ----------------- Revision 1 (dated October XXXX)
Section 3 ----------------- Revision 1 (dated October XXXX)
Section 4 ----------------- Revision 0 (dated February XXXX)

During such periods of limited HEU processing, the licensee need not follow the Plan identified in Condition SG-5.1. Whenever the possession and use limitations defined above in this condition are not applicable, the Plan identified herein shall be regarded as null and void, and the SG–5.1 Plan shall be in full force.

NRC FORM 374A	U.S. NUCLEAR REGULATORY COMMISSION		22

MATERIALS LICENSE
SUPPLEMENTARY SHEET

License Number
SNM–XXX

Docket or Reference Number
70–XX

Amendment 27
CORRECTED COPY

Section–x.0 – PHYSICAL PROTECTION REQUIREMENTS FOR STRATEGIC SPECIAL NUCLEAR MATERIAL

SG-6.1. The licensee shall follow the measures described in the physical protection plan entitled "FR Physical Safeguards Plan, Paul Plant, Revision x," dated October 27, XXXX, with replacement pages dated January 4, XXXX, and as it may be further revised in accordance with the provisions of 10 CFR 70.32(e).

SG-6.2. The licensee shall follow the safeguards contingency plan titled "FR Safeguards Contingency Plan, Revision 0," dated August 8, XXXX; and as may be further revised in accordance with the provisions of 10 CFR 70.32(g).

SG-6.3. The licensee shall follow the guard training and qualification plan titled "FR Site Security Training Plan, Revision 15," dated September XXXX; and as may be further revised in accordance with the provisions of 10 CFR 70.32(e).

SG-6.4. Notwithstanding the above Safeguards License Conditions (SG–6.1, SG–6.2, SG–6.3), upon possession of less than Category I levels of special nuclear material, the licensee shall follow the measures described in the physical protection plan titled "Physical Security Plan for the Protection of Special Nuclear Material of Moderate Strategic Significance, Revision 5" dated June 23, XXXX (letter dated June 22, XXXX), and Revision XX, dated February 6, XXXX, and as it may be further revised in accordance with the provisions of 10 CFR 70.32(e).

TRANSPORTATION CONDITIONS

Section–1.0 – TRANSPORTATION SECURITY MEASURES:

TR-1.1. The licensee shall follow the measures described in the physical security plan titled "Physical Security Plan for the Protection of Special Nuclear Material of Moderate Strategic Significance, Revision 4," dated October XXXX (letter dated December 20, XXXX), and as it may be further revised in accordance with the provisions of 10 CFR 70.32 (e).

NRC FORM 374

U.S. NUCLEAR REGULATORY COMMISSION

PAGE 1 OF 2 PAGES

MATERIALS LICENSE

Amendment No. 23

Pursuant to the Atomic Energy Act of 1954, as amended, the Energy Reorganization Act of 1974 (Public Law 93-438), and Title 10, Code of Federal Regulations, Chapter I, Parts 30, 31, 32, 33, 34, 35, 36, 39, 40, and 70, and in reliance on statements and representations heretofore made by the licensee, a license is hereby issued authorizing the licensee to receive, acquire, possess, and transfer byproduct, source, and special nuclear material designated below; to use such material for the purpose(s) and at the place(s) designated below; to deliver or transfer such material to persons authorized to receive it in accordance with the regulations of the applicable Part(s). This license shall be deemed to contain the conditions specified in Section 183 of the Atomic Energy Act of 1954, as amended, and is subject to all applicable rules, regulations, and orders of the Nuclear Regulatory Commission now or hereafter in effect and to any conditions specified below.

Licensee

In accordance with the letter dated 36552

1. BCLDP Institute

3. License number SNM–X is amended in its entirety to read as follows: its entirety as follows:

2. 505 Queen Avenue
Dayton, OH 40000

4. Expiration date December 31, 2005

5. Docket No. 070–000XX
Reference No.

6. Byproduct, source, and/or special nuclear material	7. Chemical and/or physical form	8. Maximum amount that licensee may possess at any one time under this license
A. Uranium (as defined in 10 CFR Part 150.11)	A. Any (residual material containing Special Nuclear, source, and byproduct materials)	A. As described in letter dated February 5, 1999 (*Clarification of License Possession Limits*)
B. Plutonium (as defined in 10 CFR Part 150.11)	B. Any (residual material containing Special Nuclear, source, and byproduct materials)	B. As described in letter dated February 5, 1999 (*Clarification of License Possession Limits*)

(I) Authorized places of Use:

A. and B.   Possession incident to radiological survey, Storage of waste awaiting disposal, decontamination and remediation of buildings, equipment, and materials, and outdoor areas, as described in Decommissioning Plan, BCLDP Institute, DX–92–18, Revision 4, August 3, 2000.

NRC FORM 374A	U.S. NUCLEAR REGULATORY COMMISSION	PAGE 2 of 2 PAGES

MATERIALS LICENSE
SUPPLEMENTARY SHEET

License Number
SNM–X

Docket or Reference Number
070–000XX

Amendment No. 23

CONDITIONS

10 Licensed material shall be possessed and processed at the licensee's facilities located at the BCLDP Institute's, East Adam Site, 1135 Plain City–Georgeville Road, State Route 113, Adams, Ohio.

11. The Radiation Safety Officer for this license is John G. Jensen. May B. Chance, Associate Radiation Safety Officer, may assist the Radiation Safety Officer in the management of the day-to-day oversight of the Radiation Safety Program and may act during absences of the Radiation Safety Officer.

12. Except as specifically provided otherwise in this license, the licensee shall conduct its program in accordance with the statements, representations, and procedures contained in the documents, including any enclosures, listed below. The U.S. Nuclear Regulatory Commission's regulations shall govern unless the statements, representations, and procedures in the licensee's application and correspondence are more restrictive than the regulations.

A. Application dated June 14, 2000.

B. Three letters dated August 22, 2000 with the following documents attached:
1. Response to Request for Information, *Renewal Application License SNM–X, Docket No. 070–00XX,*
2. Response to NRC Staff Review Comments, BCLDP Decommissioning Plan dated May 30, 2000, Revision 1.
3. Decommissioning Plan, BCLDP Institute, DX–93–18, Revision 4, August 8, 2000.
4. Radiation Protection Program Plan, BCLDP Institute, DX–90–03, Revision 4, August 8, 2000.

E. Letter dated February 5, 1999 (containing *Clarification of License Possession Limits*), with XP–AP–36.0, Revision 1, *Control of Revisions to Radiation Protection Documents*, and QD–XP–7.2, Revision 10, *Document Control,* attached.

FOR THE U.S. NUCLEAR REGULATORY COMMISSION

Date _____   By_____

George M. McCann
Materials Licensing Branch
Region III

Appendix K
Policy and Guidance Directive FC 94–02, Licensing Site Remediation Contractors for Work at Temporary Job Sites

UNITED STATES
NUCLEAR REGULATORY COMMISSION
WASHINGTON, D. C. 20555

JAN 2 1 1994

MEMORANDUM FOR: Those on Attached List

FROM: Carl J. Paperiello, Director
Division of Industrial and
Medical Nuclear Safety, NMSS

SUBJECT: POLICY AND GUIDANCE DIRECTIVE FC 94-02, LICENSING
SITE REMEDIATION CONTRACTORS FOR WORK AT TEMPORARY
JOB SITES

The final policy and guidance directive on licensing site remediation
contractors for work at temporary job sites is enclosed for your use.
Regional comments have been incorporated as appropriate. We have clarified
that the guidance is intended for site remediation service contractors. We
have also clarified our position that a site owner remains responsible for
eventual release of a site regardless of who the owner hires to perform
specific activities. The final guidance allows contractors to possess
calibration sources, reference standards, and contaminated equipment owned by
the contractor, and it increases the advance notification requirement to 14
days before initiating activities at a temporary job site. In addition, the
emergency response conditions were revised to clearly authorize reasonable
emergency response actions that depart from conditions in the license if NRC
is notified immediately after such action is taken.

Please note that we have requested OMB clearance for the reporting and
recordkeeping requirements in this directive, but OMB approval is still
pending. Any licensing actions involving this directive should be submitted
to Headquarters for concurrence until OMB approval is received.

If you have any questions, please contact Kevin Ramsey at (301) 504-2534.

Carl J. Paperiello, Director
Division of Industrial and
Medical Nuclear Safety, NMSS

Enclosure: As stated

POLICY AND GUIDANCE DIRECTIVE

FC 94-02

LICENSING SITE REMEDIATION CONTRACTORS
FOR WORK AT TEMPORARY JOB SITES

FC 94-02, Rev. 0

Enclosure

LICENSING SITE REMEDIATION CONTRACTORS
FOR WORK AT TEMPORARY JOB SITES

1. **Purpose:**

   The purpose of this directive is to establish the policy and guidance for authorizing service contractors to perform site remediation work under their own license at temporary job sites. This directive applies to temporary job sites owned/operated by other NRC licensees, as well as non-licensees. This directive may be used on a case-by-case basis with HQ concurrence to license other types of service contractors. However, this directive does not apply to the installation and maintenance of devices.

2. **Policy:**

   Site owners/operators may not have radiation safety programs in place that are adequate to ensure the safety of activities to be performed by a service contractor. Therefore, it is appropriate for contractors to operate under their own license at temporary job sites when they are providing the radiation safety programs under which the work is being performed. This ensures that site owners/operators do not supervise activities with which they have no experience. It also allows the NRC to authorize work without issuing a new license or amending an existing license, and it allows enforcement actions directly against contractors when violations are associated with their radiation safety programs. However, the site owner remains responsible for decommissioning financial assurance (if a licensee) and eventual release of the site regardless of who the owner hires to perform specific activities.

3. **General Guidance:**

   In general, applications for site remediation service licenses should be made in accordance with the regulations and guidance applicable to the authorized

1

use requested. For example, an application for broad authorization to handle a wide variety of radioactive materials during site remediation should be in accordance with 10 CFR Part 33 and Regulatory Guide 10.5, Applications for Type A Licenses of Broad Scope. In addition to the existing regulations and guidance, the specific provisions provided below should be addressed.

4. Specific Guidance:

4.1 A site remediation service license may authorize the use of licensed material only at temporary job sites in the United States where NRC maintains jurisdiction. Possession or use of materials at the service contractor's facilities must be authorized under a separate license. In addition, possession should be authorized only to the extent that licensed material originating from the site must be transferred to an authorized recipient or left at the site. Possession (at the temporary job site) of calibration sources, reference standards, and contaminated equipment owned by the licensee may be authorized under the service license. See example license condition 1 in the appendix.

4.2 The licensee should be required to notify the Administrator of the region issuing the license at least 14 days before initiating activities at a temporary job site. See example license condition 2 in the appendix.

4.3 If the site owner/operator (i.e., the customer) also holds a license issued by the NRC or an Agreement State, the service licensee should be required to establish a written agreement between the licensee and the customer specifying which licensee activities will be performed under the customer's license and supervision, and which licensee activities will be performed under the licensee's supervision pursuant to the service license. This agreement should include commitments by both licensees to ensure safety and it should specify whether there are any commitments by the service licensee to help the customer clean up the temporary job site if there is an accident. See example license condition 3 in the appendix.

2

FC 94-02, Rev. 0

4.4 The service licensee should maintain records of information important to decommissioning a temporary job site at the site pursuant to 10 CFR 30.35(g), 40.36(f), and 70.25(g). Customers should have access to decommissioning records throughout the decommissioning process. The service licensee should transfer these records to the customer when activities at a temporary job site are complete. See example license condition 4 in the appendix.

4.5 A service licensee may be exempted from the requirements in 10 CFR 30.35, 40.36, and 70.25 to establish decommissioning financial assurance. NMSS has made a finding that this exemption will not endanger life, or property, or the common defense and security, and is otherwise in the public interest. This exemption is based on the provision stated above in 4.1 that the service licensee is not allowed to retain possession of any licensed material originating from a temporary job site. The site owner remains responsible for eventual release of the site regardless of who monitors and supervises specific work activities. If the site owner is a licensee that has established decommissioning financial assurance or other license commitments, the site owner is responsible for ensuring that its contractors comply with those commitments. See example license condition 5 in the appendix.

4.6 An application for a service license is not required to contain an emergency plan even if the application requests authorization to use licensed material in quantities exceeding the threshold for an emergency plan. Service licensees are not in a position to establish all of the site-specific response measures necessary to execute an effective emergency plan for a temporary job site. Before handling licensed material at any one site in quantities requiring an emergency plan, the service licensee must either obtain NRC approval of an evaluation demonstrating that an emergency plan is not required, or submit written confirmation that licensee personnel have been trained and will follow an existing emergency plan for the temporary job site. See example license condition 6 in the appendix.

3

FC 94-02, Rev. 0

4.7 It is in the public interest to have site remediation service licensees who can provide immediate services in the event of a release or other incident involving uncontrolled radioactive material. However, license conditions require service licensees to establish written agreements and provide advance notification before providing services. Service licensees may be authorized to take reasonable action in an emergency that departs from conditions in the license when the action is immediately needed to protect public health and safety and no action consistent with all license conditions that can provide adequate or equivalent protection is immediately apparent. The licensee should notify the NRC before, if practicable, and in any case immediately after taking such emergency action. See example license condition 7 in the appendix.

4.8 Within 30 days of completing activities at each temporary job site, the service licensee must notify its licensing region. The notification should include the status of the temporary job site and the disposition of the material used by the service licensee. See example license condition 8 in the appendix.

4.9 Service licenses are not temporary licenses that are only in effect while work at a temporary job site is in progress. The applicant must make a clear commitment to maintain all radiation safety programs in an active status even between jobs. Service licensees may not suspend radiation programs and then attempt to re-establish them when another customer is found. This commitment should provide reasonable assurance that the licensee will remain competent to use licensed material and undertake authorized activities. This commitment should include the following:
A. Maintaining qualified personnel in key positions (i.e., RSO, etc.).
B. Holding required safety committee meetings.
C. Performing regular maintenance and calibration of safety equipment.
D. Completing required training (including periodic retraining).

4

FC 94-02. Rev. 0

Appendix

EXAMPLE LICENSE CONDITIONS FOR SERVICE LICENSES

1. Licensed materials shall be used only at temporary job sites of the licensee anywhere in the United States where the U.S. Nuclear Regulatory Commission maintains jurisdiction for regulating the use of licensed material. Except for calibration sources, reference standards, and radioactively contaminated equipment owned by the licensee, possession of licensed material at each temporary job site shall be limited to material originating from each site. This material must either be transferred to an authorized recipient or remain at the site after licensee activities are completed.

2. The licensee shall notify the Regional Administrator, NRC Region ___ in writing at least 14 days before initiating activities under this license at a temporary job site. This notification shall include:

   A. The estimated type, quantity, and physical/chemical forms of licensed material to be used,
   B. The specific site location,
   C. A description of planned activities including waste management and disposition,
   D. The estimated start date and completion date for the job, and
   E. The name and title of a point of contact for the job, including information on how to contact the individual.

3. This license does not authorize the use of licensed material at temporary job sites for uses already specifically authorized by a customer's license. If a customer also holds a license issued by the NRC or an Agreement State, the licensee shall establish a written agreement between the licensee and the customer specifying which licensee activities shall be performed under the customer's license and supervision, and which licensee activities shall be performed under the licensee's supervision pursuant to this license. The agreement shall include a commitment by the licensee and the customer to ensure safety, and any commitments by the licensee to help the customer clean up the temporary job site if there is an accident. A copy of this agreement shall be included in the notification required by license condition [example 2 above].

4. The licensee shall maintain records of information important to decommissioning each temporary job site at the applicable job site pursuant to 10 CFR 30.35(g), 40.36(f), and 70.25(g). The records shall be made available to the customer upon request. At the completion of activities at a temporary job site, the licensee shall transfer these records to the customer for retention.

A1

FC 94-02, Rev. 0

5. Pursuant to 10 CFR 30.11, 40.14, 70.14, and license condition [example 1 above], the licensee is exempted from the requirements of 10 CFR 30.35, 40.36, and 70.25 to establish decommissioning financial assurance.

6. Notwithstanding the requirements in 10 CFR 30.32(i), 40.31(j), and 70.22(i), the licensee is not required to establish an emergency plan. Before taking possession of licensed material at a temporary job site in quantities requiring an emergency plan the licensee shall either --

   A. Obtain NRC approval of an evaluation demonstrating that an emergency plan is not required pursuant to 10 CFR 30.32(i), 40.31(j), and 70.22(i), or

   B. Submit written confirmation to the Regional Administrator, NRC Region ___, that licensee personnel have been trained and will follow the provisions of an existing emergency plan approved by the NRC or an Agreement State for the temporary job site.

7. If approved by a Radiation Safety Officer specifically identified in this license, the licensee may take reasonable action in an emergency that departs from conditions in this license when the action is immediately needed to protect public health and safety and no action consistent with all license conditions that can provide adequate or equivalent protection is immediately apparent. The licensee shall notify the NRC before, if practicable, and in any case immediately after taking such emergency action using the reporting procedure specified in 10 CFR 30.50(c).

8. Within 30 days of completing activities at each job site location, the licensee shall notify the Regional Administrator, Region ___, in writing of the temporary job site status and the disposition of any licensed material used.

A2

FC 94-02, Rev. 0

Appendix L
Decommissioning Process Checklists

The following pages contain the in-process checklists for decommissioning Groups 2–7. The Group 1 checklist is in Chapter 8 because it is the sole basis for documenting decommissioning for this group. The purpose of these checklists is to provide a statement of actions to be accomplished by the licensee and by the staff during the decommissioning process.

(The next page is blank for formatting purposes only)

APPENDIX L

LICENSEE NAME: _____

LICENSE NUMBER: _____ **DOCKET NUMBER:** _____

FACILITY: _____

1. Group 2 includes the following licensees (check if applicable):

☐ Licensees that can demonstrate compliance with 10 CFR Part 20.1402 ("Radiological Criteria for Unrestricted Use") using the screening methodology discussed in Section 6.6.

☐ Licensees that possessed and used only sealed sources but cannot demonstrate current leak tight integrity.

NOTE: Group 2 licensees do not need a DP

2. Licensee Actions

☐ NRC notified as required by 10 CFR 30.36(d), 40.42(d), and 70.38(d).

☐ Licensed material disposed of in accordance with NRC requirements and cleanup performed as necessary.

☐ Obtain most recent leak tests for all sealed sources, including those no longer in licensee's possession.

☐ Decommissioning records transferred as appropriate, or affirmed that they are not required to be retained or have transferred records.

☐ NRC Form 314, submitted or equivalent information provided. Written confirmation from the recipient listed on NRC Form 314 that material has been transferred to them attached.

☐ Final Status Survey submitted demonstrating that the facility, or portion of the facility, meets NRC's criteria for unrestricted use by using the dose screening methodology.

3. NRC Actions

☐ Disposition of licensed material verified.

☐ Leak test results, the type and number of sources on the license and NRC Form 314 are in agreement and the most recent leak test results are current.

☐ Determined if Technical Assistance Control number for the decommissioning action required.

☐ Technical Assistance Control obtained, if required.

☐   If an EA is prepared, consider relying on the license termination rule GEIS, as described in Section 15.7.3 of Volume 1 of this NUREG; publish the FONSI in the *Federal Register*, offering opportunity for hearing and soliciting comments.

☐   Licensee contacted (via telephone/writing) to ascertain decommissioning schedule, and its compliance with Timeliness Rule.

☐   Based on Licensee decommissioning schedule and scope of work, determine if In-Process or Close Out Inspection is required.

☐   FSSR reviewed to ensure that it adequately demonstrates that the facility is suitable for unrestricted use. See Section 15.4 for a list of FSSR requirements.

☐   If there is an issue related to the EPA/NRC MOU (e.g., residual radioactivity levels exceed the concentrations specified in the EPA/NRC MOU), the NRC reviewer should coordinate with DWMEP for the appropriate EPA notification. The EPA/NRC MOU is provided in Appendix H of this volume.

☐   License terminated by amendment after the suitability of licensee's facility for unrestricted use verified.[1]

☐   Amendment placed in the license docket file and ADAMS, and records retired in accordance with current records management guidance (e.g., RMG 92–01 and 93–03).

---

[1]   Certain types of facilities require an additional *Federal Register* notice at issuance of license amendment (see 10 CFR 2.106(a)(2)).

LICENSEE NAME: _____

LICENSE NUMBER: _____ **DOCKET NUMBER:** _____

FACILITY: _____

1. **Group 3 includes the following licences (similar to site condition for Group 2) (check if applicable):**

☐ Licensee can demonstrate compliance with 10 CFR Part 20.1402 ("Radiological Criteria for Unrestricted Use") using the screening methodology discussed in Section 6.6.

☐ Licensees that possess and use only sealed sources but cannot demonstrate current leak tight integrity.

NOTE: Group 3 licensees do need a DP.

2. **Licensee Actions**

☐ NRC notified as required by 10 CFR 30.36(d), 40.42(d), and 70.38(d).

☐ Submit a License Amendment request with DP attached. DP addresses the program areas discussed in Section 10.2 (may reference programmatic areas already contained in the license).

☐ Licensed material disposed of in accordance with NRC requirements and cleanup performed as necessary.

☐ Most recent leak tests for all sealed sources, including those no longer in licensee's possession, demonstrate there has been no leakage.

☐ Decommissioning records transferred as appropriate, or affirmed that they are not required to be retained or have transferred records.

☐ NRC Form 314 submitted, or equivalent information provided. Written confirmation from the recipient listed on NRC Form 314 that material has been transferred to them attached.

☐ Final Status Survey submitted demonstrating that the facility, or portion of the facility, meets NRC's criteria for unrestricted use by using the dose screening methodology.

3. **NRC Actions**

☐ Issue *Federal Register* Notice of receipt of application, offering opportunity for hearing and soliciting comments. Issue notice in local paper(s) soliciting public comments. Contact State and local governments and Indian Nations soliciting comments.

☐ Disposition of licensed material verified.

☐ Leak test results, the type and number of sources on the license and NRC Form 314 are in agreement, and the most recent leak test results are current and indicate that the sources did not leak.

☐ Technical Assistance Control obtained, if required.

☐ If an EA is prepared, consider relying on the license termination rule GEIS, as described in Section 15.7.3 of Volume 1; publish FONSI in *Federal Register*.

☐ License amendment for decommissioning issued after the review of licensee's DP determined to be acceptable.[1]

OR

☐ DP deficiency letter transmitted to licensee.

☐ Based on Licensee decommissioning schedule and scope of work, determine if In-Process or Close-Out Inspections are required.

☐ FSSR reviewed to ensure that it adequately demonstrates that the facility is in compliance with approved criteria. See Section 15.4 for a list of FSSR requirements.

☐ If there is an issue related to the EPA/NRC MOU (e.g., residual radioactivity levels exceed the concentrations specified in the EPA/NRC MOU), the NRC reviewer should coordinate with DWMEP for the appropriate EPA notification. The EPA/NRC MOU is provided in Appendix H of this volume.

☐ License terminated by amendment after compliance verified.[2]

☐ Amendment placed in the license docket file, and ADAMS, and records retired in accordance with current management directives (e.g., RMG 92–01 and 93–03).

---

[1] Certain types of facilities require an additional *Federal Register* notice at issuance of license amendment (see 10 CFR 2.106(a)(2)).

[2] Ibid.

LICENSEE NAME: _____

LICENSE NUMBER: _____ **DOCKET NUMBER:** _____

FACILITY: _____

1. **Group 4 includes the following licensees (check if applicable):**

☐   Licensees that can demonstrate compliance with 10 CFR Part 20.1402 ("Radiological Criteria for Unrestricted Use").

☐   Ground water contamination does not exist.

☐   Have demonstrated residual radioactivity is ALARA.

NOTE: Group 4 licensees do need a DP.

2. **Licensee Actions**

☐   NRC notified as required by 10 CFR 30.36(d), 40.42(d), and 70.38(d).

☐   Submit a License Amendment request with DP attached. Guidance on the contents of a DP is contained in Chapters 16–18 and the checklist in Appendix D of this NUREG.

☐   Licensed material disposed of in accordance with NRC requirements.

☐   Decommissioning records transferred as appropriate, or affirmed that they are not required to be retained or have transferred records.

☐   NRC Form 314 and DOE/NRC Form 741 (if applicable) submitted, or equivalent information provided. Written confirmation from the recipient listed on NRC Form 314 that material has been transferred to them attached.

☐   Final Status Survey submitted demonstrating that the facility, or portion of the facility, meets criteria approved by the Commission.

3. **NRC Actions**

☐   Issue *Federal Register* Notice of receipt of application, offering opportunity for hearing and soliciting comments. Issue notice in local paper(s) soliciting public comments. Contact State and local governments and Indian Nations soliciting comments.

☐   Technical Assistance Control obtained, if required.

☐   EA as described in NUREG–1748 completed. Consider relying on the license termination rule GEIS, as described in Section 15.7.3 of Volume 1. Publish the FONSI in the *Federal Register*.

APPENDIX L

☐   Licensee contacted (via telephone/writing) to ascertain decommissioning schedule and its compliance with Timeliness Rule.

☐   Issue license amendment authorizing implementation of DP after the review of licensee's DP determined to be acceptable.[1]

OR

☐   DP deficiency letter transmitted to licensee.

☐   Comply with requirements of Atomic Safety and Licensing Board, if there is a hearing.

☐   Based on Licensee decommissioning schedule and scope of work, determine if In-Process or Close-Out Inspections are required.

☐   Disposition of licensed material and NMMSS update (if applicable) verified.

☐   FSSR reviewed to ensure that it adequately demonstrates that the facility is in compliance with approved criteria. See Section 15.4 for a list of FSSR requirements.

☐   If there is an issue related to the EPA/NRC MOU (e.g., residual radioactivity levels exceed the concentrations specified in the EPA/NRC MOU), the NRC reviewer should coordinate with DWMEP for the appropriate EPA notification. The EPA/NRC MOU is provided in Appendix H of this volume.

☐   License terminated by amendment after compliance verified.[2]

☐   Amendment placed in the license docket file and ADAMS, and records retired in accordance with current management directives (e.g., RMG 92–01 and 93–03).

---

[1]   Certain types of facilities require an additional *Federal Register* notice at issuance of license amendment (see 10 CFR 2.106(a)(2)).

[2]   Ibid.

LICENSEE NAME: _____

LICENSE NUMBER: _____ **DOCKET NUMBER:** _____

FACILITY: _____

1. **Group 5[1] includes the following licensees (check if applicable):**

☐   Licensees that can demonstrate compliance with 10 CFR Part 20.1402 ("Radiological Criteria for Unrestricted Use").

☐   Ground water contamination exists.

☐   Have demonstrated residual radioactivity is ALARA.

NOTE: Group 5 licensees do need a DP.

2. **Licensee Actions**

☐   NRC notified as required by 10 CFR 30.36(d), 40.42(d), and 70.38(d).

☐   Submit a License Amendment request with DP attached. Guidance on the contents of a DP is contained in Chapters 16–18 and the checklist in Appendix D of this NUREG.

☐   Licensed material disposed of in accordance with NRC requirements.

☐   Decommissioning records transferred as appropriate, or affirmed that they are not required to be retained or have transferred records.

☐   NRC Form 314 and DOE/NRC Form 741 (if applicable) submitted, or equivalent information provided. Written confirmation from the recipient listed on NRC Form 314 that material has been transferred to them attached.

☐   Final Status Survey submitted demonstrating that the facility, or portion of the facility, meets criteria approved by the Commission.

3. **NRC Actions**

☐   Issue *Federal Register* Notice of receipt of application, offering opportunity for hearing and soliciting comments. Issue notice in local paper(s) soliciting public comments. Contact State and local governments and Indian Nations soliciting comments.

☐   Technical Assistance Control obtained, if required.

---

[1]   In general, lead office responsibility for Group 5 sites will be transferred from the NRC Regional office to NRC Headquarters. Regional staff and management should discuss the decommissioning with NRC Headquarters to determine which office will assume the lead for management of the decommissioning.

☐ EA as described in NUREG–1748 completed; publish the FONSI in the *Federal Register*. If ground water is contaminated and a FONSI cannot be determined, an EIS is necessary. See Section 15.7.

☐ Licensee contacted (via telephone/writing) to ascertain decommissioning schedule and its compliance with Timeliness Rule.

☐ Issue license amendment authorizing implementation of DP after the review of licensee's DP determined to be acceptable.[2]

OR

☐ DP deficiency letter transmitted to licensee.

☐ Comply with requirements of Atomic Safety and Licensing Board, if there is a hearing.

☐ Based on Licensee decommissioning schedule and scope of work, determine if In-Process or Close-Out Inspections are required.

☐ Disposition of licensed material and NMMSS update (if applicable) verified.

☐ FSSR reviewed to ensure that it adequately demonstrates that the facility is in compliance with approved criteria. See Section 15.4 for a list of FSSR requirements.

☐ If there is an issue related to the EPA/NRC MOU (e.g., residual radioactivity levels exceed the concentrations specified in the EPA/NRC MOU), the NRC reviewer should coordinate with DWMEP for the appropriate EPA notification. The EPA/NRC MOU is provided in Appendix H of this volume.

☐ License terminated by amendment after compliance verified.[3]

☐ Amendment placed in the license docket file and ADAMS, and records retired in accordance with current management directives (e.g., RMG 92–01 and 93–03).

---

[2] Certain types of facilities require an additional *Federal Register* notice at issuance of license amendment (see 10 CFR 2.106(a)(2)).

[3] Ibid.

LICENSEE NAME: _____

LICENSE NUMBER: _____ **DOCKET NUMBER:** _____

FACILITY: _____

1.   **Group 6 includes the following licensees (check if applicable):**

☐   Licensees that can demonstrate compliance with 10 CFR Part 20.1403 ("Radiological Criteria for Unrestricted Use").

☐   Have demonstrated residual radioactivity is ALARA.

☐   Sites where Institutional Controls are required to limit dose to the public.

NOTE: Group 6 licensees do need a DP.

2.   **Licensee Actions**

☐   NRC notified as required by 10 CFR 30.36(d), 40.42(d), and 70.38(d).

☐   Submit a License Amendment request with DP attached. Guidance on the contents of a DP is contained in Chapters 16–18 and the checklist in Appendix D of this NUREG.

☐   Develop institutional controls, acquire a competent agent to implement them, and provide financial assurance to provide adequate protection of public health and safety.

☐   Obtain input from interested and affected parties, concerning the adequacy of financial assurance and institutional controls, as described in §20.1403(d). Guidance on seeking public advice is contained in Sections 17.7.5 and M.6 of this guidance.

☐   Licensed material disposed of in accordance with NRC requirements.

☐   Decommissioning records transferred as appropriate, or affirmed that they are not required to be retained or have transferred records.

☐   NRC Form 314 and DOE/NRC Form 741 (if applicable) submitted, or equivalent information provided. Written confirmation from the recipient listed on NRC Form 314 that material has been transferred to them attached.

☐   Final Status Survey submitted demonstrating that the facility, or portion of the facility, meets criteria approved by the Commission.

3.   **NRC Actions**

☐   Issue *Federal Register* Notice of receipt of application, offering opportunity for hearing and soliciting comments. Issue notice in local paper(s) soliciting public comment.

☐   Contact State and local governments and Indian Nations soliciting comment.

☐ Technical Assistance Control obtained, if required.

☐ Site-specific EIS (because the licensee plans to limit future land uses at the site) completed. See Section 15.7.4 of Volume 1.

☐ Licensee contacted (via telephone/writing) to ascertain decommissioning schedule and its compliance with Timeliness Rule.

☐ Issue license amendment authorizing implementation of DP after the review of licensee's DP determined to be acceptable.[1]

OR

☐ DP deficiency letter transmitted to licensee.

☐ Comply with requirements of Atomic Safety and Licensing Board, if there is a hearing.

☐ Based on Licensee decommissioning schedule and scope of work, determine if In-Process or Close-Out Inspections are required.

☐ Disposition of licensed material and NMMSS update (if applicable) verified.

☐ FSSR reviewed to ensure that it adequately demonstrates that the facility is in compliance with approved criteria. See Section 15.4 for a list of FSSR requirements.

☐ If there is an issue related to the EPA/NRC MOU (e.g., residual radioactivity levels exceed the concentrations specified in the EPA/NRC MOU), the NRC reviewer should coordinate with DWMEP for the appropriate EPA notification. The EPA/NRC MOU is provided in Appendix H of this volume.

☐ License terminated by amendment after compliance verified.[2]

☐ Amendment placed in the license docket file, and ADAMS, and records retired in accordance with current management directives (e.g., RMG 92–01 and 93–03).

---

[1] Certain types of facilities require an additional *Federal Register* notice at issuance of license amendment (see 10 CFR 2.106(a)(2)).

[2] Ibid.

LICENSEE NAME: _____

LICENSE NUMBER: _____ DOCKET NUMBER: _____

FACILITY: _____

1.   Group 7 includes the following licensees (check if applicable):

☐   Licensees that cannot demonstrate compliance with 10 CFR Part 20.1403 (Radiological criteria for restricted use).

☐   Have demonstrated residual radioactivity is ALARA.

☐   Have demonstrated it is unlikely dose to an average member of the critical group will exceed 100 mrem/y.

NOTE: Group 7 licensees do need a DP.

2.   Licensee Actions

☐   NRC notified as required by 10 CFR 30.36(d), 40.42(d), and 70.38(d).

☐   Submit a decommissioning plan. Guidance on the contents of a DP is contained in Chapters 16–18 and the checklist in Appendix D of this NUREG.

☐   Develop institutional controls, acquire a competent agent to implement them, and provide financial assurance to provide adequate protection of public health and safety.

☐   Obtain input from interested and affected parties, as described in §20.1404(4). Guidance on seeking public advice is contained in Sections 17.7.5 and M.6 of this guidance.

☐   Obtain approval from the Commission on the proposed residual radioactivity and doses.

☐   Licensed material disposed of in accordance with NRC requirements.

☐   Decommissioning records transferred as appropriate, or affirmed that are not required to be retained or have transferred records.

☐   NRC Form 314 and DOE/NRC Form 741 (if applicable) submitted, or equivalent information provided. Written confirmation from the recipient listed on NRC Form 314 that material has been transferred to them attached.

☐   Final Status Survey submitted demonstrating that the facility, or portion of the facility, meets criteria approved by the Commission.

3.   NRC Actions

☐   Issue *Federal Register* Notice of receipt of application, offering opportunity for hearing and soliciting comments. Issue notice in local paper(s) soliciting public comment.

Contact State and local governments and Indian Nations soliciting comments. Contact the EPA and solicit comments.

☐ Submit recommendation on proposed remediation criteria to the Commission.

☐ Technical Assistance Control obtained, if required.

☐ Site-specific EIS, as described in Section 15.7.4 of Vol. 1, completed.

☐ Licensee contacted (via telephone/writing) to ascertain decommissioning schedule, and its compliance with Timeliness Rule.

☐ Issue license amendment authorizing implementation of DP after the review of licensee's DP determined to be acceptable.[1]

OR

☐ DP deficiency letter transmitted to licensee.

☐ Comply with requirements of Atomic Safety and Licensing Board, if there is a hearing.

☐ Based on Licensee decommissioning schedule and scope of work, determine if In-Process or Close-Out Inspections are required.

☐ Disposition of licensed material and NMMSS update (if applicable) verified.

☐ FSSR reviewed to ensure that it adequately demonstrates that the facility is in compliance with approved criteria. See Section 15.4 for a list of FSSR requirements.

☐ If there is an issue related to the EPA/NRC MOU (e.g., residual radioactivity levels exceed the concentrations specified in the EPA/NRC MOU), the NRC reviewer should coordinate with DWMEP for the appropriate EPA notification. The EPA/NRC MOU is provided in Appendix H of this volume.

☐ License terminated by amendment after compliance verified.[2]

☐ Amendment placed in the license docket file and ADAMS, and records retired in accordance with current management directives (e.g., RMG 92–01 and 93–03).

---

[1] Certain types of facilities require an additional *Federal Register* notice at issuance of license amendment (see 10 CFR 2.106(a)(2)).

[2] Ibid.

**Appendix M
Overview of the Restricted Use and
Alternate Criteria Provisions of
10 CFR Part 20, Subpart E**

M.1 INTRODUCTION

The requirements of 10 CFR 20.1403 and 10 CFR 20.1404 are briefly summarized in this overview. This overview is being included in this NUREG to provide the staff with an understanding of the philosophy and approach used by the Commission in promulgating these provisions of 10 CFR Part 20, Subpart E. Staff should refer to the appropriate sections of this guidance to evaluate licensee requests for license termination under these provisions. In addition, Section 17.7.5 of this Volume and Section M.6 of this Appendix contain guidance on seeking public advice on institutional controls which should be used to evaluate a licensee's program for compliance with 10 CFR 20.1403(d)(1–2) and 10 CFR 20.1404 (a)(4).

Prior to the promulgation of the License Termination Rule (LTR) (62 FR30958), U.S. Nuclear Regulatory Commission (NRC) regulations did not contain a provision for releasing sites for other than unrestricted use. Experience with decommissioning facilities has indicated that for certain sites, achieving the unrestricted use criterion might not be appropriate because: (1) there may be net public or environmental harm in achieving unrestricted use; (2) expected future use of the site would likely preclude unrestricted use; or (3) the cost of cleanup and waste disposal to achieve the unrestricted use criterion is excessive compared with achieving the same dose criterion by restricting the use of the site and eliminating exposure pathways.

Similarly, for certain difficult sites with unique decommissioning problems, 10 CFR 20.1404 includes a provision by which the NRC may terminate a license using alternate dose criteria. The NRC expects the use of alternate criteria to be limited to rare situations. This provision was included in 10 CFR 20.1404 because the NRC believed that it was preferable to codify provisions for these difficult sites in the rule rather than require licensees to seek an exemption process outside the rule.

NRC still considers unrestricted use to be the preferable method to decommission licensed facilities and terminate radioactive materials licenses. However, in recognition that there may be a limited number of sites where license termination with restrictions may be appropriate, the NRC included provisions for terminating the licenses for these few sites in the LTR.

M.1.1 RESTRICTED USE

License termination under restricted conditions will be permitted pursuant to 10 CFR 20.1403 if all the following requirements are met:

1. The licensee can demonstrate that further reductions in residual radioactivity necessary to release the site for unrestricted use: (1) would result in net public or environmental harm; or, (2) were not being made because the residual levels are as low as is reasonably achievable (ALARA) [10 CFR 20.1403(a)].

2. The licensee has made provisions for legally enforceable institutional controls that would limit dose to the average member of the critical group to 0.25 millisieverts per year (0.25 mSv/y) [25 millirem/year (25 mrem/y)] [10 CFR 20.1403(b)].

3. The licensee has provided sufficient financial assurance to enable an independent third party to assume and carry out responsibilities for any necessary control and maintenance of the site [10 CFR 20.1403(c)].

4. The licensee has submitted a decommissioning plan or a license termination plan to the NRC that indicates the licensee's intent to release the site under restricted conditions and describes how advice from individuals and institutions in the community who may be affected by the decommissioning has been sought and incorporated, as appropriate, following analysis of that advice [10 CFR 20.1403(d)]. In seeking this advice, the licensee would have conducted the activities for seeking advice required by 10 CFR 20.1403(d)(2), including providing for participation by a broad cross-section of community interests that may be affected by decommissioning; providing an opportunity for a comprehensive collective discussion of the institutional controls and financial assurance specified in 10 CFR 20.1403(d)(1) by the affected parties; and providing a publicly available summary of all such discussions.

5. The residual radioactivity levels have been reduced so that, if the institutional controls were no longer in effect, the annual dose to the average member of the critical group would not exceed 1.0 mSv/y (100 mrem/y) or, under certain conditions, would not exceed 5.0 mSv/y (500 mrem/y). If the 5.0 mSv/y (500 mrem/y) value is used, the licensee must: (1) demonstrate that achieving 1.0 mSv/y (100 mrem/y) is prohibitively expensive, not technically achievable, or would result in net public or environmental harm, (2) make provisions for durable institutional controls, and (3) provide sufficient financial assurance to allow an independent third party to carry out rechecks of the controls and maintenance at least every 5 years and carry out any necessary controls and maintenance [10 CFR 20.1403(e)].

The NRC staff will review and evaluate the decommissioning plan and will solicit public input to determine whether the above requirements are satisfied, pursuant to 10 CFR 20.1405. Once the NRC determines that they have been met, the NRC license is terminated and the NRC no

longer regulates or oversees the site, except in the circumstances indicated in
10 CFR 20.1401(c). Specifically, 10 CFR 20.1401(c) indicates that the NRC could require
additional cleanup after license termination if it determines that, based on new information, the
criteria in Subpart E of 10 CFR Part 20 for release of a site were not met <u>and</u> residual
radioactivity remaining at the site could result in a significant threat to public health and safety.
Please note that the Commission has explicitly chosen not to define what constitutes "new
information" or "significant public risk," because this determination will be made on a case-by-
case basis.

In some instances, a licensee planning license termination with restricted conditions under an
approved decommissioning plan or license termination plan may find during remediation that the
site can be cleaned up to a level that would not require restricted conditions. Additionally, a
licensee that had planned unrestricted release may find during remediation that unrestricted
release is not practical. In these instances, the licensee should submit an amended
decommissioning plan or license termination plan to NRC as soon as possible.

The restricted conditions should be limited to the smallest portion of the site that is appropriate.
However, all areas that will be subject to restricted conditions should be contained within one or
occasionally two areas. Complicated checkerboard patterns of areas with restricted conditions
should be avoided.

M.1.2   ALTERNATE CRITERIA

Under 10 CFR 20.1404, the NRC may consider terminating a license using alternate criteria that
are greater than 0.25 mSv/y (25 mrem/y), with restrictions in place. However, licensees
requesting license termination under the alternate criteria provisions of 10 CFR 20.1404 would
still need to ensure that potential doses from residual radioactivity are less than 1.0 mSv/y
(100 mrem/y) with restrictions in place. In addition, the NRC will limit the conditions under
which a licensee could apply to the NRC for, or be granted use of, alternate criteria to unusual
site-specific circumstances subject to the following provisions:

1.   The licensee has provided assurance that public health and safety will continue to be
     protected and that it is unlikely that the dose from all man-made sources combined, other
     than medical, would be more than 1.0 mSv/y (100 mrem/y). A licensee proposing to use
     alternate criteria would have to provide a complete and comprehensive analysis of such
     possible sources of exposure.

2.   The licensee has employed, to the extent practical, restrictions on site use for minimizing
     exposure at the site, using the provisions for institutional controls and financial assurance
     in 10 CFR 20.1403, including compliance with the dose "caps" required by
     10 CFR 20.1403(e).

3.   The licensee has reduced doses to ALARA levels, based on a comprehensive analysis of
     risks and benefits of all viable alternatives.

4.   The licensee has sought advice from affected parties regarding the use of alternate criteria at the site. In seeking this advice, the licensee would have conducted the activities for seeking advice required by 10 CFR 20.1404(a)(4), including providing for participation by a broad cross-section of community interests that may be affected by decommissioning; providing an opportunity for a comprehensive collective discussion of the issues related to the alternate criteria by the affected parties; and providing a publicly available summary of all such discussions.[1] As part of this process, the licensee would submit a decommissioning plan indicating how advice of individuals and institutions in the community that may be affected by the decommissioning has been sought and addressed.

5.   The licensee has obtained the specific approval of the Commission for the use of alternate criteria. The Commission will make its decision after considering the NRC staff's recommendations that would address any comments provided by the EPA and any public comments submitted regarding the decommissioning plan pursuant to 10 CFR 20.1405.

M.1.3   INSTITUTIONAL CONTROLS

Institutional controls are used to limit intruder access to, and/or use of, the site to ensure that the exposure from the residual radioactivity does not exceed the established criteria. Institutional controls include administrative mechanisms (e.g., land use restrictions) and may include, but not be limited to, physical controls (e.g., signs, markers, landscaping, and fences) to control access to the site and minimize disturbances to engineered barriers. There must be sufficient financial assurance to ensure adequate control and maintenance of the site. As discussed below, they must be legally enforceable and the entity charged with their enforcement must have the capability, authority, and willingness to enforce the institutional controls. If the institutional control includes physical controls, they must include measures to monitor their performance and to provide for their maintenance or replacement along with sufficient financial assurance to provide for the necessary monitoring, maintenance or replacement.

Institutional controls address a variety of restrictions and need to be tailored to each site situation. Restrictions may include prohibitions on farming, industrial, recreational, or residential use. Prohibitions on excavation and water use may also be warranted. Institutional controls are usually characterized as "proprietary" or "governmental." Generally, a layering of different restrictions and mechanisms are needed to provide for durable and effective institutional controls.

Institutional controls based on property rights involve a party that owns rights that restrict the use of, or access to, the property and are referred to as "proprietary institutional controls."

---

[1]   Licensees are required by 10 CFR 20.1403 to obtain advice from institutions and individuals that may be affected by the decommissioning on specific issues related to institutional controls and financial assurance. However, 10 CFR 20.1404 provides for a much broader discussion of the issues associated with the use of alternate criteria, and, as such, licensees must obtain advice on essentially any issue associated with the use of alternate criteria.

Institutional controls based on property rights apply to land owned by individuals or private institutions and land owned by governments.

One example of such an institutional control is an environmental covenant. The Uniform Environmental Covenants Act (UECA), is a model law that was approved by the National Conference of Commissioners on Uniform State Laws in 2003. If enacted in a State, the UECA would establish requirements for an environmental covenant, a new valid real estate document that can control the future use of a site when real estate is transferred from one person to another. UECA includes provisions absent from most existing State statutes, which may help to overcome obstacles that lead to the ineffectiveness of other land use controls. The environmental covenants created under UECA would be based upon traditional property law principles and would be recorded in the local land records and thereby bind successive owners of the property. State and local governments would have clear rights to enforce the land use restrictions. UECA ensures that a covenant will survive despite tax lien foreclosure, adverse possession, marketable title statutes, and other common law doctrines which might otherwise inadvertently extinguish an intended land use control. UECA also provides detailed provisions on dealing with recorded interests that have priority over the new covenant.

Institutional controls that involve a government using its sovereign or police powers to impose restrictions on citizens or sites under its jurisdiction to limit the use and occupation of privately owned lands are referred to as "governmental institutional controls." Among the more common governmental institutional controls are zoning, well-use restrictions, and building permit requirements.

Zoning is a legal designation placed on land by a local government that restricts the types of uses on a particular property. Overlay zoning consists of zones drawn on a municipality's existing zoning map that provides protection not explicitly stated under existing zoning requirements. Since zoning is subject to change, zoning generally should be used in combination with other restrictions.

Governments, most often local, can place restrictions on private property prohibiting or limiting use. Such government-imposed restrictions could include prohibiting construction of wells for water use, restricting the use of other potential water supplies, issuing permits for certain activities including use of wells for drinking water or construction or use of buildings, and establishing county or State ordinances and property law regulations.

At some sites, institutional controls may include physical controls (e.g., fences, markers, earthen covers, radiological monitoring, and the maintenance of those controls). Physical controls alone do not meet the requirement in 10 CFR 20.1403(b) for legally enforceable institutional controls because they lack a mechanism for legal enforcement. Physical controls and their maintenance can be used to meet the requirement in 10 CFR 20.1403(b) only when they are used in combination with an instrument that permits legal enforcement of the physical control.

In addition to requiring that the institutional controls function to limit the dose to 0.25 mSv/y (25 mrem/y) in 10 CFR 20.1403(b), Subpart E also contains (in 10 CFR 20.1403(e)) two levels of protection based on potential exposure if the institutional controls become ineffective. Based on those two levels, the institutional controls and the parties enforcing the controls need to meet the following criteria:

1. If the annual dose to the average member of the critical group would not exceed 1.0 mSv/y (100 mrem/y) if the institutional controls were no longer in effect, a private individual, organization, or Federal, State or local government may be acceptable as the entity responsible for enforcing the institutional control[2] depending on the circumstances at the site; or

2. If the annual dose could exceed 1.0 mSv/y (100 mrem/y) but be less than 5.0 mSv/y (500 mrem/y), if the institutional controls were no longer in effect, 10 CFR 20.1403(e) requires that a more durable institutional control be used. To meet the requirement in 10 CFR 20.1403(e), an institutional control that involves government ownership of land would be generally acceptable. On privately owned land, a Federal, State or local government as the entity responsible for enforcing the restriction could also be acceptable, depending on the circumstances at the site.

Finally, restrictions will need to remain in place for the duration that they are needed, up to 1000 years. The duration may be a definite specified duration or an indefinite duration. Definite durations are for a specified number of years (for example, the number of years until radiological decay or other processes have reduced the concentration to a level corresponding to an annual dose to the average member of the critical group of less than 0.25 mSv/y (25 mrem/y) without the restrictions). Indefinite durations might end when some measurable event has occurred (for example, when natural processes have adequately reduced the risk of exposure to the residual radioactivity).

The NRC staff will review and evaluate the decommissioning plan and will solicit public input to determine whether the above requirements are satisfied, pursuant to 10 CFR 20.1405.

M.1.4   INSTITUTIONAL CONTROL IMPLEMENTATION ISSUES

NRC and licensees have had difficulty implementing the LTR requirements for license termination under restricted conditions. For example, States have not been agreeable to becoming the independent third party to act as a backup to an owner and often oppose the restricted use approach. Similarly, NRC's efforts to make arrangements for DOE to take

---

[2]   The Commission has stated (see Section B.3.3 of the "Statements of Consideration" for 10 CFR Part 20, Subpart E, "Radiological Criteria for License Termination") that stringent institutional controls would be needed for sites involving large quantities of uranium and thorium contamination. Typically, these would involve legally enforceable deed restrictions and/or controls backed up by State and local government control or ownership, engineered barriers, and as appropriate, Federal ownership.

ownership of commercial sites and provide the necessary access and land use controls or maintenance under the provisions of Section 151(b) of the Nuclear Waste Policy Act of 1982 have not been successful. Finally, for sites with long half-life radionuclides such as uranium and thorium, long-term effectiveness of institutional controls is recognized as a significant challenge given many examples of institutional control failure even after short periods of time.

In response to these difficulties that caused decommissioning delays, NRC evaluated the institutional control issues and developed new policies that should help resolve the issues. These evaluations and new policies are described in the LTR Analysis (SECY-03-0069), and a May 2004 Regulatory Issue Summary, RIS-2004-08.

M.2   RISK-INFORMED GRADED APPROACH TO INSTITUTIONAL CONTROLS

The first of the three new policies is a risk-informed graded approach to selecting institutional controls under the LTR, so that licensees can have flexibility to arrange the appropriate level of controls. The risk-informed graded approach consists of risk framework and associated grades of institutional controls. The general risk framework is defined by the hazard level and likelihood of hazard occurrence. The hazard level is established in the LTR (10 CFR 20.1403(e)(ii)) as the dose level of 1.0 mSv/y (100 mrem/y), calculated assuming institutional controls are not in effect. This dose level is the public dose limit. Sites with calculated doses above the public dose limit, but below 5.0 mSv/y (500 mrem/y), are considered higher risk sites. Those sites below the public dose limit are considered lower risk sites. In addition, higher risk sites are those with longer hazard duration (e.g., longer dose persistence or longer half-life, greater than 100 years). The LTR also defines the general grades of controls: sites below the 1.0 mSv/y (100 mrem/y) dose level require legally enforceable institutional controls, and sites above the 1.0 mSv/y (100 mrem/y) dose level require both legally enforceable and durable institutional controls. Thus, the LTR requires that institutional controls provide more reliable or sustainable protection over the time period needed (i.e., durable) for higher risk sites that could exceed the public dose limit assuming no restrictions. Durable institutional controls are also appropriate for long-lived radionuclides regardless of the dose limit. Therefore, the hazard magnitude and duration criteria for a higher risk site define when durable institutional controls are needed. Table M.1 illustrates this risk-informed graded approach and gives examples.

Specific grading of institutional controls can be selected within the two general grades defined above. This approach recognizes that the site-specific factors affecting risk can be highly variable from site to site. As a result, specific grading recognizes the need for flexibility to tailor institutional controls to achieve the desired effectiveness. Specific grading involves evaluating and balancing numerous site-specific factors such as: (a) physical characteristics of the site that limit future land use; (b) land uses that could be adverse to performance/compliance and therefore should be prohibited; (c) land uses that are acceptable and could result in productive reuse of the site; (d) dose assessment results; e) engineered barriers and related maintenance; (f) monitoring controls and maintenance; (g) jurisdictional limitations on enforceability and long-term effectiveness of institutional controls; and h) advice from affected parties, such as local governments and the public.

The graded approach has important benefits. For the public, protection is increased, especially over the long term. The approach clearly identifies when durable controls might be needed and specific controls would be designed to mitigate site-specific risks that are significant to maintaining safety. For licensees and NRC, clearer guidance is provided for licensees to select institutional controls and NRC to review licensees' proposed controls. Licensees also have the flexibility to select appropriate controls that could be less costly and easier to arrange.

Table M.1 NRC'S Risk-Informed Graded Approach for Institutional Controls to Restrict Site Use

Lower Risk	Higher Risk
Lower Hazard Level [0.25–1.0 mSv/y (25–100 mrem/y)]	**Higher Hazard Level [1.0–5.0 mSv/y (100–500 mrem/y)]**
Shorter Hazard Duration – Lower Likelihood of IC Failure	**Longer Hazard Duration – Higher Likelihood of IC Failure**
Shorter Dose Persistence or Half-Life (less than 100 years)	**Longer Dose Persistence or Half-Life (greater than 100 years) ***
General Grade Legally enforceable institutional controls Specific Grade Tailor specific type of institutional controls and land use restrictions to site-specific circumstances using scenario analyses from dose assessments Examples Single conventional "deed restriction," such as a restrictive covenant (less control) Layered/redundant controls such as restrictive covenant, deed notice, and State registry (more control) (Note that either the NRC Long-Term Control license or legal agreement/restrictive covenant could be used if other types of conventional institutional controls cannot be established.)	General Grade Durable and legally enforceable institutional controls with 5-year review Specific Grade Tailor specific type of institutional controls and land use restrictions to site-specific circumstances using scenario analyses from dose assessments Examples Layered/redundant controls that include State government control (durable) Conventional institutional control with NRC monitoring and enforcement after license termination using legal agreement and restrictive covenant (durable) Conventional institutional control with NRC monitoring and enforcement after license termination using regulatory authority under 10 CFR 20.1401(c) (more durable) State or Federal government ownership and control (NWPA §151(b)) (most durable) NRC Long-Term Control license (most durable)
* It may be appropriate to treat sites with longer half-life radionuclide contamination, but with doses close to 0.25 mSv/y (25 mrem/y) assuming no controls, as "Lower Risk" sites.	

M.3   LONG-TERM CONTROL LICENSE OPTION

NRC staff recommended to the Commission, in SECY-03-0069, that a new type of possession-only specific license for long-term control be established as one option for resolving the LTR institutional control issue at sites where restricted use or alternate criteria could be used. This option should not be considered a guaranteed option, but would be used as a last resort (see Section 17.7 of this volume). This new type of possession-only license is referred to in this guidance as a long-term control (LTC) license to clearly distinguish it from the NRC's existing possession-only licenses for storage. The existing possession-only license is typically used at NRC licensed sites in the operating or decommissioning phases. In contrast, the LTC license is for use as an institutional control in the long-term control phase after completion of decommissioning. A licensee may propose use of an LTC license only if the licensee cannot otherwise establish acceptable institutional controls or independent third party arrangements. Attachment 1 of SECY-03-0069 provides a description and evaluation of the staff's recommended option of possession-only license for long-term control. On November 17, 2003, the Commission approved this LTR recommendation (SRM-SECY-03-0069).

M.3.1   PURPOSE OF LTC LICENSE

The primary purpose of NRC's LTC license is to provide the legally enforceable and durable institutional controls required by 10 CFR 20.1403(b) to ensure the long-term protection of the public health, safety, and the environment. Therefore, the LTC license is for long-term control of a restricted use site after decommissioning is completed. The LTC license is not for the purpose of storage of radioactive materials. It also should not be considered a guaranteed option, but would be a last resort under the criteria in 10 CFR 20.1403(b). With use of an LTC license, the licensee must still meet all the restricted use requirements of the LTR, to ensure protection of the public health and safety.

The conditions of the LTC license would specify the necessary controls to limit site access and land use that the licensee must monitor and maintain and that NRC would inspect and enforce, if necessary. The LTC license also would specify other required long-term control activities to be conducted by the licensee, such as surveillance, maintenance, reporting, records retention, and stakeholder involvement (see guidance below). Detailed plans to implement the LTC license conditions would be given in a Long-Term Control and Maintenance Plan that the licensee would prepare and NRC would approve during decommissioning and before the LTC license is established.

M.3.2   ROLES AND RESPONSIBILITIES

The licensee has the primary responsibility for long-term protection of the public health, safety, and the environment by implementing and then maintaining the effectiveness of the controls required by the LTC license. The licensee would maintain the required site access and land use controls, as well as engineered barriers, using periodic surveillance, maintenance, and monitoring, if needed. The licensee also would provide an annual report to NRC, with copies to State and local governments. Finally, licensing records would be maintained by the licensee.

NRC is responsible for assuring that the licensee's controls and maintenance remain effective by conducting oversight reviews, making periodic inspections, conducting five-year license renewals, issuing a new LTC license when ownership changes in the future, enforcing the license, if needed, and maintaining licensing records for the duration of the LTC license.

Oversight reviews could include reviewing licensee annual reports and other reports (e.g., corrective action reports or requests for NRC approval of the sale of the site) and obtaining advice from stakeholders. NRC's inspection role might include an annual inspection for the first five years and then once every five years thereafter as part of the license renewal process. Periodic inspections might also be needed to address specific adverse events, allegations, and licensee corrective actions. NRC inspections could involve seeking advice and information from stakeholders. A license renewal process also would be conducted every five years, considering licensee reports, NRC inspections, and stakeholder advice. The five-year renewal process would also provide the five-year rechecks, as needed per 10 CFR 20.1403(e)(2). License renewal is a regulatory mechanism to evaluate the sustainability of the LTC license over the long term including: effectiveness of site access and land use controls, licensee performance, new site information, and sufficiency of funding. These evaluations could result in revised license conditions necessary to ensure long-term effectiveness of controls. Enforcement actions may be taken if the conditions of the license are not met.

Stakeholders have a role under the LTR during the licensee's preparation of the decommissioning plan for a restricted use site. For these sites, the licensee is required by 10 CFR 20.1403(d) to seek advice from affected parties regarding a number of matters, including the plans for enforceable institutional controls, sufficient financial assurance, and undue burdens on the local community or other affected parties. The licensee shall document in the decommissioning plan how the advice was sought and incorporated, as appropriate, following analysis of that advice. Similarly, under 10 CFR 20.1405, NRC shall notify and solicit comments from the public upon receipt of the decommissioning plan.

In addition to the State and EPA, other stakeholders may have an ongoing interest in the site after decommissioning has been completed and the LTC license is in place. From time to time, it might be appropriate to schedule public meetings, such as during the five-year license renewal process, to obtain information about the site and to maintain a local awareness of the site and the restrictions on site access and use.

M.3.3 REQUIREMENTS FOR LICENSEES PROPOSING RESTRICTED USE WITH THE LTC LICENSE

The decommissioning goal for a site proposing the LTC license is the same as any other decommissioning site proposing restricted use — safe site decommissioning that complies with the LTR. However, for such sites, the license is not terminated after remediation; it is amended to become an LTC license. Nevertheless, a licensee proposing to use the LTC license needs to comply with all the criteria of 10 CFR 20.1403, even though the license will not be terminated. These restricted use requirements for licensees are:

- 10 CFR 20.1403(a): Eligibility for restricted use (ALARA or public/environmental harm)

- 10 CFR 20.1403(b): Legally enforceable institutional controls and 0.25 mSv/y (25 mrem/y) dose criterion. (The institutional control requirements would be met with the LTC license conditions.)

- 10 CFR 20.1403(c): Sufficient financial assurance for control and maintenance

- 10 CFR 20.1403(d): A decommissioning plan or a license termination plan for restricted use that includes how advice from affected parties has been sought and incorporated

- 10 CFR 20.1403(e): 1.0 mSv/y (100 mrem/y) and 5.0 mSv/y (500 mrem/y) dose "cap" requirements if institutional controls were no longer in effect

In addition, because the NRC license would be amended and not terminated, other NRC requirements for NRC licensees would continue, such as recordkeeping.

M.3.4 INITIAL ELIGIBILITY FOR RESTRICTED RELEASE AND THE LTC LICENSE OPTION

In the Statements of Consideration for the LTR, the Commission noted that it allows restricted use as an appropriate method of decommissioning while maintaining the philosophy that "…in general, termination of a license for unrestricted use is preferable because it requires no additional precautions or limitations on use of the site after licensing control ceases, in particular for those sites with long-lived nuclides."

As a result, sites considering restricted use must first comply with the existing "initial eligibility" requirements of 10 CFR 20.1403(a) that further reductions in residual radioactivity to comply with unrestricted use criteria would result in net public or environmental harm or are not being made because the levels associated with restrictions are as low as reasonably achievable (ALARA).

In addition, the selection of the LTC license option would be an appropriate option if:

a. Durable institutional controls are required because the site is considered higher risk under the staff's graded approach to institutional controls in Table M.1.

b. The licensee can demonstrate to NRC satisfaction that it was unable to establish other types of acceptable institutional controls and independent third party arrangements (e.g., letter from the State rejecting responsibility for ownership, control, or independent third party oversight).

c.   The site would need long-term monitoring or maintenance requiring technical skills to conduct.

M.3.5   PARTIAL RESTRICTED RELEASE UNDER AN LTC LICENSE AND MAINTAINING SINGLE OWNERSHIP OF THE SITE

Under the LTC license option, there is flexibility in the approach to subdivide the existing site and allow license termination or release of unrestricted use portions of the site while maintaining the restricted use portion(s) under the LTC license.

For government owned sites, the site could be subdivided and unrestricted use portions could be released from the license for reuse. The remaining restricted use portion(s) would remain under the license and sustained government ownership would be assumed.

Similarly, for private sites, where the restricted use portion would clearly have resale value to sustain future ownership (even with restrictions on use), the site could be subdivided and unrestricted use portions could be released for reuse. For example, if the restricted use portion has soil contamination or buried contaminated slag but an industrial use would be safe (e.g., warehouse or parking lot), then it would likely have a resale value. For this case, sustaining ownership might be more likely because of the value of the property, and the potential for an "abandoned" site in the future would be reduced.

A different approach could be beneficial for privately owned sites where long-term controls are needed and where the restricted use portion has little or no resale value, but the unrestricted use portion has valuable reuse that would sustain future ownership, both at the present time and in the future. For this case, there could be a benefit to maintain the current license boundaries, including both the restricted and unrestricted use portions together. Even under the LTC license, the unrestricted use portion would be available for any use consistent with local zoning constraints. The only conditions on these portions of the site would be to: (1) conduct monitoring, if needed; and (2) prohibit the sale separately from the restricted use portion containing the residual contamination. The staff recognizes that this approach is a challenging issue with pros and cons, with the possibility that undue burdens could result. Thus, both potential benefits and burdens should be evaluated, on a case-by-case basis, considering the views of affected parties. Some considerations are given below.

- Potential benefits:
  - The unrestricted use portion of the site could have resale value that balances the lack of resale value or even perception of liability associated with the restricted use portion. Prospective buyers would clearly understand the permitted uses on the unrestricted use portion. Permitted uses should enhance future resale of the site (with both restricted and

unrestricted use portions) as a whole. Overall, this approach would help ensure sustainability of owner/licensee controls, and thus protection of public health and safety, for sites that need long-term control and where numerous ownership changes are expected over the long-term. This approach minimizes the possibility of the restricted portion of the site being abandoned in the future, if there is a gap in ownership.

— This approach is intended to allow reuse of the site while enhancing the long-term protection.

— Maintaining ownership is the most effective and efficient approach to long-term protection. While NRC would have some options in the event of a gap in ownership, these options could be difficult and time consuming to establish.

— Maintaining ownership of the complete site would help ensure monitoring over the long-term, if needed.

- Potential burdens:

— Under this approach, the sensitivity to an NRC license might discourage future reuse of the unrestricted portion of the site and impact productive future use and revenue for the local community.

It is important to note that the licensee must provide for sufficient financial assurance (an independent trust fund) to allow a third party to monitor and maintain the site for the long-term; therefore, future entities would not be responsible for any costs associated with the monitoring or maintenance or routine costs. This approach would eliminate the liability of future owners, as long as they abided by the provisions of the LTC license. Clearly communicating this in the LTC license could help reduce this sensitivity.

— There could be a perception that this approach is inconsistent with NRC's existing policy for partial site release for operating sites, for example a power reactor that could maintain an independent spent fuel storage installation (ISFSI) on a portion of the original site and release the remaining portion for unrestricted use.

While the two approaches might seem inconsistent, they have different purposes and time periods, and therefore are not the same. The partial site release example for an ISFSI is for the purpose of short-term storage of up to a few decades, and therefore maintaining ownership should not be a concern. In contrast, the purpose of the LTC license approach for a restricted use site needing restrictions for a long time period is to sustain ownership and a license over a long time period where ownership is expected to change many times.

— This approach could be legally challenged.

M.3.6 MINIMIZING THE SIZE OF THE RESTRICTED AREA

The licensee should minimize the size of the restricted portion of the site, while also considering dose assessments and the need for monitoring in its determination of what area needs to be restricted. Such an approach would contribute to demonstrating ALARA. It also would result in a smaller area to control, which may make access limitations like fencing and surveillance simpler and more effective.

M.3.7 FLEXIBILITY TO SEEK UNRESTRICTED RELEASE OR ANOTHER TYPE OF INSTITUTIONAL CONTROL IN THE FUTURE

The LTC license is not necessarily permanent, but would be in place as long as needed to protect public health and safety and the environment based the half-life of the radionuclides and other factors. Similarly, the LTC license would not preclude a licensee from removing residual radioactivity in the future and seeking unrestricted release, for example, if a new inexpensive disposal option becomes available. NRC would not require removal because the LTR finality provision in 10 CFR 20.1401(c) applies. However, this provision does not preclude a licensee from making a business decision to request license termination with unrestricted release under the LTR. For this case, a licensee would submit a decommissioning plan for NRC review and then conduct the approved remediation and final status survey as is currently required by the LTR. For example, at a site with contaminated slag, if reuse of the slag becomes viable, the licensee could submit a license amendment request and decommissioning plan for NRC approval. After NRC approval of the DP and license amendment for decommissioning, the material could be removed for reuse and the LTC license terminated with unrestricted release under the LTR. Thus, unrestricted use would not be precluded by the LTC license. The LTC license could explain this process so that future licensees understand this flexibility.

NRC would also allow a request to terminate the NRC LTC license and release the site with restrictions using another acceptable type of legally enforceable and, if needed, durable institutional control and independent third party arrangement, if approved by NRC.

M.3.8 TRANSFER OF CONTROL/OWNERSHIP AND DEED NOTICE

Transfers of site ownership are expected over the long-term, and the new owner(s) will need to become the licensee and provide the controls as specified in the conditions of the LTC license. Thus, the required control and maintenance under the LTC license would continue to be effective over the long-term even when ownership transfers as a condition of the license. The licensee must notify NRC of a potential sale and obtain NRC approval of the new owner by amending the license prior to the effective date of the sale of the licensed property. The licensee and potential new owner would submit to NRC a request for ownership transfer that would include information about the new owner such as: financial viability; willingness to accept their new responsibilities under the LTC license; technical capability to conduct monitoring, maintenance, and potential corrective actions; and willingness to become an NRC licensee. The prospective owner must become an NRC licensee effective at the time of the sale. The licensee also must establish and maintain/re-record a deed notice, approved by NRC, as a condition of the license. This will provide additional assurance that potential future owners will be informed that an NRC LTC license is required as well as the conditions of the license.

M.3.9   SUFFICIENT FINANCIAL ASSURANCE AND TRUST

The licensee must establish a trust and place sufficient funds into it to produce annual income that is sufficient to cover the (1) annual average costs of licensee surveillance, control, radiological monitoring of surface and groundwater if needed, and routine maintenance, (2) NRC oversight costs, and (3) trustee fees and expenses. The licensee should assume 1 % return on investment (consistent with 10 CFR Part 40, Appendix A). This trust fund is independent of the licensee and, therefore, not affected in the event of licensee bankruptcy and would continue independent of site ownership. The NRC would be the beneficiary of the trust. The licensee would request, and the independent trustee would pay, in accordance with the instrument, for the costs of surveillance, control, maintenance, and NRC oversight costs, most likely on an annual basis. Because the fund would produce income sufficient to hire a contractor to perform the surveillance and control tasks, the licensee could hire a contractor to perform the duties, and be reimbursed for the full cost, rather than performing the work itself.

In the event the licensee does not perform its duties, or goes bankrupt, NRC could take enforcement action, as necessary, to ensure that control activities are maintained. Alternatively, the trustee could be directed by NRC to provide funds to a contractor to work on behalf of the licensee. NRC could seek a court to appoint a custodial trustee to continue the long-term control activities using funds from the trust in the event that no licensee exists.

M.3.10   NRC FEES FOR LTC OVERSIGHT ACTIVITIES

No annual fees (10 CFR Part 171) are required for the LTC license. However, fees for NRC services would be recovered (10 CFR Part 170). Therefore, the licensee would be charged for NRC activities during the year, expected to be review of one annual report, annual inspections during the first five years, license renewal activities every five years, enforcement actions if needed, and responses to events and licensee corrective actions as needed. The licensee should assume an NRC fee of $10,000 for one report review and one inspection each year. Also assume a fee of $20,000 once every five years for the five-year license renewal, expanded inspection, and report review. These fees are in 2005 dollars and should be adjusted for inflation. To adjust for inflation, use the ratio of the cost of professional staff hours found in 10 CFR 170.20. In 2005, the staff-hour cost for NMSS was $197 per hour.

M.3.11   COMPLETION OF DECOMMISSIONING AND FINALITY OF DECOMMISSIONING DECISIONS

NRC recognizes the importance of the finality of its decommissioning decisions. Under 10 CFR 20.1401(c), the Commission could require additional cleanup in the future, based on new information, if it determines that the criteria in the LTR were not met and residual radioactivity remaining at the site could result in significant threat to public health and safety. This requirement also would apply to a site with the LTC license and may be particularly important to potential future owners/licensees who may be concerned about future liabilities should they purchase the site.

The definition of "decommission" in 10 CFR Part 20 states, "Decommission means to remove a facility or site safely from service and reduce residual radioactivity to a level that permits–(1) release of the property for unrestricted use and termination of the license; or (2) release of the property under restricted conditions and the termination of the license." As the Part 20 definition notes that decommissioning includes reducing residual radioactivity to a level that *permits* release and termination of the license, the staff considers a site with an LTC license to be decommissioned (even though the license is amended and not actually terminated), given all of the applicable restricted use requirements in the LTR have been met.

Although an existing license could be terminated and a new LTC license established at the end of the decommissioning process, the staff believes that amending the license is administratively more efficient and helps preserve a single Agency record for the site. The staff informed the Commission of this and other public comments on this issue (see SECY-06-0143). The Commission confirmed (in SRM-SECY-06-0143) that the LTC license could be put into place by amendment, rather than termination of the decommissioning license and establishment of a new LTC license, and that, for this circumstance, decommissioning would be considered complete and the finality provision of the LTR in 10 CFR 20.1401(c) would be relevant to the site (i.e., NRC would require additional cleanup only if, based on new information, it determined that the LTR criteria were not met and residual radioactivity could result in a significant threat to public health and safety).

M.3.12 LONG-TERM RECORD RETENTION AND AVAILABILITY

The licensee will be required to maintain the decommissioning records that are necessary for maintaining effective long-term protection. In addition, new LTC records must be maintained for the duration of the LTC license. The purpose of recordkeeping is to support those LTC licensee activities necessary for effective long-term protection. In the event of ownership and license transfer in the future, there are existing NRC requirements for records transfer to ensure that important records remain available.

In addition, NRC intends to continue maintaining the LTC licensing records in the same docket file used for operations and decommissioning. This approach should result in a continuous and completely documented history of the site operations, decommissioning, and long-term control available in a single file that will improve the efficiency and effectiveness of future search and retrieval of site information. These records are expected to be available to the public in the future. Finally, NRC currently maintains the site decommissioning database, which includes restricted use sites. This publicly available database provides web-based access to general site information about all NRC decommissioned sites.

NRC recognizes that maintaining records and making them publicly available over the long term is one of the important elements to ensure protection for long periods of time so that knowledge of the site will not be forgotten. Retention of duplicate records in different locations by the licensee and NRC enhances long-term record retention.

M.3.13 CONTENT OF THE LTC POSSESSION-ONLY LICENSE AND LTC PLAN

LTC license conditions specify requirements for: prohibited site access and land use, permitted site access and land use, physical controls (fences, signs, monuments), surveillance, groundwater monitoring (if needed), corrective actions, maintenance, reporting, records retention and availability.

The LTC Plan provides site information and implementation activities and procedures for each license condition (similar to the Long-Term Surveillance Plan for uranium mill tailings sites required by 10 CFR Part 40, Appendix A. See Appendix D of NUREG-1620, Rev. 1 for guidance). The LTC Plan would include the following information:

• Legal description and ownership of the land

• Final condition of the site, residual contamination, engineered barriers, and physical controls

• LTC license conditions and implementing activities and procedures (e.g., restrictions on land use, monitoring requirements, and reporting requirements)

M.4 LEGAL AGREEMENT AND RESTRICTIVE COVENANT (LA/RC)

M.4.1 PURPOSE OF LA/RC

The primary purpose of NRC's legal agreement and restrictive covenant (LA/RC) option is to provide the legally enforceable and durable institutional controls required by 10 CFR 20.1403(b) to ensure the long-term protection of the public health, safety, and the environment. The combination of a legal agreement and restrictive covenant is an institutional control option with the NRC having an enforcing role after license termination under restricted conditions. A licensee may propose use of a LA/RC only if the licensee cannot otherwise establish acceptable institutional controls or independent third party arrangements.

The current licensee or site owner and NRC enter into a legal agreement on the restrictions and controls needed for restricted use. The legal agreement includes using a restrictive covenant, which outlines the restrictions on site use and any necessary maintenance, monitoring, or reporting. Monitoring could include the owner agreeing to provide a response annually or at other frequency to an NRC inquiry on effectiveness of controls or land uses. The LA/RC is not for the purpose of storage of radioactive materials. It also should not be considered a guaranteed option, but would be a last resort under the criteria in 10 CFR 20.1403(b). With use of a

LA/RC, the licensee must still meet all the restricted use requirements of the LTR, to ensure protection of the public health and safety.

In the legal agreement, the licensee or site owner agrees to abide by the restrictive covenant, to record the restrictive covenant in the deed, and to not withdraw the restrictive covenant from the deed. In the legal agreement, NRC agrees to monitor and enforce the restrictions, under the authority written into the legal agreement and restrictive covenant. The legal agreement is only between the NRC and the present owner (owner at time of license termination or completion of decommissioning) and includes language that requires the present owner to record the restrictive covenant with the proper recordation body (e.g., Registrar of Deeds) and to not modify or rescind the restrictive covenant. This will help assure that the restrictive covenant will run with the property and will be binding upon all subsequent owners. NRC would need to approve any modifications to or recision of the restrictive covenant. The legal agreement and restrictive covenant should also contain a legal description of the property, and if there are both unrestricted and restricted portions, a clear delineation of these portions of the site.

Under the LA/RC option, after the licensee demonstrates compliance with all of the requirements for restricted use under the LTR, the license is terminated, and the legal agreement and restrictive covenant become the legal tools for maintaining needed restrictions on the site. The LA/RC option has not been implemented by the NRC or legally tested, and NRC's ability to enforce the LA/RC depends on the laws of the jurisdiction where the site is located. Therefore, the licensee must demonstrate that the LA/RC is an effective and legally enforceable institutional control in the jurisdiction where the site is located.

If complex monitoring or maintenance of a site is needed (where the site owner needs to have knowledge, expertise, or the means to carry out these activities), the LA/RC option should not be used, as NRC would not have the authority to approve the sale of the site, and, thus assess the capability of a new site owner to perform any complex monitoring or maintenance activities. In

this case, it is more appropriate to use the LTC license, as NRC would need to review and approve any transfer of the site to a new owner, as the new owner would need to become an NRC licensee. Section 17.7.3 discusses the criteria for determining whether the LA/RC or the LTC license option should be used.

M.4.2   ENFORCEABILITY OF THE LA/RC

It is the licensee's responsibility to demonstrate that the LA/RC option is a legally enforceable institutional control, given that the enforceability of this option depends on the property laws in the jurisdiction where the site is located. The licensee should obtain an independent legal opinion or analysis of the property laws in the jurisdiction where the site is located, demonstrating that the restrictive covenant would transfer to each subsequent owner of the property through the deed and "run with the land." The independent legal opinion should address the following issues:

- how future owners or operators of the property will have notice of the restrictive covenant;

- whether the restriction on land use can pass to the next owner when the property is sold;

- whether the restrictive covenant can be enforced by governmental entities or agencies (NRC, as well as local and State governments, who may be enforcing parties), which do not own any property in close proximity to the restricted property;

- whether the restrictive covenant/deed restriction is enforceable in perpetuity, or whether it would become unenforceable after the elapse of some period of time by operation of law;

- if the property is rezoned for a use which is restricted per the restrictive covenant, whether such rezoning would void or nullify the restrictive covenant; and

- if there are restricted and unrestricted portions of the property, whether the restrictive covenant would continue with the new deed/title of the restricted portion of the property, if, in the future, the property is divided.

Under the LA/RC option for institutional controls, NRC would enforce the restrictions under authority written into the legal agreement and restrictive covenant. If there was a breach of the legal agreement or restrictive covenant, NRC would address this by taking legal action in a court of the appropriate jurisdiction. NRC also could exercise its authority under the Atomic Energy Act, as amended, and take appropriate actions to assure that the use of the site or site conditions are protective of public health and safety.

The legal agreement and restrictive covenant would outline the methods and frequency in which NRC (as the enforcing party) would verify that the site owners were complying with the restrictive covenant. For example, NRC may correspond with the site owner annually, inquiring as to how the site is being used, or may conduct an inspection at the site at least every five years to verify that restrictions remain in place and that the site is being used appropriately. NRC also could review the property laws in the jurisdiction where the site is located at the time site ownership changes (or at least every five years), to assure that the laws still support the enforceability of the restrictive covenant. The LA/RC would outline how the NRC would enforce the restrictions, and if the legal agreement or restrictive covenant was breached, what steps NRC would take to restore the restrictive covenant (and the land use restrictions, monitoring, or reporting actions it contains). For example, if NRC determined that the site was being used inappropriately, it could take legal action in the courts to order the owner to restore the site to proper use.

M.4.3   ROLES AND RESPONSIBILITIES

The site owner has the primary responsibility for long-term protection of the public health, safety, and the environment by implementing and then maintaining the effectiveness of the controls required by the restrictive covenant. The site owner would maintain the required site access and land use controls and would conduct any necessary maintenance and monitoring. The site owner would also maintain any necessary records related to complying with the restrictive covenant and for any monitoring and maintenance needed.

NRC is responsible for assuring that the controls and restrictions on the site continue and that the use of the property is consistent with the restrictive covenant. NRC should periodically verify the effectiveness of controls, by sending inquiries to the site owner and conducting periodic inspections. NRC is responsible for taking actions to enforce the legal agreement and restrictive covenant, if the conditions of these legal instruments are not met. NRC also is responsible for maintaining the records associated with the restricted use of the site.

Stakeholders also have a role under the LTR during the licensee's preparation of the decommissioning plan for a restricted use site. For these sites, the licensee is required by 10 CFR 20.1403(d) to seek advice from affected parties regarding a number of matters, including the plans for enforceable institutional controls, sufficient financial assurance, and undue burdens on the local community or other affected parties. The licensee shall document in the decommissioning plan how the advice was sought and incorporated, as appropriate, following analysis of that advice. Section 17.7.5 of this volume and Section M.6 of this Appendix discusses obtaining public advice on institutional controls in more detail. Similarly, under 10 CFR 20.1405, NRC shall notify and solicit comments from the public upon receipt of the decommissioning plan.

In addition to local and State governments and the EPA, other stakeholders may have an ongoing interest in the site after decommissioning has been completed and the restrictive covenant is in place. From time to time, it might be appropriate to schedule public meetings or solicit input from the local community or affected parties about the site, to maintain a local awareness of the site and knowledge of the restrictions on site access and use.

M.4.4 TRANSFER OF CONTROL/OWNERSHIP

The LA/RC option should only be used at sites where there are no monitoring or maintenance activities that would require the site owner to have special expertise or knowledge to carry them out. Therefore, it would not be necessary for NRC to approve the transfer of site ownership, as the property is sold. Instead, NRC would continue to monitor the site, to ensure that the restrictive covenant was still in effect and that the site owner was abiding by the provisions of the restrictive covenant. A provision could be included in the LA/RC, which provides that the site owner must notify NRC when it is planning to sell the property. NRC, at that time, may wish to review the property laws in the jurisdiction where the site is located to make sure that the laws continue to support the enforceability of the restrictive covenant by NRC. As previously noted, NRC also may also periodically review (e.g., every five years) the property laws in the jurisdiction where the site is located, to assure that the laws still support the enforceability of the restrictive covenant.

M.4.5 SUFFICIENT FINANCIAL ASSURANCE AND TRUST

As required in 10 CFR 20.1403(c), the licensee or site owner must establish sufficient financial assurance for the long-term cost of any necessary monitoring, maintenance and control of the site, and the cost of NRC monitoring and enforcing the controls (as the independent third party). The licensee, as part of the license termination and the establishment of the legal agreement and restrictive covenant, should provide a single payment (in an independent fund) for future maintenance and monitoring costs.

At the time of establishing the LA/RC, the licensee must establish a trust and place sufficient funds into it to produce annual income that is sufficient to cover the (1) annual average costs of any surveillance, control, monitoring or routine maintenance, (2) NRC oversight costs, and (3) trustee fees and expenses. The licensee should assume a 1% return on investment (consistent with 10 CFR Part 40, Appendix A). This trust fund is independent of the licensee or site owner and, therefore, not affected in the event of site owner bankruptcy and would continue independent of site ownership. The NRC would be the beneficiary of the trust. The site owner would request, and the independent trustee would pay, in accordance with the instrument, for the costs of surveillance, control, maintenance, and NRC oversight costs, most likely on an annual basis. Because the fund would produce income sufficient to hire a contractor to perform the surveillance and control tasks, the site owner could hire a contractor to perform the duties and be reimbursed for the full cost, rather than performing the work itself.

M.4.6 FINALITY OF DECOMMISSIONING DECISIONS

NRC recognizes the importance of the finality of its decommissioning decisions. Under 10 CFR 20.1401(c), the Commission could require additional cleanup in the future, based on new information, if it determines that the criteria in the LTR were not met and residual activity remaining at the site could result in significant threat to public health and safety. This requirement also would apply to a site with a restrictive covenant and may be particularly

important to potential future owners/licensees who may be concerned about future liabilities should they purchase the site.

It is also important to note that if a future site owner wished to decommission the property to unrestricted use, it would be able to propose this to NRC. The LA/RC option for restricted use does not preclude a future decommissioning to meet the radiological criteria for unrestricted use. In that case, the site owner would submit a decommissioning plan for NRC review and decommission the site in accordance with NRC's decommissioning regulations, and the NRC would assure that the site is properly decommissioned and suitable for unrestricted use, before taking actions to terminate or rescind the restrictive covenant (remove it from the deed).

M.4.7 LONG-TERM RECORD RETENTION AND AVAILABILITY

There is concern that over time, knowledge of the site conditions and appropriate uses of the site may be lost. NRC recognizes that maintaining records and making them publicly available over the long term will help to preserve public knowledge of the site and the restrictions on its use. Preserving this knowledge and ensuring that pertinent information related to the site is easy to find and is readily accessible, is one of the important elements to ensure protection of public health and safety for long periods of time. One approach to managing this information is to duplicate the responsibilities of different agencies and groups for retaining records, as well as maintain the records in different locations, to better assure that the records will be preserved and made available to those who use the site in the future.

Under the LA/RC option, NRC would have the primary responsibility for maintaining records and making those available to the public. The NRC would continue to maintain any pertinent records related to the decommissioning and license termination in the same docket file that was used for the operating site. This approach should result in a continuous and completely documented history of the site operations, decommissioning, and license termination in a single file that will improve the efficiency and effectiveness of future search and retrieval of site information. These records are expected to be available to the public in the future. Also, NRC currently maintains (and expects to maintain in the future), a site decommissioning database, which includes restricted use sites. This publicly available database provides web-based access to general site information about all NRC decommissioned sites.

The site owner also would maintain any necessary records related to complying with the restrictive covenant and any monitoring and maintenance needed. Such records include the decommissioning plan, final status survey report, legal agreement (site owner at time of license termination), restrictive covenant, and correspondence between NRC and the site owner.

In accordance with the legal agreement, the licensee or site owner would be required to record the restrictive covenant with the appropriate recordation body responsible for maintaining records related to land ownership (e.g., Registrar of Deeds) in the jurisdiction where the site is located. In addition to the site owner, NRC, and the recordation body, other local and State groups or agencies [e.g., the State Department of Environmental Protection, local government agencies, civic groups, libraries (either physical locations or web-based information collection and storage systems)] could also preserve records. From time to time, it might be appropriate to hold public meetings or solicit input from the local community or affected parties about the site, to maintain a local awareness of the site and knowledge of the restrictions on site access and use. These recordkeeping responsibilities and knowledge management activities should be outlined in the LA/RC.

M.5 TOTAL SYSTEM APPROACH TO SUSTAIN SITE PROTECTION AT RESTRICTED USE SITES

Long-term effectiveness of institutional controls that would be required to restrict future site use is an important part of the overall LTR issue for institutional controls. Currently, all of NRC's decommissioning sites that are considering restricted use have radionuclides with long half lives such as uranium or thorium, and, therefore, would need long-term controls. Therefore, the purpose of this section is to explain NRC's approach for long-term protection. This section integrates many of the approaches described in the previous sections and presents them as a system to sustain protection.

In the Statement of Considerations for the LTR, the Commission recognized that requiring absolute proof that institutional controls would endure over long periods of time would be difficult, and the Commission did not intend to require this of licensees. Rather, the Statement of Considerations explained that institutional controls should be established with the objective of lasting 1000 years to be consistent with the time-frame used for calculations, and these controls would be expected to remain effective into the foreseeable future. However, the LTR also included added assurances that the public would be protected. Therefore, protection of public health and safety is provided by a total system of controls and assurances that is durable and provides defense-in-depth. The total system described below is based on the requirements of the LTR, descriptions in the Statement of Considerations for the LTR, new policy options for institutional controls described in the LTR Analysis (SECY-03-0069) and the Regulatory Issue Summary (RIS) for the LTR Analysis in RIS 2004-08, and decommissioning guidance in NUREG-1757.

NRC's total system for protection consists of six elements: (1) legally enforceable institutional controls; (2) engineered barriers; (3) monitoring and maintenance; (4) independent third party oversight; (5) sufficient funding; and (6) upper limits on dose (i.e., "dose caps") if institutional controls fail. In addition, potential involvement by State and local governments and the community can add to the process. Each of these elements is described below, including how it contributes to protection, how it sustains protection for the duration needed, and what entity is responsible.

M.5.1 LEGALLY ENFORCEABLE INSTITUTIONAL CONTROLS

Legally enforceable institutional controls are required by the LTR. Institutional controls are administrative/legal mechanisms such as deed restrictions, permits, zoning, government ownership, or even an NRC LTC license. Institutional controls also can include physical controls such as fences, signs, markers, or vegetation.

Institutional controls are intended to protect the public health and safety by preventing adverse site access and land uses so that the LTR dose criterion of 0.25 mSv/y (25 mrem/y) is not exceeded. Limiting exposure time or preventing groundwater or agricultural uses can prevent adverse exposure pathways to people.

NRC's risk-informed graded approach is used to select the appropriate grade or type of institutional control, based on duration and magnitude of the hazard, so that restrictions are appropriately targeted using risk insights. Dose assessments are used to tailor site-specific restrictions to avoid adverse land uses.

Durable institutional controls, such as government ownership or control, could be used for higher risk sites with longer duration or higher magnitude hazards, to provide additional assurance of sustaining protection over the time period needed. Under new NRC policy, two options are available to provide durable institutional controls using either a NRC LTC license or a legal agreement and restrictive covenant, where NRC would have a monitoring and enforcing role.

Maintaining institutional controls is the responsibility of the owner or contractor to the owner, referred to as the custodian. The custodian also is responsible for conducting five-year reviews for higher risk sites to ensure the institutional controls are in place and continue to function. These reviews would include onsite inspections to verify that prohibited adverse activities are not being conducted. The custodian would also maintain records and make them available to the public.

Institutional controls also are required by the LTR to be legally enforceable by an entity other than the custodian (e.g., local government, courts) that has the authority to enforce the particular type of institutional control. This entity would need to be identified and potential corrective actions described in the event the controls fail. Sustaining protection is also addressed by having legal opinions of the State or locality submitted to NRC to demonstrate that the institutional controls can be enforced and will be binding on future owners.

M.5.2   ENGINEERED BARRIERS

Engineered barriers are man-made structures and can be a variety of types such as disposal cells, erosion protection covers, or cover layers to prevent or divert infiltration. These barriers are typically used to control adverse natural processes, such as erosion (that might expose contamination) or infiltration of water (that could cause release and migration of contaminants). Engineered barriers also can be designed to inhibit adverse human intrusion such as excavation and removal of cover material or contaminants.

The LTR does not require use of engineered barriers or specific designs that should be used, but the Statement of Considerations for the LTR recognizes that engineered barriers might be needed for sites with long-lived radionuclides. The LTR's performance-based approach allows flexibility for a licensee to determine if engineered barriers are needed to meet the LTR dose criteria and what contribution to performance might be needed considering how the barriers might degrade over time.

Although engineered barriers are not institutional controls, they can be used to supplement institutional controls and contribute to protection. In some cases, protection can be sustained for long time periods by using robust designs that do not rely on ongoing active maintenance. For example, erosion protection covers designed for up to 1000 years that have been used for uranium mill tailings sites may also have use at some decommissioning sites.

M.5.3   MONITORING AND MAINTENANCE

The site would be maintained by the custodian in accordance with the institutional controls. Monitoring and maintenance consists of identifying potential problems with institutional controls or engineered barriers and taking appropriate corrective actions to maintain the performance of the institutional controls or engineered barriers. Typically, monitoring could include a variety of activities such as visual surveillance or using instruments for radiological monitoring of surface or groundwater. Monitoring also could be used to detect indicators of potential future problems or measuring natural processes that could eventually impact the performance of the total system, unless corrected. Maintenance would include corrective actions to prevent disruptive processes that could result in non-compliance such as intrusion of covers by plants or burrowing animals, or repair of fences and signs. A risk-informed graded approach is taken for both monitoring and maintenance to focus on disruptive processes and engineered barrier performance important to compliance. However, in such cases, the licensee should justify claims for long-term performance and consider in his justification the interaction between multiple barriers if more than one system is employed.

M.5.4   INDEPENDENT THIRD PARTY OVERSIGHT

The LTR requires an independent third party to provide oversight to assure that the custodians' controls are performed and corrective actions are taken, as needed, to sustain the controls and maintenance. The independent third party also would act as a backup to the custodian to assume and carry out the responsibilities for control and maintenance, if needed. The independent third party could be a government entity, or even NRC (under its new policy for the LTC license or legal agreement/restrictive covenant) if other government entities do not accept this responsibility.

M.5.5   SUFFICIENT FUNDING

The LTR requires that sufficient financial assurance be established to enable an independent third party, including a governmental custodian of a site, to assume and carry out responsibilities for any necessary control and maintenance of the site. A trust fund, or other financial assurance mechanism, would be established independent from the custodian and managed by a trustee. Sufficient funds would need to be placed into the trust fund to produce an annual income that is sufficient to cover (1) the annual average costs of controls, maintenance, and monitoring, if needed; (2) independent third party oversight costs; and (3) trustee fees and expenses. Thus, the fund balance would be sustained over time and not depleted because the annual costs of controls and maintenance are provided by the annual interest income.

M.5.6   MAXIMUM LIMITS ON DOSE IF INSTITUTIONAL CONTROLS FAIL

Because it is not possible to preclude the failure of controls, the LTR also requires that remediation be conducted so that there would be a maximum value, or "cap" on the dose if the institutional controls are no longer in effect. Compliance with the dose cap would prevent exposures in excess of the public dose limit of 1.0 mSv/y (100 mrem/y) or 5.0 mSv/y (500 mrem/y) under certain rare circumstances. These dose caps act as a safety net if institutional controls fail and, therefore, sustain protection by providing defense-in-depth.

M.6   OBTAINING PUBLIC ADVICE ON INSTITUTIONAL CONTROLS

Subpart E of 10 CFR Part 20 requires that public input on the institutional controls proposed by the licensee be sought during the decommissioning process. Licensees, as part of their planning for restricted use, are required by 10 CFR 20.1403(d) to seek advice from individuals and institutions in the community that may be affected by the decommissioning. The rationale for this requirement is that the licensee's direct involvement regarding diverse community concerns and interests can be useful in developing effective institutional controls, and this information should be considered and incorporated, as appropriate, into the DP or License Termination Plan (LTP) before it is submitted to NRC for review. This section provides guidance on complying with 10 CFR 20.1403(d).

To comply with 10 CFR 20.1403(d) and to ensure that the fundamental performance objectives of institutional controls are met, licensees who plan to release a site under restricted conditions must satisfy the following:

- Seek advice on whether the provisions for institutional controls will:

  — provide reasonable assurance that annual doses will not exceed 0.25 mSv/y (25 mrem/y);

  — be enforceable; and

  — not impose undue burden on the local community or other affected parties.

- Seek advice on whether the licensee has provided sufficient financial assurance for any necessary control and maintenance of the site by an independent third party.

- Seek advice from representatives of a broad cross-section of individuals and institutions in the community that may be affected by the decommissioning (affected parties).

- Provide an opportunity for a comprehensive, collective discussion on the issues.

- Provide a publicly available summary of the results of all such discussions, including a description of the individual viewpoints of the participants on the issues and the extent of agreement and disagreement among the participants on the issues.

- Describe, in the DP or LTP, how advice from the affected parties has been sought and incorporated, as appropriate, following analysis of that advice. The licensee is not required to reach consensus with the affected parties on the various aspects of the proposed institutional controls.

Once the DP or LTP is submitted to NRC, NRC reviews the licensee's plans for license termination, including the institutional controls proposed to restrict site use. NRC also evaluates the public comments gathered by the licensee and the licensee's consideration of comments from affected parties. As part of NRC's review process for all submitted DPs and LTPs, NRC also must notify and solicit comments from the public regarding the proposed licensee action, under 10 CFR 20.1405. Significant and appropriate public involvement in NRC's review process will take place at this time. It is NRC's, not the licensee's, responsibility to carry out these actions required under 10 CFR 20.1405.

Identifying Affected Parties

The licensee should first identify the individuals and institutions in the community who may be affected by the decommissioning (affected parties). According to 10 CFR 20.1403(d)(2), the licensee must provide for participation by representatives of a broad cross-section of community interests who may be affected by the decommissioning. The affected parties may vary for each specific site and may include the following:

- any State, local, or Federal government agency, other than NRC, that has jurisdiction, responsibilities, knowledge, or expertise (e.g., zoning, community land use planning, public health, environmental protection, or radiological, decommissioning or regulatory issues) with respect to the site to be decommissioned;

- local community, civic, labor, or environmental organizations with an interest in the decommissioning, and whose members would be affected by the decommissioning;

- adjacent landowners whose properties abut the site or portions of the site to be released under restricted conditions; and/or

- any Indian tribe or other indigenous people who have relevant treaty or statutory rights that may be affected by the decommissioning of the site.

Methods of Seeking Advice

The licensee should establish a method for seeking advice, from the affected parties, on the adequacy of the institutional controls and the sufficiency of financial assurance. The type of process a licensee uses should be tailored to its site (based on site-specific considerations and the stakeholders at a site) and can include a variety of approaches. The licensee is encouraged to meet with NRC staff to discuss its intended methods for seeking advice from affected parties, prior to beginning this activity, in order to ensure that the proposed method will be acceptable to NRC staff.

Appropriate mechanisms for seeking advice from affected parties could include a public meeting or series of meetings, meetings with individual groups/organizations to discuss the licensee's decommissioning plans and obtain stakeholder input, a specific process for obtaining written or electronic public comments by e-mail or website means, or formation of a site-specific advisory board (SSAB) (i.e., a group representing a broad cross-section of the community that may be affected by the decommissioning).

Convening a Site-Specific Advisory Board

In general, NRC considers that convening a SSAB should be the starting point in providing for public involvement because a SSAB is the most effective way to ensure that the licensee considers the diversity of views in the community. Small group discussions can be a more effective mechanism than written comments or large public meetings for articulating the exact nature of community concerns, determining how much agreement or disagreement there is on a particular issue, and facilitating the development of acceptable solutions to issues. Also, the type of close interaction resulting from a small group discussion could help the licensee develop a credible relationship with the community in which it is operating.

It is important to note that the SSAB does not have to be a new group formed specifically for the decommissioning. Any group that can perform the functions of a SSAB may be considered to be a SSAB. Thus, if an existing or established group in the community has enough participation by the affected parties and can effectively perform the functions of the SSAB, that group may be used by the licensee as the SSAB.

Licensees should use the following guidance in establishing and convening a SSAB:

- The licensee should solicit members to serve on the SSAB. Membership should reflect the full range of the affected parties' interests by selecting representatives from the affected parties to present the views of the organization or interest that they represent. Government agencies and other organizations should be able to nominate their own representatives to the SSAB. Invited participants should be informed of the objectives of the SSAB. The SSAB normally consists of about 8 to 10 members.

- Members of the SSAB should agree to meet their responsibilities as a condition of membership. In general, NRC regulations require that the DP be submitted within 12 months after notifying NRC that the site will be decommissioned. The licensee is responsible for meeting this requirement. Therefore, the licensee is responsible for ensuring that the SSAB is meeting a schedule that will allow the licensee to submit the plan within the required time. If the board does not meet its responsibilities, the licensee should evaluate and discuss with the SSAB any problem and how to resolve it.

- The SSAB members should be selected as soon as practical after the licensee notifies NRC of its intention to decommission and terminate the license.

- The licensee should provide reasonable administrative support for SSAB activities and access to licensee studies and analyses that are pertinent to the proposed decommissioning.

- To avoid the appearance of a conflict of interest, members of the SSAB usually are not paid by the licensee. However, reimbursement for expenses incurred is acceptable.

- The licensee should establish a schedule for the work of the SSAB that allows the licensee to obtain advice from the SSAB, incorporate the advice into the DP or LTP as appropriate, and submit the DP or LTP within the time required by NRC regulations. The schedule should include submittal of the SSAB's advice, allowing sufficient time for the licensee to analyze the advice and describe in the DP or LTP how the advice was incorporated, as appropriate.

- The licensee should propose a charter and operating procedures for the SSAB's consideration. The charter and operating procedures should address the advice to be sought and the characteristics of a SSAB.

- The SSAB should:

  — select a chairperson;

  — adopt a charter and operating procedures;

  — work with the licensee to identify and obtain information needed in its evaluation process;

  — hold meetings open to the public, provide for a comprehensive, collective discussion of the issues, and allow the opportunity for public comment at the meetings;

  — respond to concerns and questions raised by the public, making the results publicly available;

  — provide advice to the licensee on the topics listed above and on any other topics the licensee wants discussed;

  — to the extent feasible, abide by the schedule established by the licensee to meet NRC requirements; and

  — ensure that a publicly available summary of the results of all discussions, including descriptions of the individual viewpoints of the participants on the issues and the extent of agreement and disagreement among the participants on the issues, is developed to support the meeting.

- SSAB meetings should be open to the public with adequate public notice (at least two weeks in advance) of the location, time, date, and agenda for the meetings. Consideration should be given to using print, electronic, and website notification methods. The licensee should inform NRC of SSAB meetings and distribution of information made at SSAB meetings, as these meetings and distributions may cause the public to contact NRC.

- A summary of the results of all collective discussions, including a description of the individual viewpoints of the participants on the issues and the extent of agreement and disagreement among the participants on the issues, should be made publicly available.

Other Appropriate Methods for Seeking Advice

The use of a SSAB may not be appropriate in all situations, such as, if a broad cross-section of the community clearly has insufficient interest or wishes to defer its involvement to a State or local governing body. If a licensee determines that a SSAB is not appropriate or feasible, the licensee should determine the best methods to allow for a comprehensive collective discussion of the issues associated with restricted use decommissioning and an opportunity for the affected parties to provide advice on the institutional controls. The licensee should develop a public involvement process, which has the following characteristics:

- The affected parties should be informed of the decommissioning and informed that their advice is being sought. The methods and efforts that can be used initially to inform the public can include, as appropriate for the specific site:

  — information in mass media, such as articles, advertisements, and public service announcements in newspapers, television, and radio;

  — websites or other related technologies;

  — flyers or brochures distributed in the neighborhood or mailings to individual residents close to the site;

  — letters or telephone contacts with government agencies and local community, civic, and labor organizations; or

  — presentations at public meetings.

- The licensee should clearly state to the affected parties the matters on which advice is being sought with sufficient clarity to obtain meaningful input.

- The initial information provided to interested affected parties should describe the decommissioning process, characterize in basic terms the nature and extent of residual radioactivity at the site, and provide pertinent information about the licensee's request for license termination under restricted conditions.

  The licensee should present information on the provisions for restricting uses of the site to meet the dose criteria of the LTR, the nature of the institutional controls expected to restrict use over extended time periods, how the restrictions would be enforced, the effect on the community, and the adequacy of the level of financial assurance.

- Information should be provided early enough to allow sufficient time for review by the affected parties. The initial information and any subsequent long, complex studies should be provided at least 30 days before any public meeting(s). Although there should be as much time provided as practical, it is acceptable for short simple supplemental information to be provided with very little time for review.

- The licensee should establish a method for receiving advice from the affected parties. There should always be a method to receive written comments. The licensee should also hold public meetings to obtain oral comments, and could obtain comments electronically, such as by e-mail or through a website. Comments received should be available for public inspection.

- The licensee should inform NRC of any public meetings and the information distributed at the meetings, as these meetings and distributions may cause the public to contact NRC.

- A summary of the results of all collective discussions, including a description of the individual viewpoints of the participants on the issues and the extent of agreement and disagreement among the participants on the issues, is to be made publicly available.

Following solicitation of advice from affected parties, the licensee will document in its DP or LTP how the advice of affected parties has been sought and incorporated, as appropriate, following analysis of that advice, but the licensee is not required to reach a consensus with the affected parties on the various aspects of the proposed institutional controls.

Suggestions for Effective Public Involvement

In relation to the information the licensee provides to the affected parties, it is important for the licensee to provide necessary background information to promote understanding of institutional controls and sufficient time to allow discussion on the institutional control issues. The licensee also could clearly identify what the permitted uses of the site are, as well as the adverse uses that must be restricted to meet the dose criteria of the LTR, as this information could assist the affected parties in providing advice on whether the proposed institutional controls impose an undue burden on them or the local community. The definition of an "undue burden" is specific to the site and its stakeholders or affected parties.

In public meetings with the affected parties, the licensee could use a facilitator to promote discussion and dialogue on the institutional control issues. The licensee also could host a roundtable discussion with a smaller group of the representatives of the broad spectrum of stakeholder interests to ensure that there is a dialogue among those key interests on the specific institutional control issues.

As the licensee develops specific plans and analyses, the discussions with the affected parties could become more detailed. These discussions could focus on topics such as the cost of maintenance and monitoring that could be needed in the future, preliminary results of dose assessments, and other special provisions, such as periodic rechecks of the restricted area and the continued effectiveness of institutional controls.

The licensee should refer to "Best Practices for Effective Public Involvement in Restricted-Use Decommissioning of NRC-Licensed Facilities" (June 2002), which was prepared for the NRC by the U.S. Institute for Environmental Conflict Resolution. This report offers guidance and advice on the best practices for achieving effective public involvement in NRC's decommissioning program, specifically for restricted use decommissioning.

Appendix N
MOU Between NRC and USACE

MEMORANDUM OF UNDERSTANDING

BETWEEN

THE U.S. NUCLEAR REGULATORY COMMISSION

AND

THE U.S. ARMY CORPS OF ENGINEERS

FOR COORDINATION ON CLEANUP & DECOMMISSIONING OF THE FORMERLY UTILIZED SITES REMEDIAL ACTION PROGRAM (FUSRAP) SITES WITH NRC-LICENSED FACILITIES

ARTICLE I - PURPOSE AND AUTHORITY

A. This Memorandum of Understanding (MOU) is entered into by and between the U.S. Nuclear Regulatory Commission (NRC) and the U.S. Army Corps of Engineers (USACE), ("The Parties") for the purpose of minimizing dual regulation and duplication of regulatory requirements at FUSRAP sites with NRC-licensed facilities. For activities where a potential for dual regulation could exist, the two agencies agree to cooperate, share information, and/or coordinate activities in their respective programs. This MOU applies to USACE response actions meeting the decommissioning requirements of 10 C.F.R. 20.1402, "Radiological Criteria for Unrestricted Use." USACE Response actions meeting the restricted release requirements of 10 C.F.R. 20.1403, are outside the scope of this MOU.

B. The NRC has the statutory responsibility for the protection of the public health and safety related to the possession and use of source, byproduct, and special nuclear material under the Atomic Energy Act of 1954, as amended (Public Law 83-703, 68 Stat. 919). This includes ensuring the decommissioning of the nuclear facilities that it licenses. The Commission's licenses and regulations set out conditions to provide for the protection of the public health and safety and the environment. To terminate such licenses, NRC must ensure that licensees meet the Commission's decommissioning requirements including the provisions of 10 CFR 20 Subpart E – Radiation Criteria for License Termination.

C. USACE is administering and executing cleanup at FUSRAP sites pursuant to a March 1999, MOU with the Department of Energy and the provisions of the Energy and Water Development Appropriations Acts for Fiscal Years 1998-2001 (Public Laws 105-62, 105-245, 106-60 and 106-377, respectively). Section 611 of Pub. L. 106-60 requires the USACE to remediate FUSRAP sites, in accordance with, and subject to the Comprehensive Environmental Response, Compensation, and Liability Act of 1980, as amended (CERCLA), 42 U.S.C. 9601 et seq., and the National Oil and Hazardous Substances Pollution Contingency Plan (NCP), 40 C.F.R., Chapter 1, Part 300. Section 611 also confers lead agency status on the USACE for remedy selection. USACE, as provided for in section 121(e) of CERCLA and 40 C.F.R. 300.400(e), is not required to obtain a NRC license for its on-site remediation activities conducted under its CERCLA authority. However, if a response action is required, CERCLA requires the remedy to be protective of human health and the environment.

D. This MOU describes how the two agencies will work together to meet their existing statutory responsibilities. It neither creates nor removes any agency responsibility or authority. This MOU is not an admission of responsibility or liability on the part of the United States with regard to any hazardous substances or operations at a licensed site; does not relieve a license holder of its responsibilities and liabilities under any law; and does not create rights in any third party against USACE, NRC, or the United States.

E. CERCLA obligations imposed on the USACE may duplicate the obligations established by NRC regulations and licenses, resulting in duplicate regulatory requirements at NRC-licensed FUSRAP sites that will impose an added regulatory burden without an added safety benefit. To avoid unnecessary duplication of regulatory requirements and effort, this MOU sets out the conditions, consistent with the protection of the public health and safety, that will permit NRC to exercise its discretion to suspend NRC issued licenses at FUSRAP sites so that NRC requirements do not hinder USACE in its remediation of sites under CERCLA.

F. Each agency will bear its own costs for actions consistent with this MOU, but this does not preclude each agency from recovering costs, based on it's statutory authority, from the licensee or responsible parties.

G. USE OF TERMS.

1. The term "response action" means response actions as defined in CERCLA at 42 U.S.C. 9601(25) including removal and remedial actions and related CERCLA enforcement actions.

2. The term "closeout" means that all construction activities and reports are complete, the cleanup goals specified in the final ROD are achieved, coordination with regulatory agencies, and publication of notice in accordance with the provisions of CERCLA, the National Contingency Plan (NCP) and USACE procedures have been completed.

3. The term "completed response action" means that all construction activities are complete; for components other than ground or surface water, the cleanup goals specified in the ROD are achieved; any ground and/or surface water restoration remedies are operating as designed; and a remedial or removal action report is complete.

4. The term "FUSRAP site" means any geographic area certified by the Department of Energy (DOE) to have been used for activities in support of the Nation's early atomic energy program, and determined by USACE to require a response action pursuant to CERCLA or placed into the FUSRAP program pursuant to Congressional direction. A FUSRAP site may overlap all, or any part, of an NRC-licensed site.

5. The term "possession" means physical control of the property or materials for purposes of environmental restoration and protection of the health and safety of the public. Possession does not require ownership nor is USACE assuming responsibility for the operations and activities of the NRC licensee or owner of the materials. The USACE will take control only of the FUSRAP-related materials on the licensed site as provided in paragraph III. B. Non-FUSRAP materials, unless the responsibility of the USACE under CERCLA, remain under control of the licensee.

6. The term "licensed site" means that a NRC license has been issued, and remains active or suspended, to possess and use material licensed under the Atomic Energy Act at the site.

ARTICLE II - INTERAGENCY COMMUNICATION

To provide for consistent and effective communication between NRC and USACE, each agency shall appoint a Principal Representative to serve as its headquarters-level point of contact on matters relating to this MOU. Written notices required by the MOU shall be sent to the USACE's and NRC's Principal representatives. The Principal Representatives are:

Chief, Decommissioning Branch
Division of Waste Management
U.S. Nuclear Regulatory Commission
Washington, D.C. 20555

Chief, Environmental Division
Directorate of Military Programs
U.S. Army Corps of Engineers
441 G Street, N.W.
Washington, D.C. 20314-1000

ARTICLE III – AGREEMENT

A. At the request of USACE, NRC will initiate action for the suspension of the NRC license or portions of the license for a FUSRAP site to be remediated by USACE under CERCLA authority contingent upon USACE notifying the NRC in writing that:

1) USACE is prepared to take physical possession of all or part of the licensed site for purposes of control of radiation from FUSRAP materials subject to NRC jurisdiction and be responsible for the protection of the public health and safety from those materials consistent with 10 CFR Part 20 "Standards For Protection Against Radiation" and other requirements consistent with CERCLA;

2) USACE will conduct a response action at the licensed site under its FUSRAP and CERCLA authority, with regard to FUSRAP materials subject to NRC jurisdiction, to meet at least the standards required under 10 C.F.R. 20.1402, and

3) USACE has no objection to, and will facilitate, NRC observing USACE in-process remediation activities.

Such written notification to the NRC should be provided after the final Record of Decision (ROD), or its equivalent, is issued, if one is prepared, and at least 90 calendar days prior to USACE's expected date of initiation of a site response action so that the NRC can initiate the process for suspension of the license. Prior to submitting the notification, USACE will make a reasonable attempt to obtain the licensee's consent to USACE's proposed action and document the results of this effort in the notification.

B. Depending on the extent of FUSRAP materials and their separability from other hazardous substances on the site, USACE's responsibility may encompass the entire site, portions of the site, all the radioactive materials or just the FUSRAP and commingled materials, as specified in the final ROD. USACE will notify NRC of its findings regarding the type and extent of hazardous substance on a licensed site prior to requesting license suspension. Prior to USACE submitting a request for license suspension on a site where the NRC license suspension will not encompass the entire site, USACE and NRC will meet to agree on the scope of the suspension. The licensee may be involved in these discussions.

C. NRC licensing action for the suspension of the license, or portions of the license, will be effective, subject to:

1) written notification from USACE to the NRC that USACE has taken physical possession of the licensed site for purposes of radiation control and is now responsible for the protection of the public health and safety consistent with the requirements of 10 CFR Part 20 and

2) the effectiveness rules of the NRC hearing process pursuant to 10 CFR Part 2, "Rules Of Practice For Domestic Licensing Proceedings And Issuance Of Orders."

Prior to license suspension, the licensee retains responsibility for meeting the Commission's requirements for protecting the environment and the health and safety of the public.

D. NRC may observe, as it deems warranted, remediation activities being conducted by USACE. For the purpose of scheduling in-process activity observation, USACE shall provide the NRC with the schedule of major activities, regular progress reports on sites' activities, studies, and/or remediation, and planned work stoppages.

E. The NRC shall keep USACE apprised in writing of questions, comments or concerns arising from any NRC observations of USACE response action activities and shall immediately notify the USACE of any conditions having a potential to adversely affect the environment or the health and safety of the public.

F. USACE shall be responsible for the protection of the health and safety of the public consistent with the requirements of CERCLA and 10 CFR Part 20 during the time it is in physical possession of the licensed site or portions thereof which are suspended in accordance with the agreement at the time of license suspension.

G. USACE shall remediate the licensed site to meet at least the requirements of CERCLA and of 10 CFR 20.1402. The Applicable or Relevant and Appropriate Requirement (ARAR) in the final executed ROD will include 10 CFR 20.1402 or a more stringent requirement.

H. USACE shall manage all activities and prepare program estimates, funding requirements, and budget justifications for all FUSRAP activities for which it has been given responsibility as provided by the annual Energy and Water Development Appropriations Act, and the terms of this MOU. USACE shall request FUSRAP appropriations in the annual Energy and Water Development Appropriations Act for these activities. USACE shall respond to inquiries from public officials, Congressional interests, stakeholders, and members of the press regarding USACE activities under FUSRAP.

I. USACE shall consult with NRC if USACE surveys, investigations, and data analyses are inconsistent with the NRC description of the potential radioactive and/or chemical contaminants and processes involved in the historical activities at a licensed site at which the USACE is conducting a FUSRAP investigation or response action under CERCLA. USACE shall immediately notify NRC if, as a result of its Preliminary Assessments, Remedial Investigations, or other surveys prior to production of a ROD, conditions warrant a time-critical removal action, and the agencies will identify an appropriate response that protects the environment and the health and safety of the public.

J. USACE shall notify NRC in writing if there is a need for a radiological response action under FUSRAP on any property not covered by the license suspended or to be suspended (whether or not owned by the licensee) as a result of radioactive contamination from a licensed site undergoing a FUSRAP investigation or response action.

K. Following completion of the response action at a FUSRAP site with an NRC-licensed facility, USACE shall provide the NRC with a copy of the CERCLA Administrative Record for the NRC historical public record. At the time of close out USACE will provide NRC with copies of any additional information that has been placed in the CERCLA Administrative Record.

L. USACE shall notify the NRC in writing if there are NRC-licensed facilities on FUSRAP sites that may require coordination with the NRC in addition to the four known sites:

Maywood Site (Stepan), Maywood, NJ; CE-Windsor Site, Windsor, CT; St. Louis Downtown Site (Mallinkrodt), St. Louis, MO; and the Shallow Land Disposal Area, Parks Township, PA.

M. USACE shall keep NRC apprised in writing of progress toward completion of Preliminary Assessments and/or Site Investigations at licensed sites to determine:

1) Whether FUSRAP and commingled materials at the site are a threat or potential threat to public health and safety or the environment as a result of the licensed materials there; and

2) Whether the release requires a response under CERCLA.

N. The NRC will reinstate the license or portions of the license put into suspension due to USACE's remediation if USACE:

1) is no longer controlling the FUSRAP-related portion of the licensed site for radiation protection purposes,

2) is no longer proceeding with a response action at the licensed site under CERCLA, or

3) has otherwise completed its response action.

At least 90 calendar days prior to USACE terminating its physical possession of the licensed site for purpose of control of radiation, USACE will notify the NRC in writing so that the NRC can initiate the process for reinstating the license. USACE shall promptly notify NRC in writing if annual funding for the FUSRAP response action at an NRC-licensed site does not appear to be sufficient to complete the response action.

O. NRC shall be responsible for appropriate regulatory action, including requiring any further decommissioning if necessary, following license reinstatement.

P. As may be necessary, NRC and USACE will develop working procedures to implement this MOU. Such procedures will be approved by the Principal Representatives.

ARTICLE IV – FURTHER ASSISTANCE

NRC and USACE shall provide such information as may be reasonably necessary or required, which are not inconsistent with applicable laws and regulations, and the provisions of this MOU, in order to give full effect to this MOU and to carry out its intent.

ARTICLE V- DISPUTE RESOLUTION

Every effort will be made to resolve issues between NRC and USACE by the staff directly involved in the activities at issue, through consultation and communication. If a mutually acceptable resolution cannot be reached, the dispute will be elevated to successively higher levels of management up to the signers of this MOU. If resolution cannot be reached, NRC may in its discretion reinstate the licenses involved after providing a written 30 calendar day advance notice to the USACE. Upon license reinstatement, USACE's obligations under this MOU for the particular site shall cease and the licensee becomes responsible for control of radioactive materials on the licensed site, as well as protecting the environment and the health and safety of the public, subject to NRC regulation and other applicable law. Upon determining that the licensee has established control of the site and hazardous substances, USACE will relinquish possession of the site and hazardous substances, will cease remediation activities, and will vacate the site. License reinstatement constitutes notice of the shift in responsibility for control of the site and its hazardous substances.

ARTICLE VI- AMENDMENT AND TERMINATION

This MOU may be modified or amended in writing by the mutual agreement of the parties. Either party may terminate the MOU by providing written notice to the other party. The termination shall be effective 60 calendar days following notice, unless the parties agree to a later date. Termination of this MOU does not relieve USACE of its statutory responsibility for protecting the environment or the health and safety of the public until NRC has reinstated the license and the licensee has taken control of the site and its hazardous substances.

ARTICLE VII - EFFECTIVE DATE

This MOU shall become effective when signed by authorized officials of NRC and USACE.

U.S. Nuclear Regulatory Commission   **U.S. Army Corps of Engineers**

Martin J. Virgilio   **M.G. Hans A. Van Winkle**
Director,   Major General, U.S. Army
Office of Nuclear Materials Safety   Director, Civil Works
and Safeguards   U.S. Army Corps of Engineers
U.S. Nuclear Regulatory Commission

_____/RA/_____   _____/RA/_____
Signature   Signature

Date: July 2, 2001   Date: July 5, 2001

NRC FORM 335 (9-2004) NRCMD 3.7	U.S. NUCLEAR REGULATORY COMMISSION	1. REPORT NUMBER (Assigned by NRC, Add Vol., Supp., Rev., and Addendum Numbers, if any.)
BIBLIOGRAPHIC DATA SHEET *(See instructions on the reverse)*		NUREG-1757, Vol. 1, Rev. 2

2. TITLE AND SUBTITLE	3. DATE REPORT PUBLISHED	
Consolidated Decommissioning Guidance: Decommissioning Process for Materials Licensees	MONTH	YEAR
	September	2006
Final Report	4. F N OR GRANT NUMBER	

5. AUTHOR(S)	6. TYPE OF REPORT
K.L. Banovac, J.T. Buckley, R.L. Johnson, G.M. McCann, J.D. Parrott, D.W. Schmidt, J.C. Shepherd, T.B. Smith, P.A. Sobel, B.A. Watson, D. A. Widmayer, and T.H. Youngblood	Technical
	7. PERIOD COVERED *(Inclusive Dates)*

8. PERFORMING ORGANIZATION - NAME AND ADDRESS *(If NRC, provide Division, Office or Region, U.S. Nuclear Regulatory Commission, and mailing address; if contractor, provide name and mailing address.)*

Division of Waste Management and Environmental Protection

Office of Nuclear Material Safety and Safeguards

U.S. Nuclear Regulatory Commission

Washington, DC 20555–0001

9. SPONSORING ORGANIZATION - NAME AND ADDRESS *(If NRC, type "Same as above"; if contractor, provide NRC Division, Office or Region, U.S. Nuclear Regulatory Commission, and mailing address.)*

Same as above

10. SUPPLEMENTARY NOTES

K.L. Banovac and D.W. Schmidt, NRC Project Managers

11. ABSTRACT *(200 words or less)*

As part of its redesign of the materials license program, the U.S. Nuclear Regulatory Commission (NRC), Office of Nuclear Material Safety and Safeguards has consolidated and updated numerous decommissioning guidance documents into a three-volume NUREG.

Volume 1 of this NUREG series, entitled "Consolidated Decommissioning Guidance: Decommissioning Process for Materials Licensees," takes a risk-informed, performance-based approach to the information needed to support an application for decommissioning a materials license and compliance with the radiological criteria for license termination in 10 CFR Part 20, Subpart E. The approaches to license termination descr bed in this guidance will help to identify the information (subject matter and level of detail) needed to terminate a license by considering the specific circumstances of the wide range of radioactive materials users licensed by NRC. Licensees should use this guidance in preparing license amendment requests. NRC staff will use this guidance in reviewing these amendment requests.

Volume 1 is intended to be applicable only to the decommissioning of materials facilities licensed under 10 CFR Parts 30, 40, 70, and 72 and to the ancillary surface facilities that support radioactive waste disposal activities licensed under 10 CFR Parts 60, 61, and 63. However, parts of this volume are applicable to reactor licensees, as described in the Foreword to this volume.

12. KEY WORDS/DESCRIPTORS *(List words or phrases that will assist researchers in locating the report.)*	13. AVA LAB LITY STATEMENT
consolidated decommissioning guidance risk-informed license termination LTR	unlimited
	14. SECURITY CLASSIFICATION
	(This Page)
	unclassified
	(This Report)
	unclassified
	15. NUMBER OF PAGES
	16. PRICE

NRC FORM 335 (9-2004)   PRINTED ON RECYCLED PAPER

www.ingramcontent.com/pod-product-compliance
Lightning Source LLC
Chambersburg PA
CBHW081427170526
45166CB00008B/2123